METHODS IN MOLECULAR BIOLOGY

Series Editor
John M. Walker
School of Life and Medical Sciences
University of Hertfordshire
Hatfield, Hertfordshire, AL10 9AB, UK

For further volumes:
http://www.springer.com/series/7651

TGF-β Signaling

Methods and Protocols

Edited by

Xin-Hua Feng

Life Sciences Institute and Innovation Center for Cell Signaling Network, Zhejiang University, Hangzhou, Zhejiang Province, China; Departments of Surgery, and Molecular & Cellular Biology, Baylor College of Medicine, Houston, Texas, USA

Pinglong Xu

Life Sciences Institute, Zhejiang University, Hangzhou, Zhejiang Proviince, China

Xia Lin

Department of Surgery, Baylor College of Medicine, Houston, USA

Humana Press

Editors
Xin-Hua Feng
Life Sciences Institute and Innovation
Center for Cell Signaling Network
Zhejiang University, Hangzhou
Zhejiang Province, China

Departments of Surgery
and Molecular & Cellular Biology
Baylor College of Medicine
Houston, Texas, USA

Pinglong Xu
Life Sciences Institute
Zhejiang University
Hangzhou, Zhejiang Proviince, China

Xia Lin
Department of Surgery
Baylor College of Medicine
Houston, USA

ISSN 1064-3745 ISSN 1940-6029 (electronic)
Methods in Molecular Biology
ISBN 978-1-4939-2965-8 ISBN 978-1-4939-2966-5 (eBook)
DOI 10.1007/978-1-4939-2966-5

Library of Congress Control Number: 2015951954

Springer New York Heidelberg Dordrecht London

Printed on acid-free paper

Humana Press is a brand of Springer
Springer Science+Business Media LLC New York is part of Springer Science+Business Media (www.springer.com)

Preface

Cells that respond to environmental cues through the complex and dynamic network of signaling pathways maintain a critical balance between cellular proliferation, differentiation, and death. Since the original discovery of TGF-β about more than 30 years ago, the molecule only represents a prototype of a large family that consists of at least 33 members encoded by the human genome. The TGF-β superfamily members are secreted proteins, including TGF-β, activins, bone morphogenetic proteins, growth/differentiation factors, and Müllerian inhibiting substance, and can regulate many developmental processes in a range of organisms from worms to humans. At cellular level, they control a wide range of cellular functions such as proliferation, death, differentiation, and other functions in many cell types. Dysfunctions of the TGF-β family members often result in the pathogenesis of cancer, autoimmune diseases, diabetes, heart disease, hereditary hemorrhagic telangiectasia, Marfan syndrome, Vascular Ehlers-Danlos syndrome, Loeys–Dietz syndrome, and neurodegenerative diseases. The characterization of the TGF-β family underscores its importance in physiological and pathophysiological functions.

The research on TGF-β biology, including its regulation, signaling, and physiological functions, has developed rapidly into a large field with thousands of publications per year. In the last 20 years, the components of the canonical signaling pathways—the TGF-β receptors and downstream intracellular effectors Smad proteins—have been identified, and the concept for context-dependent TGF-β actions has been well established. Many conventional methodologies or state-of-the-art technologies have been employed to elucidate how the TGF-β pathways are regulated and what functions they have in the context from single cells to complexed tissues to the whole organism. The rapidly evolving nature of TGF-β signaling research also necessitates a continuous updating of methods used. *TGF-β Signaling: Methods and Protocols* brings together a comprehensive collection of methods and techniques in TGF-β signaling research that are scientifically grounded within the cancer and development fields. Thus, this volume grows out of the necessity that a comprehensive method book covering biochemical, molecular, and biological description of TGF-β ligands, receptors, and intracellular events is needed for researchers who are already in the TGF-β field and for those who wish to enter the field.

This volume provides the reader with up-to-date information in this continuing evolving field and attempts to take the reader into the exciting realm of TGF-β from the basic principles to the practical applications. All the chapters are provided by leading researchers in the TGF-β field. The first chapter by Budi and Derynck gives a basic introduction of TGF-β receptor signaling at the cell surface. Subsequent chapters are generally concerned with methods and techniques for the investigation of TGF-β signaling mechanism including receptors, intracellular kinases, microRNA, epigenetic regulation, post-translational regulations, non-Smad pathway; the physiological implications including those in epithelial-mesenchymal transition, endothelial cells, adipogenesis, Th differentiation, stem cell, bone remodeling, ovary, zebrafish development, and frog animal capping; and the methodologies including metastasis imaging, 3D morphogenesis, membrane receptor quantification, conditional knockout, bone remodeling, kinase and phosphatase assays, BiFC interaction assays, and genome-wide siRNA screen.

This book would not be possible were it not for all the contributors who devoted their precious time and considerable energy to bring this volume into reality and provide such clear and detailed accounts of their experimental protocols and useful hints. We are greatly indebted to them for their excellent contributions and for their patience in dealing with the editors. I also wish to thank Dr. P.J. Higgins and their overall help and patience in keeping the book on track. We hope this volume will prove valuable to all researchers, rookie or veteran, in the TGF-β signaling field and serve as a useful reference for many years to come.

Hangzhou, China and Houston, TX, USA ***Xin-Hua Feng***
Hangzhou, China ***Pinglong Xu***
Houston, TX, USA ***Xia Lin***

Contents

Contributors

HIROYUKI ABURATANI • *Genome Science Division, Research Center for Advanced Science and Technology (RCAST), The University of Tokyo, Tokyo, Japan*

LINDSAY ALCARAZ • *Inserm U1052, Centre de Recherche en Cancérologie de Lyon, Lyon, France; CNRS UMR 5286, Centre de Recherche en Cancérologie de Lyon, Lyon, France; Université de Lyon, Lyon, France; Université Lyon 1, Lyon, France; Centre Léon Bérard, Lyon, France*

LAURENT BARTHOLIN • *Inserm U1052, Centre de Recherche en Cancérologie de Lyon, Lyon, France; CNRS UMR 5286, Centre de Recherche en Cancérologie de Lyon, Lyon, France; Université de Lyon, Lyon, France; Université Lyon 1, Lyon, France; Centre Léon Bérard, Lyon, France*

JUAN C. BOURNAT • *Department of Pediatrics, Baylor College of Medicine, Houston, TX, USA*

CHESTER W. BROWN • *Department of Molecular and Human Genetics, Baylor College of Medicine, Houston, TX, USA; Department of Pediatrics, Baylor College of Medicine, Houston, TX, USA; Texas Children's Hospital, Houston, TX, USA*

ERINE H. BUDI • *Department of Cell and Tissue Biology, Broad Center, University of California, San Francisco, CA, USA*

XU CAO • *Department of Orthopaedic Surgery, Johns Hopkins University School of Medicine, Baltimore, MD, USA*

JONATHON CARTHY • *Ludwig Institute for Cancer Research, Science for Life Laboratory, Uppsala University, Uppsala, Sweden*

CHENBEI CHANG • *Department of Cell, Developmental and Integrative Biology, University of Alabama at Birmingham, Birmingham, AL, USA*

XIAOCHU CHEN • *Program in Molecular Medicine, University of Massachusetts Medical School, Worcester, MA, USA*

YAN CHEN • *Key Laboratory of Nutrition and Metabolism, Institute for Nutritional Sciences, Shanghai Institutes for Biological Sciences, Chinese Academy of Sciences, Shanghai, China*

YE-GUANG CHEN • *State Key Laboratory of Membrane Biology, Tsinghua-Peking Center for Life Sciences, School of Life Sciences, Tsinghua University, Beijing, China*

JANET L. CRANE • *Department of Orthopaedic Surgery, Johns Hopkins University School of Medicine, Baltimore, MD, USA; Department of Pediatrics, Johns Hopkins University School of Medicine, Baltimore, MD, USA*

RIK DERYNCK • *Department of Cell and Tissue Biology, Eli and Edythe Broad Center of Regeneration Medicine and Stem Cell Research, Programs in Cell Biology, and Developmental and Stem Cell Biology, University of California, San Francisco, CA, USA*

PETER TEN DIJKE • *Department of Molecular Cell Biology, Cancer Genomics Centre Netherlands, Centre for Biomedical Genetics, Leiden University Medical Center, Leiden, The Netherlands*

SHENG DING • *Department of Pharmaceutical Chemistry, Gladstone Institute of Cardiovascular Disease, University of California, San Francisco, CA, USA*

MAARTEN VAN DINTHER • *Department of Molecular Cell Biology, Leiden University Medical Center, Leiden, The Netherlands*
HENG-YU FAN • *Life Sciences Institute and Innovation Center for Cell Signaling Network, Zhejiang University, Hangzhou, China*
ZIPEI FENG • *Department of Chemistry and Biochemistry, University of Colorado Boulder, Boulder, CO, USA; Otto-Warburg Laboratory, Max Planck Institute for Molecular Genetics, Berlin, Germany*
MAI FUJIMOTO • *Genome Science Division, Research Center for Advanced Science and Technology (RCAST), The University of Tokyo, Tokyo, Japan; Department of Molecular Oncology, Graduate School of Medicine, Chiba University, Chiba, Japan*
TAKANORI FUJITA • *Genome Science Division, Research Center for Advanced Science and Technology (RCAST), The University of Tokyo, Tokyo, Japan*
M.J. GOUMANS • *Department of Molecular Cell Biology, Cancer Genomics Centre Netherlands, Centre for Biomedical Genetics, Leiden University Medical Center, Leiden, The Netherlands*
SHYAM KUMAR GUDEY • *Department of Medical Biosciences, Umeå University, Umeå, Sweden*
XING GUO • *Department of Pharmacology and Cancer Biology, Duke University Medical Center, Durham, NC, USA*
AKIKO HATA • *Cardiovascular Research Institute, University of California, San Francisco, CA, USA*
ANDREW P. HINCK • *Department of Structural Biology, University of Pittsburgh, Pittsburgh, PA, USA*
TAO HUANG • *Protein Chemistry, Novo Nordisk Research Center China, Beijing, China*
ATSUSHI KANEDA • *Genome Science Division, Research Center for Advanced Science and Technology (RCAST), The University of Tokyo, Tokyo, Japan; Department of Molecular Oncology, Graduate School of Medicine, Chiba University, Chiba, Japan; CREST, Japan Science and Technology Agency, Saitama, Japan*
YIBIN KANG • *Lewis Thomas Laboratory 255, Department of Molecular Biology, Princeton University, Princeton, NJ, USA*
HARA KANG • *Division of Life Sciences, College of Life Sciences and Bioengineering, Incheon National University, Incheon, Republic of Korea; Cardiovascular Research Institute, University of California, San Francisco, CA, USA*
MITSUYASU KATO • *Department of Experimental Pathology, Graduate School of Comprehensive Human Sciences, University of Tsukuba, Tsukuba, Ibaraki, Japan*
MARENE LANDSTRÖM • *Department of Medical Biosciences, Umeå University, Umeå, Sweden*
WENLIN LI • *Department of Cell Biology, Second Military Medical University, Shanghai, China*
XIA LIN • *Michael E. DeBakey Department of Surgery, Baylor College of Medicine, Houston, TX, USA*
XUEDONG LIU • *Department of Chemistry and Biochemistry, University of Colorado Boulder, Boulder, CO, USA*
FANG LIU • *Center for Advanced Biotechnology and Medicine, Rutgers, The State University of New Jersey, Piscataway, NJ, USA; Susan Lehman Cullman Laboratory for Cancer Research, Department of Chemical Biology, Ernest Mario School of Pharmacy, Rutgers, The State University of New Jersey, Piscataway, NJ, USA; Rutgers Cancer Institute of New Jersey, Rutgers, The State University of New Jersey, New Brunswick, NJ, USA*

Xingfeng Liu • *State-key Laboratory of Biomembrane and Membrane Engineering, Tsinghua-Peking Center for Life Sciences, School of Life Sciences, Tsinghua University, Beijing, China*

Kunxin Luo, Ph.D. • *Department of Molecular and Cell Biology (MCB), University of California, Berkeley, CA, USA*

J.A. Maring • *Department of Molecular Cell Biology, Cancer Genomics Centre Netherlands, Centre for Biomedical Genetics, Leiden University Medical Center, Leiden, The Netherlands*

Geoffrey J. Markowitz • *Department of Pharmacology and Cancer Biology, Duke University Medical Center, Duke University, Durham, NC, USA*

Graziano Martello • *Department of Molecular Medicine, University of Padua School of Medicine, Padua, Italy*

Isao Matsuura • *Division of Molecular Genomics and Medicine, National Health Research Institutes, Zhunan Town, Miaoli County, Taiwan*

L.A. van Meeteren • *Department of Molecular Cell Biology, Cancer Genomics Centre Netherlands, Centre for Biomedical Genetics, Leiden University Medical Center, Leiden, The Netherlands*

Anming Meng • *State-key Laboratory of Biomembrane and Membrane Engineering, Tsinghua-Peking Center for Life Sciences, School of Life Sciences, Tsinghua University, Beijing, China; Institute of Zoology, Chinese Academy of Sciences, Beijing, China*

Kohei Miyazono • *Department of Molecular Pathology, Graduate School of Medicine, The University of Tokyo, Tokyo, Japan*

Marco Montagner • *Department of Molecular Medicine, University of Padua School of Medicine, Padua, Italy*

Aristidis Moustakas • *Ludwig Institute for Cancer Research, Science for Life Laboratory, Uppsala University, Uppsala, Sweden; Department of Medical Biochemistry and Microbiology, Science for Life Laboratory, Uppsala University, Uppsala, Sweden*

Maria Namwanje • *Department of Molecular and Human Genetics, Baylor College of Medicine, Houston, TX, USA*

Aya Nonaka • *Genome Science Division, Research Center for Advanced Science and Technology (RCAST), The University of Tokyo, Tokyo, Japan*

Yukari Okita • *Ludwig Institute for Cancer Research, Science for Life Laboratory, Uppsala University, Uppsala, Sweden; Department of Experimental Pathology, Graduate School of Comprehensive Human Sciences, University of Tsukuba, Tsukuba, Ibaraki, Japan*

Yi Pan • *Key Laboratory of Nutrition and Metabolism, Institute for Nutritional Sciences, Shanghai Institutes for Biological Sciences, Chinese Academy of Sciences, Shanghai, China*

Stefano Piccolo • *Department of Molecular Medicine, University of Padua School of Medicine, Padua, Italy*

Lan Qin • *Michael E. DeBakey Department of Surgery, Baylor College of Medicine, Houston, TX, USA*

Juliet Rashidian • *Department of Molecular and Cell Biology (MCB), University of California, Berkeley, CA, USA*

Takashi Sekiya • *Department of Microbiology and Immunology, Keio University School of Medicine, Tokyo, Japan*

Tao Shen • *Department of Molecular and Cellular Biology, Baylor College of Medicine, Houston, TX, USA; Michael E. DeBakey Department of Surgery, Baylor College of Medicine, Houston, TX, USA*

SYLVIE THUAULT • *Faculty of Pharmacy, INSERM UMR 911 CRO2, Marseille, France*
ULRICH VALCOURT • *Inserm U1052, Centre de Recherche en Cancérologie de Lyon, Lyon, France; ; CNRS UMR 5286, Centre de Recherche en Cancérologie de Lyon, France; Université de Lyon, Lyon, France; Université Lyon 1, Lyon, France; Centre Léon Bérard, Lyon, France*
QIANG WANG • *Institute of Zoology, Chinese Academy of Sciences, Beijing, China*
XIAO-FAN WANG • *Department of Pharmacology and Cancer Biology, Duke University Medical Center, Durham, NC, USA*
WANGUO WEI • *Stem Cell and Regenerative Medicine Center, Chinese Academy of Science, Shanghai Advanced Research Institute, Shanghai, China*
LINGLING XIAN • *Department of Orthopaedic Surgery, Johns Hopkins University School of Medicine, Baltimore, MD, USA*
JIAN XU • *Center for Craniofacial Molecular Biology, Ostrow School of Dentistry of USC, University of Southern California, Los Angeles, CA, USA*
LAN XU • *Blueprint Medicines, Cambridge, MA, USA*
PINGLONG XU • *Life Sciences Institute and Innovation Center for Cell Signaling Network, Zhejiang University, Hangzhou, China*
RYOTA YAMANAKA • *Genome Science Division, Research Center for Advanced Science and Technology (RCAST), The University of Tokyo, Tokyo, Japan*
XIAOHUA YAN • *State Key Laboratory of Membrane Biology, Tsinghua-Peking Center for Life Sciences, School of Life Sciences, Tsinghua University, Beijing, China*
PENGYUAN YANG • *CAS Key Laboratory of Infection and Immunity, Institute of Biophysics, Chinese Academy of Sciences, Beijing, China*
AKIHIKO YOSHIMURA • *Department of Microbiology and Immunology, Keio University School of Medicine, Tokyo, Japan*
CHAO YU • *Life Sciences Institute and Innovation Center for Cell Signaling Network, Zhejiang University, Hangzhou, China*
LONG ZHANG • *Life Sciences Institute and Innovation Center for Cell Signaling Network, Zhejiang University, Hangzhou, China*
YUN ZHANG • *Department of Pharmacology and Cancer Biology, Duke University Medical Center, Durham, NC, USA*
FANGFANG ZHOU • *Institutes of Biology and Medical Sciences, Soochow University, Suzhou, China*
JIAN-JIE ZHOU • *Life Sciences Institute and Innovation Center for Cell Signaling Network, Zhejiang University, Hangzhou, China*
ZHIKE ZI • *Otto-Warburg Laboratory, Max Planck Institute for Molecular Genetics, Berlin, Germany*

Chapter 1

Regulation of TGF-β Receptors

Erine H. Budi, Jian Xu, and Rik Derynck

Abstract

In cells responding to extracellular polypeptide ligands, regulatory mechanisms at the level of cell surface receptors are increasingly seen to define the nature of the ligand-induced signaling responses. Processes that govern the levels of receptors at the plasma membrane, including posttranslational modifications, are crucial to ensure receptor function and specify the downstream signals. Indeed, extracellular post-translational modifications of the receptors help define stability and ligand binding, while intracellular modifications mediate interactions with signaling mediators and accessory proteins that help define the nature of the signaling response. The use of various molecular biology and biochemistry techniques, based on chemical crosslinking, e.g., biotin or radioactive labeling, immunofluorescence to label membrane receptors and flow cytometry, allows for quantification of changes of cell surface receptor presentation. Here, we discuss recent progress in our understanding of the regulation of TGF-β receptors, i.e., the type I (TβRI) and type II (TβRII) TGF-β receptors, and describe basic methods to identify and quantify TGF-β cell surface receptors.

Key words TβRI, TβRII, Posttranslational modifications, Cell surface receptor, Clathrin, Caveolin, Biotin labeling, Affinity labeling, Immunoprecipitation, Neutravidin precipitation, Cell sorting, Chemical crosslinking, Immunofluorescence

1 Introduction

The transforming growth factor-β (TGF-β) family includes TGF-βs, activins, and nodal, and a large number of bone morphogenetic proteins (BMPs). These secreted proteins regulate a wide range of biological processes in embryonic and adult development, as well as tissue homeostasis, and deregulation of TGF-β family signaling is at the basis of the pathogeneses of various diseases [1]. TGF-β proteins act as disulfide-linked dimers and activate cell responses following binding to heteromeric cell surface receptor complexes [2–4]. TGF-β family receptors are tetramers consisting of two type II and two type I receptors. Both receptor types are transmembrane kinases with dual specificity, able to phosphorylate substrates not only on serine (Ser) and threonine (Thr), but also on tyrosine (Tyr) residues [5, 6]. Upon ligand binding to the receptor complex

Xin-Hua Feng et al. (eds.), *TGF-β Signaling: Methods and Protocols*, Methods in Molecular Biology, vol. 1344, DOI 10.1007/978-1-4939-2966-5_1, © Springer Science+Business Media New York 2016

in the extracellular domains, the type II receptors phosphorylate the type I receptors in their characteristic GS domain in the juxtamembrane domain, which then results in activation of the type I receptor kinases [4]. Phosphorylation of the GS domain enables the TβRI receptors to recruit receptor-regulated Smads (R-Smads), and to phosphorylate them at the C termini, resulting in R-Smad activation. TGF-β and activin induce activation of Smad2 and Smad3, whereas BMP type I receptors activate Smad1, Smad5, and Smad8 [2–4]. The specificity of the ligand-induced responses is defined by the nature of the heteromeric type II/type I receptor combinations, of which the receptor ectodomain combinations provide ligand binding specificity and the type I receptor kinases act as primary effectors in initiating the signaling responses [2–4].

With the identification of TGF-β1 as the first member of the TGF-β family [7], TGF-β and the TGF-β receptors have largely served as prototype system to study ligand-induced receptor activation of this class of growth and differentiation factors. Binding of TGF-β to the heteromeric receptor complexes is largely specified by the ectodomain of TβRII, which is not known to act as type II receptor for any other ligand [4]. The TGF-β receptor complex is composed of two TβRII polypeptides and two type I TGF-β receptor polypeptides, typically TβRI, also known as ALK-5 [3, 4]. Some TGF-β responses have been attributed to TGF-β binding to complexes of TβRII with other type I receptors, i.e., ALK-1 or ALK-2, which more commonly act as BMP type I receptors. TβRI also functions as type I receptor for the TGF-β-related myostatin, but does so in combination with another type II receptor, i.e., ActRIIB [8, 9].

In the absence of TGF-β, the TβRII and TβRI receptors are found as dynamic dimers, in equilibrium with monomers, at the cell surface [10]. The TβRII receptors are thought to act as constitutively active kinases, and TGF-β binding stabilizes receptor dimer formation, and induces TβRII autophosphorylation and transphosphorylation of the type I receptors [11]. Phosphorylation of the TβRI receptors in their GS domains in turn induces activation of type I receptor kinases, enabling TβRI autophosphorylation and possibly TβRII phosphorylation by TβRI [4]. TβRI phosphorylates Smad2 and Smad3 C-terminally at two C-terminal serines in response to TGF-β. The resulting activation of these R-Smads by TβRI is thought to occur in a 1:1 stoichiometry, allowing activation of two R-Smads by the activated receptor complex [4, 12]. Following their dissociation from the receptors, two receptor-activated Smads form trimers with one Smad4 [2–4]. These translocate into the nucleus and activate or repress target gene expression through association with DNA sequence-specific transcription factors, and transcription coactivators and/or corepressors [3]. TGF-β also activates non-Smad responses, most notably the Erk, p38, and JNK MAP kinase pathways, and the PI3-kinase-Akt-TOR pathway [13, 14].

Activation of the signaling responses induced by TGF-β through the type II/type I receptor complexes is tightly regulated at the receptor level. Although much remains to be learned, our current knowledge suggests important roles for posttranslational modifications of the receptors in defining the level of functional TGF-β receptors at the cell surface. Furthermore, receptor interactions with accessory proteins define the intracellular routing of the activated receptor complexes and the signaling response. This short review addresses our current view of how TβRI and TβRII presentation is regulated at the cell surface, and the mechanisms and signaling mediators that help define the signaling specificity at the receptor level.

1.1 Posttranslational Modifications of the TGF-β Receptors

Similarly to other cell surface receptors, TGF-β receptors are targets of posttranslational modifications in both their extracellular and cytoplasmic domains (Fig. 1). These modifications involve actions of various enzymes and interacting proteins, and provide mechanisms that define the functions of the receptors as initiators of signaling.

1.1.1 TGF-β Receptor Phosphorylation

TGF-β binding to the receptor complex leads to Smad activation and non-Smad signaling pathways. Accordingly, cytoplasmic

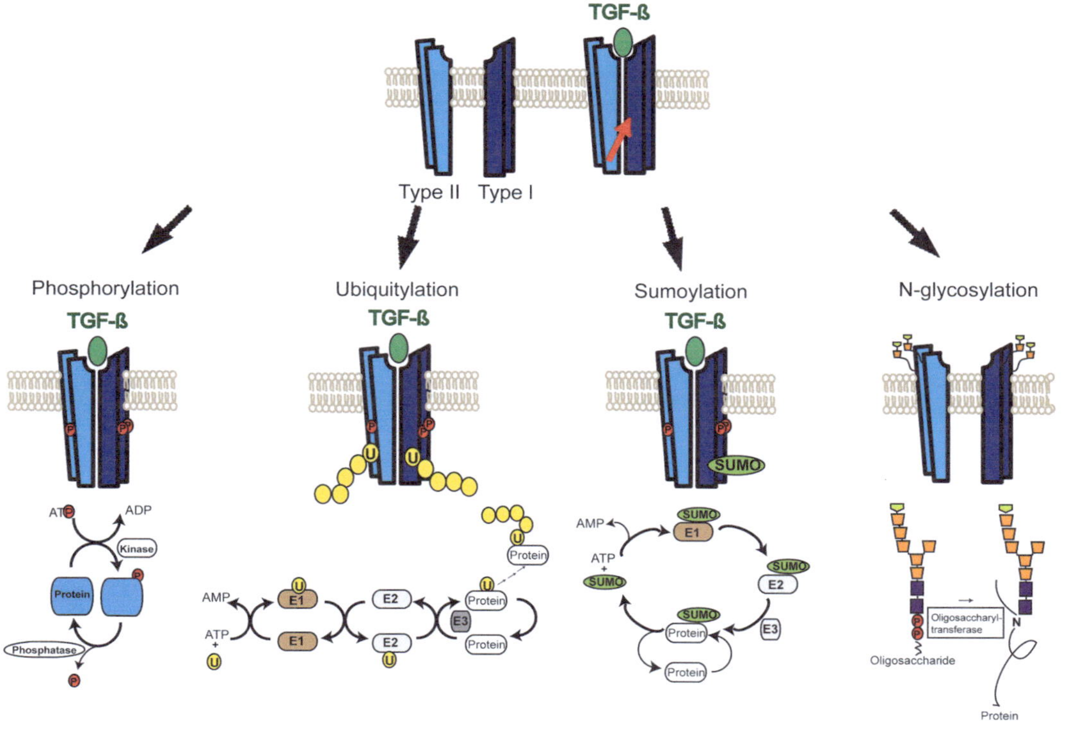

Fig. 1 Posttranslational modifications of TβRI and TβRII receptors. The underlying enzymatic reactions of receptor phosphorylation, ubiquitylation, sumoylation, and *N*-glycosylation are shown

receptor phosphorylation plays a key role in the activities of the receptors. The type II receptors are thought to be constitutively catalytically active, with ligand binding inducing dimer stabilization and autophosphorylation in trans [5, 11]. However, while TβRII was shown to be phosphorylated [5, 11], the identities and roles of the phosphorylated Ser and Thr residues have not been characterized, nor is it clear to what extent other kinases contribute to Ser/Thr phosphorylation of TβRII. Consistent with its dual specificity kinase activity, TβRII is also able to autophosphorylate on Tyr, and is found to be Tyr phosphorylated [5]. By analogy with MAP kinases, TβRII phosphorylation on Tyr may be autoregulatory [5]. Additionally, Src phosphorylates TβRII on Tyr, and the phosphorylated Tyr residue may provide a docking site for Grb2 and Shc, which has been proposed to facilitate TGF-β-induced, TβRII-mediated activation of the p38 MAPK pathway [15].

TGF-β binding to the receptor complex stabilizes the TβRII/TβRI interactions and induces conformational changes that activate TβRI. Central in this activation process is the ligand-induced phosphorylation of TβRI's GS domain by TβRII on Ser and Thr, which results in a conformational change that fully activates the TβRI kinase, enabling it to phosphorylate and activate Smad2 and Smad3 [4]. TGF-β binding also triggers phosphorylation of Ser and Thr residues outside the GS domain [16], although the identities of these amino acids and their roles remain largely unknown. Additionally, by analogy with receptor tyrosine kinases, it is likely that the activated type I receptors autophosphorylate in trans, and that the type I receptors phosphorylate type II receptors, but also these mechanisms and the phosphorylated residues remain to be defined at the molecular level. Finally, the TβRI kinase also has dual specificity, and TβRI is Tyr phosphorylated in response to TGF-β, raising the possibility of TGF-β-induced TβRI autophosphorylation on Tyr residues. These sites in turn may mediate TGF-β-induced Shc binding to TβRI, thus initiating Erk MAPK pathway activation [6].

Consistent with the dynamic nature of phosphorylation and the role of TβRI in initiating TGF-β-induced signaling, TβRI is also targeted by phosphatases. Indeed, the protein phosphatase PP1c was shown to dephosphorylate TβRI, which may be facilitated through interaction of PP1c with SARA, a scaffolding protein that stabilizes the interactions of Smad2/3 with TβRI [17, 18]. Introducing a mutant SARA that is incapable of PP1c binding results in hyperphosphorylation of TβRI and hyperactivation of TGF-β signaling [18]. In addition, the multimeric serine-threonine protein phosphatase 2A (PP2A) has been implicated in modulating TGF-β signaling at the receptor level [19]. Regulatory subunits of PP2A were shown to interact with activated TβRI and to be phosphorylated by TβRI in response to TGF-β [20]. Depending on which regulatory subunit associates with TGF-β receptors, Bα or Bδ, the net effects of these interactions in signaling activated by TGF-β, activin or nodal are opposite [19, 20]. No phosphatases have as yet been implicated in the control of TβRII phosphorylation.

The significance of TGF-β receptor phosphorylation in cancer is supported by the occurrence of mutations in the kinase domains of TβRII or TβRI that attenuate or inactivate their kinase activities [21, 22]. These mutations were found to occur primarily in carcinomas that benefit from loss of growth inhibition by autocrine TGF-β. Additionally, impaired kinase activities of these receptors attenuate TGF-β's ability to promote epithelial–mesenchymal transition, invasion, and cancer dissemination [15, 23].

1.1.2 TGF-β Receptor Ubiquitylation

The stability of TβRI is regulated by poly-ubiquitylation, a process that results in the covalent addition of a chain of ubiquitin polypeptides to lysine residues. Poly-ubiquitylation targets TβRI for degradation. The ubiquitylation process is catalyzed and tightly regulated by the consecutive activities of a ubiquitin activating E1 enzyme, a ubiquitin-conjugating E2 enzyme, and a ubiquitin E3 ligase that covalently attaches ubiquitin to its target. The E3 ligase provides target selectivity [24, 25]; accordingly a large number of ubiquitin E3 ligases have been identified, with several of them targeting TβRI [26, 27]. Ubiquitylation of TβRII has not been demonstrated.

Smad7, an inhibitory Smad that interacts with TβRI thereby preventing Smad2/3 activation [28, 29], appears to play a key role in TβRI ubiquitylation [30]. Receptor-bound Smad7 can recruit one of several E3 ubiquitin ligases to TβRI, resulting in TβRI ubiquitylation and degradation, thus attenuating TGF-β signaling. Several E3 ligases have been shown to target TβRI, with Smurf1 and Smurf2 thought to be the most prevalent ones [26, 27]. Their Hect domain is required for the transfer of ubiquitin from the E2 ligase to the substrate [31]. Two other Hect domain E3 ligases, NEDD4-2 and WWP1, have also been shown to associate with TβRI via Smad7 and to induce receptor degradation [32, 33].

Poly-ubiquitylation also controls protein localization [34]. The ubiquitin ligase TRAF6 (tumor necrosis factor receptor-associated factor 6) can poly-ubiquitylate TβRI at different lysine positions, and plays an important role as intermediate in the activation of p38 MAPK and JNK pathways in response to TGF-β [35]. TRAF6 also plays a role in ectodomain shedding of TβRI and the release of the intracellular domain of TβRI [36]. Posttranslational ectodomain shedding is discussed later.

Conversely, the deubiquitylating enzyme USP15 regulates TGF-β signaling at the receptor level [37, 38]. USP15 does not bind TβRI directly, but deubiquitylates and stabilizes activated TβRI by binding to the Smad7-Smurf2 complex [38]. In this case, Smad7 acts as a scaffold that recruits both the ubiquitin ligase

Smurf2 and the deubiquitylating enzyme USP15 to the activated TβRI [37, 38], setting up a scenario of competitive ubiquitylation and de-ubiquitylation. De-ubiquitylation of TβRI enhances the levels of TβRI, leading to higher levels of TGF-β-induced Smad activation. The deubiquitylase USP4 has also been shown to target TβRI, but without requirement of Smad7 as the scaffold [39]. Increased USP4 expression translates into enhanced Smad2 phosphorylation and enhanced TGF-β target gene expression [39].

Aberrant ubiquitylation of TβRI has been implicated in the initiation or progression of certain cancer types. USP15 expression is increased in a fraction of glioblastomas, breast, and ovarian cancers, and this increase correlates with elevated TGF-β activity [38]. Increased USP4 levels, as seen in several types of human cancers, e.g., urinary tract and prostate carcinomas [40], enhances TGF-β-induced epithelial–mesenchymal transition and cell migration of breast cancer cells, and cancer cell dissemination [37]. In addition, pharmacological inhibition of deubiquitylases decreases TGF-β signaling and reduces the tumor sizes of human gliomas in an orthotopic mouse model [38].

1.1.3 TGF-β Receptor Sumoylation and Neddylation

The TβRI receptor is also targeted for sumoylation, a process that results in covalent attachment of a single SUMO polypeptide to a targeted lysine of a substrate. The process of sumoylation shows similarities with ubiquitylation. Not only does SUMO structurally resemble ubiquitin, but sumoylation also requires three-step enzymatic actions of an E1, E2, and E3 SUMO ligase [41, 42]. Most known targets for sumoylation are transcription factors, perinuclear and nuclear proteins [43]. Similar to ubiquitylation, only a small fraction of the targeted proteins are sumoylated at any given time, consistent with the reversible nature of sumoylation [42, 44, 45].

In response to TGF-β binding, TβRI is sumoylated at a single lysine residue [46]. TβRI sumoylation depends on the kinase activities of both TβRI and TβRII, strongly suggesting that TβRI sumoylation requires TβRI phosphorylation, and effectively illustrating functional cross talk between two types of posttranslational modifications [46]. Sumoylation of the TβRI cytoplasmic domain enhances the efficiency of Smad2/3 recruitment to TβRI, thus increasing the level of Smad2/3 activation and Smad-mediated signaling in response to TGF-β [46].

Although TβRII is not known to be ubiquitylated or sumoylated, it is modified by neddylation, i.e., addition of the ubiquitin-like protein Nedd8. The ubiquitin E3 ligase encoded by the proto-oncogene c-Cbl was shown to mediate covalent linkage of Nedd8 to TβRII at two lysine positions, thus promoting enhanced stability of TβRII [47]. Targeted inactivation of c-Cbl decreases the TβRII levels and desensitizes hematopoietic stem/progenitor cells to TGF-β signaling [47]. Inactivating c-Cbl mutations have been identified in leukemias [47], suggesting a link between aberrant TβRII neddylation and predisposition to some leukemias.

1.1.4 TGF-β Receptor N-glycosylation

Both TβRII and TβRI are *N*-glycosylated [48–50]. *N*-glycosylation of the extracellular domains of transmembrane proteins and secreted proteins is thought to contribute to protein stability and solubility and may facilitate intracellular transport through the Golgi to the plasma membrane [49, 51–53]. In TβRII, N-linked glycosylation occurs on two asparagine residues in the extracellular domain [53]. Mutation of these two sites, which prevents *N*-glycosylation, blocks transport of the TβRII receptor to the plasma membrane and leads to receptor accumulation in the ER, while mutants with a single residue mutation are still able to localize TβRII at the cell surface [53]. The defect in *N*-glycosylation results in impaired TGF-β signaling [53]. The role of glycosylation in the type I receptor in cell surface transport and signaling is unknown. Whether N-linked glycosylation affects TβRI-TβRII complex formation is similarly unknown.

MED12, a component of transcription Mediator complex, is found to interact with a partially and unglycosylated form of TβRII [54]. MED12 overexpression causes a decrease in cell surface TβRII by interfering with proper glycosylation of TβRII, hence blocking the appearance of TβRII at the cell surface. Conversely, attenuation of MED12 expression confers resistance to a range of chemotherapy drugs that are used in treating colon cancer, melanoma, and liver cancer, through activation of TGF-β signaling [54].

1.1.5 Ectodomain Shedding of TGF-β Receptors

The levels of functional TGF-β receptor at the cell surface are also regulated by ectodomain shedding, i.e., proteolytic release of ectodomains by cell surface metalloproteases [55]. The transmembrane metalloprotease TACE, also known as ADAM17, cleaves and thus removes the extracellular domain of TβRI, but not TβRII, resulting in TβRI inactivation without affecting TGF-β binding to TβRII. TACE-mediated ectodomain shedding is activated by the Erk MAPK pathway, e.g., in response to growth factors that act through receptor tyrosine kinases, and by p38 MAPK signaling in response to inflammatory stimuli, thus decreasing functional receptor availability under these conditions, and attenuating TGFβ-induced Smad responses [56, 57]. Ectodomain cleavage of TβRI not only leads to the release of its extracellular domain, but also leads to the release of the intracellular domain of TβRI into the cytosol. This domain can then translocate into the nucleus and act as a transcription regulator [36]. Another metalloprotease, ADAM12, also known as meltrin-α, was found to interact with the extracellular domain of TβRII [58]. ADAM12 does not impair TGF-β signaling, as one would expect from its protease activity, but instead enhances TGF-β-induced Smad signaling by stabilizing the receptors and facilitating their localization from the plasma membrane to early endosomes [58].

1.2 TGF-β Receptor-Interacting Proteins

In addition to roles of posttranslational modifications in regulating TGF-β receptor presentation and activities, proteins with diverse functions have been identified to interact with the TGF-β receptor complexes and modulate TGF-β signaling [59]. Some proteins are required for efficient activation of TGF-β-induced Smad signaling, some repress TGF-β responses, and others participate in non-Smad TGF-β signaling. At the cell surface, several transmembrane proteins associate with TβRII-TβRI heteromeric complexes. Most notably, betaglycan and endoglin enhance TGF-β binding efficiency to the TβRII-TβRI complex, and thus serve as coreceptors to facilitate TGF-β signaling [60–62]. Additionally, the integrin dimer α5β3, the cell adhesion proteins E-cadherin, and endothelial VE-cadherin bind TβRII/TβRI complex, thus enhancing their stability and TGF-β signaling efficiency [63–66]. Other cell surface proteins, however, interfere with TGF-β receptor complex formation and consequently attenuate TGF-β signaling. For example, the neurotrophin receptor TrkC and a related fusion protein ETV6-NTRK3, found in congenital fibrosarcomas, interact with TβRII, thus interfering with the formation of TβRI/TβRII complexes [67, 68]. A glycosyl phosphatidylinositol (GPI)-anchored protein, CD109, which is discussed later, has also been shown to interact with TβRI [69] and negatively regulate TGF-β signaling [70, 71].

An increasing number of cytoplasmic proteins were found in association with either TβRII or TβRI and regulate TGF-β signaling. Among these, the immunophilin FKBP12, an abundant and ubiquitously expressed protein, binds the cytoplasmic domain of TβRI, and silences leaky TGF-β signaling in the absence of ligand [72, 73]. Also the ubiquitous dynein light chain Tctex-type family protein, Tctex2β, also known as TctexD4 (Tctex1 domain-containing protein 4), interacts with TβRII and inhibits TGF-β-induced signaling in association with endoglin [74]. STRAP, a WD-40 domain protein stabilizes the association of Smad7 with TβRI, consequently attenuating Smad2/3-mediated transcription activation [75]. Other cytoplasmic proteins that associate with TGF-β receptors enhance TGF-β signaling. The scaffold protein SARA that is associated with the plasma membrane through its FYVE domain interacts with TβRI and stabilizes the TGF-β-induced recruitment of Smad2 and Smad3 to TβRI, thus enhancing Smad2/3 activation and signaling [76]. Another FYVE domain protein, Hrs/Hgs has similarly been implicated in enhancing Smad2/3 recruitment to the activated TβRI [77]. The abilities of SARA and Hrs/Hgs to enhance Smad signaling may result from their roles in directing the receptors to clathrin-mediated endocytosis, as discussed further.

Additionally, the heat shock protein Hsp90 interacts with both TβRII and TβRI, thus acting as their chaperone. Interfering with Hsp90 function enhances TGF-β receptor ubiquitylation dependent on Smurf2, and degradation. Hsp90 may therefore aid in determining the TGF-β receptor cell surface levels and amplitude of TGF-β signaling [78].

While some proteins facilitate Smad2/3 recruitment and activation in response to TGF-β, others interfere with Smad2/3 recruitment, thus attenuating TGF-β signaling. Most prominent among these are the inhibitory Smads (I-Smads), i.e., Smad6 and Smad7, which interact with TβRI similarly to R-Smads, thus effectively competing with Smad2/3 recruitment. Consequently, the balance of Smad2/3 and inhibitory Smad recruitment to TβRI defines the amplitude of Smad2/3 activation and signaling in response to TGF-β [79]. As mentioned, the inhibitory Smads also recruit E3 ubiquitin ligases that target TβRI for poly-ubiquitylation and degradation, and decrease the cell surface levels of activated and signaling competent TβRI receptors [26, 27, 30, 32, 33]. Additionally, Smad7 also recruits the phosphatase complex GADD34-PP1c to dephosphorylate, and therefore inactivate TβRI [18]. Competition between R-Smads and I-Smads for receptor availability is further integrated into distinct endocytic routing of the activated receptor complexes, which is discussed in the next section.

Other receptor-associated proteins play key roles in the activation of non-Smad signaling pathways in response to TGF-β. Of these, the adaptor protein Shc associates with the activated TβRI similarly to receptor tyrosine kinases, is phosphorylated by TβRI, and mediates assembly of Grb2-SOS complexes, leading to activation of the Erk-MAPK pathway [6]. As mentioned, MED12, a component of the transcription Mediator complex, associates with TβRII and represses its signaling, thus repressing TGF-β-induced Erk-MAPK signaling [54]. Another example is TRAF6, a mediator of TNF receptor and Toll/IL-1 receptor signaling. TRAF6 interacts with TβRI to transduce signals through the TAK1 kinase, thus initiating p38 and JNK MAPK signaling in response to TGF-β [35, 80]. Finally, XIAP also associates with TβRI, ubiquitylates TAK1 and links TGF-β receptor activation to activation of NF-κB signaling [81].

Using different methods and approaches, a large number of diverse proteins, including the ones already mentioned, were found to interact with TGF-β receptors. Regulating intracellular routing, cell surface presentation, or signaling functions, these proteins should often been seen as integral regulators of the life cycles of TβRII and TβRI, and of TGF-β receptor function. A curated list of proteins that have been identified to interact with the type I and type II TGF-β receptors are shown in Table 1.

Table 1
Proteins interacting with TGF-ß receptors

Interacting proteins	Regulatory mechanism	References
Axin	Facilitates Smad3 activation by TβRI and interacts with Smad7 and Arkadia to promote Smad7 degradation	[103–105]
BAT3	Interacts with TβRI and TβRII and enhances TGF-β-induced transcription	[106]
c-Ski	Associates with activated TβRI and stabilizes a non-functional TβRI/R-Smad/Smad4 complex	[107]
CD109	Enhances TGF-β binding to its receptors, promotes caveolae-mediated receptor internalization and Smad7-Smurf mediated receptor degradation	[70, 71]
CD44	Associates with TβRI to promote Smad2/3 activation and PTH-RP expression	[108]
cPML	Stabilizes R-Smad/SARA complex and enriches TGF-β and SARA in early endosomes	[109]
Dab2	Associates with TGF-β receptors and R-Smads to facilitate R-Smad activation and promotes clathrin-mediated recycling of TβRII to decrease degradation	[94, 95]
Dpr2	Binds to TβRI receptor and promote lysosomal degradation of the receptor	[97, 98]
ETV6-NTRK3	Binds directly to TβRII and prevents TβRII from interacting with TβRI	[68]
FKBP12	Binds to the GS domain of TβRI and prevents leaky TGF-β signaling, interacts with Smad7, Smurf1, and TβRI to target TβRI for degradation	[72, 110, 111]
Hgs/Hrs	Cooperates with SARA to stabilize TGF-β receptor/R-Smad complexes	[77, 112]
HSP90	Binds to TβRI and TβRII and protects receptors from Smurf2-mediated ubiquitylation and degradation	[78, 113]
Integrin αvβ3	Interacts with TβRII to enhance TGF-β1 activation	[63]
Itch/AIP4	Enhances Smad2 or Smad7 association with TβRI and modulates R-Smad activation	[114, 115]
Km23-1	Associates with TGF-β receptors and Smad2 in early endosomes to enhance Smad activation	[116]
MED12	Interacts with TβRII and inhibits Smad2 and Erk MAPK activation	[54]
Mgat5	Modifies TGF-β receptor *N*-glycosylation and delays internalization	[117]
Occludin	Regulates TβRI localization and facilitates TGF-β-dependent tight junction dissolution	[118]
PDK1	Facilitates STRAP-induced stabilization of Smad7-receptor complexes	[119]
Rock2	Binds to TβRI receptor and promotes lysosomal degradation of the receptor	[120]

(continued)

Table 1 (continued)

Interacting proteins	Regulatory mechanism	References
SARA	Stabilizes the association of TGF-β receptors and R-Smads, and directs activated receptors towards clathrin-mediated endocytic pathways	[76, 89, 121]
Shc	Interacts with and is phosphorylated by TβRI, activates ERK-MAPK signaling	[6]
SIK	Associates with Smad7, TβRI, and Smurf2 and promotes proteasomal degradation of the receptor	[122, 123]
Smad7	Competes with R-Smads for TβRI receptor binding to prevent R-Smad activation, recruits ubiquitin E3 ligases to TβRI receptor for receptor degradation, and GADD34-PP1c to TβRI for receptor dephosphorylation	[18, 27, 30, 33, 122]
STRAP	Interacts with TβRI and TβRII receptors and stabilizes Smad7-receptor complexes	[75]
Tctex2β	Interacts with TβRII and inhibits TGF-β signaling when overexpressed	[74]
TLP	Associates with TβRI and TβRII to regulate the balance between Smad2 and Smad3 signaling	[124]
TMEPAI	Competes with TβRI receptor for R-Smad binding	[125]
Tollip	Interacts with ubiquitylated TβRI and promotes receptor degradation	[126]
TRAF6	Associates with activated TGF-β receptors, inhibits TGF-β-induced R-Smad activation and IL-2 blockage in T cells, and mediates p38 and JNK activation in response to TGF-β	[35, 80]
TrkC	Binds to TβRII, suppresses TβRII interaction with TβRI, and TGF-β signaling	[67]
VE-cadherin	Promotes active TGF-β receptor complex formation and enhances R-Smad phosphorylation	[65]
xIAP	Associates with TβRI and ubiquitylates TAK1 to activate NF-κB	[81]
YAP-65	Enhances Smad7 association to activated TβRI and inhibits R-Smad activation	[127]

1.3 Endocytic Trafficking of TGF-β Receptors

Endocytosis, which is often thought of as a mechanism to switch off signaling, plays an important role in regulating TGF-β signaling. In contrast to receptor tyrosine kinase receptors [82, 83], TGF-β receptors constitutively recycle in the absence or presence of ligand [84, 85]. Two major pathways mediate internalization of the TGF-β cell surface receptors: clathrin-mediated and non-clathrin-mediated endocytosis, with the latter mode being mediated by caveolin-1 and lipid rafts (Fig. 2). The dynamic partitioning of the receptors between these two internalization routes helps define the signaling outcome.

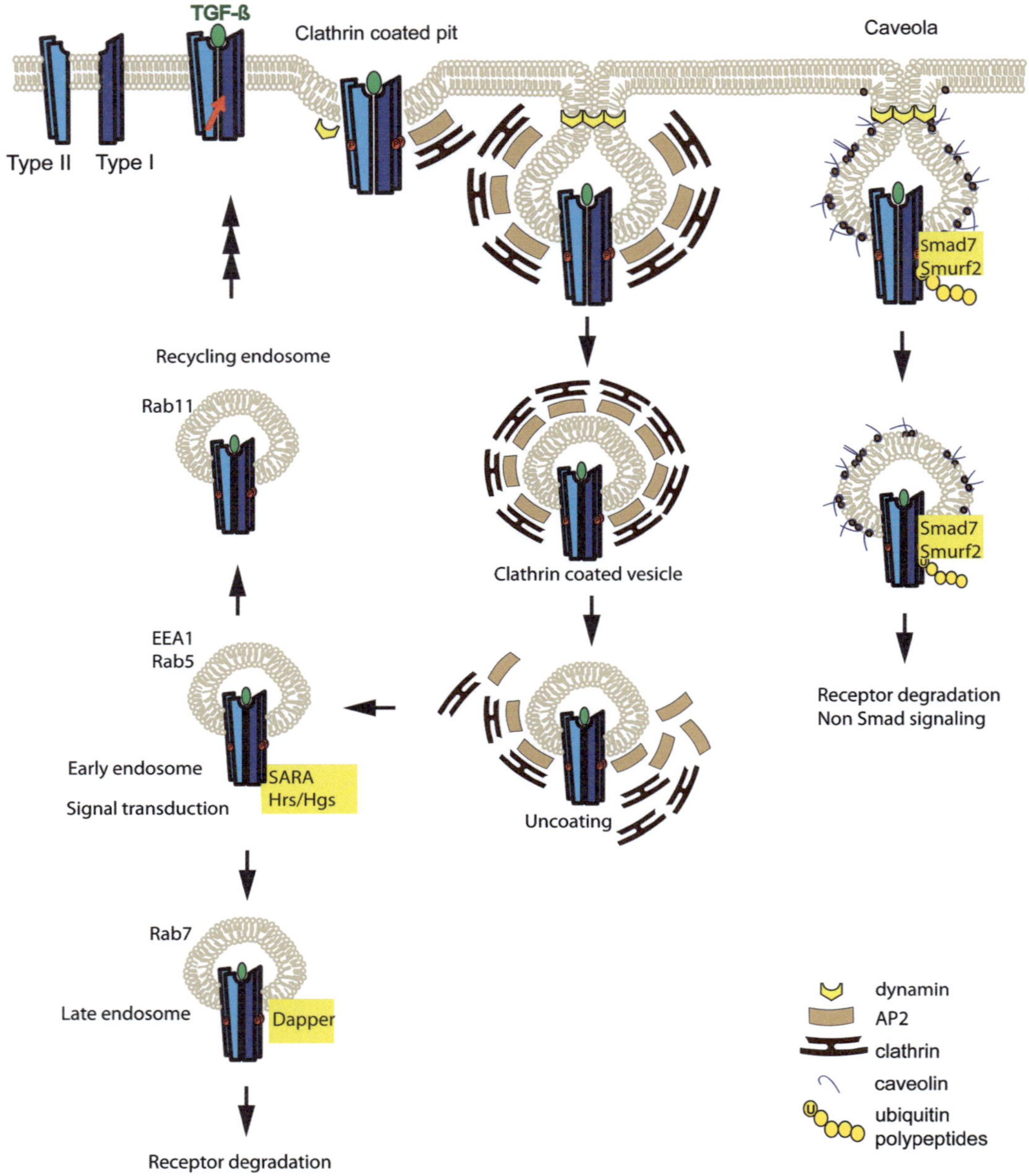

Fig. 2 Endocytic routing of TGF-β receptors. At the plasma membrane, tetrameric TGF-β receptor complexes are internalized by clathrin-mediated endocytosis or non-clathrin-mediated endocytosis mediated by caveolin-1 and lipid-rafts. In clathrin-mediated endocytosis, the ligand-bound TGF-β receptor complexes internalize through clathrin-coated pits and localize in EEA-1- and Rab5-positive early endosomes [85]. Clathrin is recruited to the endosomal invaginations and clathrin triskelial-AP2 protein complexes are formed enabling interactions with other endosomal mediators at the plasma membrane. The coated pits bud and pinch off from the membrane in a dynamin-dependent manner and give rise to clathrin-coated vesicles that then uncoat and fuse with EEA1- and Rab5-positive early endosomes. TGF-β receptors are recycled back to the plasma membrane, dependent on Rab11 [84], or are degraded using a Rab7-dependent mechanism [97, 98]. In lipid raft-mediated caveolar endocytosis, caveolin-1-dependent caveolae are formed in association with lipid rafts. The ligand-bound TGF-β receptor complexes are localized in invaginating caveolae that internalize dependent on dynamin. This route of internalization has been associated with TGF-β receptor degradation and non-Smad signaling [92]

1.3.1 Clathrin-Mediated Endocytosis

Clathrin-mediated endocytosis of the TGF-β receptors is mediated by a di-leucine motif in TβRI [86], and has been linked to TGF-β-induced Smad activation and transcription responses [85]. Additionally, receptor internalization in clathrin-coated endosomes also functions in TGF-β receptor recycling and degradation [84]. Inhibition of clathrin-mediated endocytosis, through targeted gene silencing or inactivation, prevents or decreases TGF-β-induced Smad activation [85, 87]. The process of clathrin-dependent TGF-β receptor internalization involves many proteins. The integral membrane protein AP2, a component of the clathrin adaptor complex, was shown to interact with the cytoplasmic domains of TβRII and TβRI [88]. TGF-β receptor binding to AP2 promotes clathrin cage assembly and dynamin-dependent endosome invagination [88]. Functional inactivation of dynamin's GTPase activity prevents clathrin-mediated endocytosis of the TGF-β receptors [85]. Subsequent entry of the TGF-β receptors in EEA1-positive and Rab5-positive endosomes further reinforces Smad signaling [85, 89]. In these endosomes, the adaptor proteins SARA and/or Hgs/Hrs promote Smad activation by the receptors [77]. Furthermore, neddylation of TβRII has also been shown to promote internalization of TβRII in EEA1-positive endosomes, and thus facilitates TGF-β-induced Smad signaling [47].

1.3.2 Caveolin- and Lipid Raft-Mediated Endocytosis

TGF-β receptors are also found in lipid raft microdomains and caveolin-positive endocytic compartments, i.e., caveolae [85, 90, 91]. TβRI interacts with caveolin-1, a key integral membrane protein in caveolae [90]. This interaction promotes TGF-β receptor turnover via Smad7-mediated recruitment of Smurfl or Smurf2 [85, 90]. Interfering with caveolin-1 expression using antisense RNA results in an enhanced TGF-β-induced Smad activation [90]. Similar results are seen with suppression of caveolae-mediated internalization by nystatin [85]. In contrast, expression of caveolin-1 in caveolin-1-deficient cells increases the turnover of TGF-β receptors [85]. The GPI-anchored cell surface protein CD109 may play a role in partitioning the receptor complexes between clathrin- and caveolin-mediated endocytosis. CD109 associates with caveolin and TβRI, and colocalizes with activated-TGF-β receptors complex in caveolae [69, 70]. CD109 enhances the association of ligand-activated TβRI with Smad7-Smurf2-complex and thus inhibits TGF-β signaling and responses [70, 71].

In addition to its role in promoting receptor degradation, lipid raft-caveolar endocytosis has also been linked to TGF-β-induced Erk MAPK signaling [92]. Whether activation of non-Smad signaling pathways requires differential endosomal receptor routing requires further studies.

1.3.3 Roles of GTPases and Other Proteins in TGF-β Receptor Trafficking

The differential routing of the internalized receptors is regulated by different Rab GTPases that localize in specific intracellular membrane structures, where they function as regulators of distinct steps in vesicle or membrane trafficking. Rab5 promotes internalization of the activated TGF-β receptors into early EEA-1-positive endosomes [85]. Like other GTPases, Rab5 is active in its GTP-bound form, and the generation of the activated form is promoted by a Rab-specific guanine nucleotide-exchange factor (GEF) such as RIN1 that promotes exchange of GDP for GTP. Consistent with the role of Rab5, RIN1 promotes TGF-β-induced Smad signaling [93].

Another small GTPase, Rab11, is involved in recycling the TGF-β receptors back to the plasma membrane [84]. The clathrin adaptor Dab2 (disabled-2) has been shown to associate with the type I and type II TGF-β receptors [94] and regulates TGF-β receptor trafficking between EEA1-positive early endosomes and Rab11-positive recycling endosomes [95]. Rap2, a Ras family GTPase, has also been implicated in enhancing the TGF-β response by promoting recycling of the receptors, preventing their degradation, and thus maintaining their levels on the cell surface [96]. Rap2 competes with Smad7 for binding to activated TβRI to prevent their degradation, and therefore contributes to the upregulation of Smad activation [96].

Lastly, Rab7 is known to direct the TGF-β receptors to lysosomes for degradation via a Dapper2 (Dpr2) binding complex. Dpr2, first identified as a Dishevelled (Dsh)-interacting protein, preferentially forms a complex with activated TβRI [97]. It negatively regulates TGF-β/nodal signaling by facilitating the transport of the activated receptors to lysosomes [97, 98]. Consistent with this notion, TGF-β-induced TβRI degradation is enhanced by Dpr2 and inhibited by lysosomal inhibitors, e.g., bafilomycin A or chloroquine, but not by the proteasome inhibitor MG132 [97].

1.4 Concluding Remarks

TGF-β signaling is regulated at the receptor level by different molecular events. Although posttranslational modifications determine receptor function, the exact functions and interplay of these modifications remain to be better defined. Also, the regulation of intracellular receptor routing by posttranslational modifications and direct or indirect protein associations requires much additional study. Understanding of how the TGF-β receptors are regulated is fundamental to gain valuable insights into the nature and complexity of the TGF-β response, and may aid in the development of TGF-β-based therapeutic approaches toward human diseases.

2 Materials

2.1 Biotin Labeling of Cell Surface TGF-β Receptors

2.1.1 Chemicals

TGF-β1 is from Humanzyme (Product No. HZ-1011). The chemical crosslinkers EZ-link sulfo NHS-LC-biotin (Product No. 21335), EZ-Link-NHS-SS-Biotin (Product No. 21441), and the NeutrAvidin beads (Product No. 29201) are from Thermo Scientific. See Blue Plus2 protein marker (Product No. LC5925), 4× LDS sample buffer (Product No. NP0008), and NuPAGE 10× MOPS-SDS running buffer (Product No. NP0001) are from Invitrogen. The Bio-Rad Protein dye assay (Product No. 500-0006) and β-mercaptoethanol (BME) electrophoresis purity reagent (Product No. 161-0710) are from Bio-Rad. Complete protease inhibitor cocktail tablets (Product No. 11 836 170 001) are from Roche. Bovine Serum Albumin (BSA) and Tween 20 are from Sigma.

2.1.2 Antibodies

Anti-TGF-β receptor type I and II are from Abcam (Product No. ab31013 and ab61213).

Anti-phospho-Smad2 and phospho-Smad3 antibodies are from Cell Signaling Technology (Product No. 3108 and No. 9520).

2.1.3 Solutions

PBS (sterile filtered-tissue culture grade): 0.1 g/L $CaCl_2$, 0.1 g/L $MgCl_2 \cdot 6H_2O$, 0.2 g/L KCl, 0.2 g/L KH_2PO_4, 8 g/L NaCl, 2.16 g/L $Na_2HPO_4 \cdot 7H_2O$.

MLB lysis buffer: 20 mM Tris–HCl pH 8.0, 200 mM NaCl, 10 mM NaF, 1 mM Na_3VO_4, 1 % NP-40, and Complete Protease inhibitor (Roche); MLB lysis buffer can be made in advance, aliquotted and stored at −20 °C.

MLB wash buffer: 20 mM Tris–HCl pH 8.0, 200 mM NaCl, 10 mM NaF, 1 mM Na_3VO_4, 1 % NP-40; MLB wash buffer can be made in advance and stored at 4 °C.

RIPA lysis buffer: 20 mM Tris–HCl pH. 8.0, 150 mM NaCl, 10 mM NaF, 1 mM Na_3VO_4, 1 % NP-40, 0.1 % SDS, 0.25 % $C_{24}H_{39}O_4Na$ (sodium deoxycholate), and complete Protease inhibitor (Roche); RIPA lysis buffer can be made in advance, aliquotted, and stored at −20 °C.

Stop buffer: 0.1 M Glycine in PBS.

Western blot sample buffer: 4× LDS from Invitrogen.

Western blot transfer buffer (4 L): 57.8 g Glycine, 12.0 g Tris base, 750 mL Methanol, top up to 4 L.

10× TBS (1 L): 24.23 g Tris base, 80.06 g NaCl, pH to 7.6 with HCl, top up to 1 L with ultrapure water.

1× TBS (1 L): 100 mL of 10× TBS, 900 mL of water.

1× TBST (1 L): 100 mL of 10× TBS, 900 mL of water, 1 mL of Tween 20.

Western blot membrane wash buffer: 1× TBST.

Western blot blocking: 5 % nonfat dry milk in 1× TBST, or 3 % sterile filtered BSA in 1× TBST.

NuPAGE 10× MOPS-SDS running buffer for Bis-Tris gel (Invitrogen).

2.2 Affinity Crosslinking of Cell Surface TGF-β Receptors Using Radiolabeled TGF-β

2.2.1 Chemicals

TGF-β is from Humanzyme (Product No. HZ-1011). Sephadex G25 is from GE Healthcare. Disuccinimidyl suberate (DSS) is from Thermo Scientific (Product No. 21555). Acetonitrile, CF_3COOH (trifluoroacetic acid), Na_3PO_4 (sodium phosphate), chloramine-T, *N*-acetyl-tyrosine, NaI (sodium iodide), NH_2CONH_2 (urea), CH_3CO_2H (acetic acid), suramin, $C_{12}H_{22}O_{11}$ (sucrose), $C_7H_7FO_2S$ (phenylmethylsulfonyl fluoride), EDTA, HEPES, bovine serum albumin (BSA) are from Sigma.

2.2.2 Radioactive

$Na^{125}I$, consult with your EH&S office for purchasing radioactive materials.

2.2.3 Solutions

Cold binding buffer: 128 mM NaCl, 5 mM KCl, 1.2 mM $CaCl_2$, 5 mM $MgSO_4$, 50 mM HEPES pH 7.4.

Elution buffer: 4 mM HCl, 100 mM NaCl, 0.1 mg/mL BSA.

Detachment buffer: 0.25 M sucrose, 10 mM Tris–HCl pH 7.4, 1 mM EDTA, 1 mM $C_7H_7FO_2S$ (phenylmethylsulfonyl fluoride).

2.3 Fluorescence-Based Detection of Cell Surface TGF-β Receptors

2.3.1 Antibodies

HA.11 Clone 16B12 is from Covance. Mouse or rabbit IgG is from Upstate Biotechnology. Alexa Fluor® 488 Goat Anti-Rabbit IgG (H+L) or Alexa Fluor® 568 Goat Anti-Rabbit IgG (H+L) are from Invitrogen-Molecular Probes (Product No. A-11034 and A11036). Anti-Flag M2 monoclonal antibody is from Sigma.

2.3.2 Solutions

PBS (sterile filtered, cell culture grade): 0.1 g/L $CaCl_2$, 0.1 g/L $MgCl_2 \cdot 6H_2O$, 0.2 g/L KCl, 0.2 g/L KH_2PO_4, 8 g/L NaCl, 2.16 g/L $Na_2HPO_4 \cdot 7H_2O$.

PBS-IHC (sterile filtered, cell culture grade): 0.2 g/L KH_2PO_4, 2.16 g/L $Na_2HPO_4 \cdot 7H_2O$, 0.2 g/L KCl, 8.0 g/L NaCl; this PBS contains no Mg and Ca and is used for immunohistochemistry.

FC (Flow Cytometry) staining buffer: PBS-IHC containing 3 % BSA, and 0.5 % sodium azide.

IHC fixation: 3.7 % paraformaldehyde in PBS-IHC (PFA) pH 7.4.

IHC blocking: 3 % BSA in PBS-IHC, filter through a standard 0.2 μm filter before use, and store at 4 °C.

PBT: PBS-IHC with 0.2 % Triton X.

2.4 Immuno-precipitation

2.4.1 Chemicals and Antibodies

TGF-β1 is from Humanzyme (Product No. HZ-1011). Dithiobis [succinimidylpropionate] (DSP) and EZ-link sulfo NHS-LC-biotin are from Thermo Scientific (Products No 22585 and 21335, respectively). Protein A or Protein G Sepharose beads are from GE Healthcare (Products No. 17-0618-01 and 10069350, respectively). 4× LDS is from Invitrogen (Product No. NP0008).

Anti-HA.11 clone 16B12 is from Covance (Product No MMS-101P). DMSO, Streptavidin-HRP antibody, rabbit anti-Flag (M2) monoclonal antibody, and anti-Myc (9E10) monoclonal antibody are from Sigma. Mouse or rabbit IgG are from Upstate NY Biotechnology.

2.4.2 Solutions

PBS (sterile filtered-tissue culture grade): 0.1 g/L $CaCl_2$, 0.1 g/L $MgCl_2{\cdot}6H_2O$, 0.2 g/L KCl, 0.2 g/L KH_2PO_4, 8 g/L NaCl, 2.16 g/L $Na_2HPO_4{\cdot}7H_2O$.

IP lysis buffer: 50 mM Tris–HCl pH 7.5, 150 mM NaCl, 2 mM EDTA, 0.1 % NP-40, Complete protease inhibitor tablet (Roche).

Wash IP lysis buffer: 50 mM Tris–HCl pH 7.5, 150 mM NaCl, 2 mM EDTA, 0.1 % NP-40.

Stop buffer: 0.1 M glycine in PBS.

Sample loading buffer: 4× LDS (Invitrogen).

3 Methods

This section assumes that the readers are familiar with common molecular biology practices, including cell culture, RT-PCR, SDS-PAGE, Western blot analysis, and basic immunohistochemistry (IHC). Prepare all solutions using ultrapure pyrogen-free cell culture grade water. Prepare and store all reagents at 4 °C, unless indicated otherwise. Follow waste disposal regulations when disposing biological and chemical wastes.

3.1 Biotin Labeling of Cell Surface TGF-β Receptors

Extracellular domains of cell surface proteins of intact cells can be modified by covalent attachment of a small (244 Da) molecule, biotin, conjugated with a linker arm. Avidin-based isolation of biotin-labeled proteins has found widespread use because it is consistent and reliable. We use this method to detect the level of TGF-β receptors at the cell surface [57, 99].

We use EZ-Link-Sulfo-NHS-LC-Biotin (with a 22.4 Å spacer arm), a water-soluble, cell impermeable reagent that enables easy and efficient biotinylation of proteins and any other primary amine-containing macromolecules in solution. The stable covalent bond is formed by *N*-hydroxysuccinimide (NHS) esters and the primary amine in the side chain of lysine (Lys, K) residues. Additional information accompanies the product protocol from Thermo Scientific. Chemical crosslinking using EZ-Link-Sulfo-NHS-LC-Biotin is not reversible. As an alternative crosslinker with the same spacer arm as EZ-Link Sulfo-NHS-LC-Biotin, EZ-Link-NHS-SS-Biotin provides cleavable and reversible crosslinking for cell surface protein biotinylation. The reversibility of the crosslinking is provided by a built-in disulfide bond that can be cleaved under reducing condition, thus releasing the biotin tag. Cleavable biotin is often used for internalization assays, or cell surface labeling of proteins with multiple transmembrane domains, e.g., the glucose transporters [100], that cannot be boiled. Other crosslinkers that are cell impermeable may also be used, but are not discussed.

Cell Surface Protein Biotinylation for Adherent Monolayer Cells

1. Cells at 80–90 % confluence in 60 mm dish are treated with TGF-β (0.5–1 ng/mL) or any soluble ligand or regulator for the desired amount of time. Following treatment, transfer and keep cells on ice for the duration of the biotinylation reaction (*see* **Note 1**).
2. Prepare a fresh dilution of EZ-Link-Sulfo-NHS-LC-Biotin or EZ-Link-NHS-SS-Biotin by weighing using an analytical balance, and dissolve at a final concentration of 0.5 mg/mL in PBS (*see* **Note 2**).
3. Rinse monolayer cell samples twice with ice-cold PBS on ice to remove media and any contaminating proteins. On the final rinse, aspirate as much PBS as possible. Remove remaining PBS by tilting the dish, allowing the buffer to collect to one side for convenient aspiration (*see* **Note 3**).
4. Add 0.5 mg/mL EZ-Link-Biotin solution to each dish and incubate for 30 min with gentle shaking on ice or at 4 °C. For 60 mm dish, add 4 mL of 0.5 mg/mL EZ-Link-Biotin solution.
5. Wash cells with ice-cold PBS to remove non-reacted biotin and incubate samples in stop buffer, 0.1 M Glycine in PBS for 20–40 min on ice to sequester the unbound biotin reagent.
6. Wash cells with PBS two or three times and remove all PBS using the tilting method.

Preparation of Biotin-Labeled Cell Lysates

7. Add 200 μL of MLB lysis buffer onto the cells (60 mm dish), scrape to collect cells, and transfer cell lysate to 1.5 mL microcentrifuge tube. Pipet up and down several times to break cell clumps. Repeat with each of the remaining sample dishes. Rotate the microcentrifuge tubes using a rotator for 15 min in cold room.
8. Spin samples at 10,000 × *g* or at max speed in a tabletop centrifuge for 15 min at 4 °C to remove cell debris. Carefully remove the supernatants without disturbing the pellet into a prechilled microcentrifuge tube.
9. Measure protein concentrations using Bio-Rad Protein Dye Assay and adjust the samples of lysates with MLB buffer to ensure the same amount of protein input.

Affinity Purification and Examination of Biotin-Labeled Cell Surface Proteins

10. Prepare the NeutrAvidin beads slurry at 20 μL per sample, by washing three times in lysis buffer. Spin down beads by centrifugation using low speed at 2500×*g* or 5000 rpm using a tabletop centrifuge at 4 °C (*see* **Note 4**).
11. Incubate a sample of 250 μg of protein with 20 μL of NeutrAvidin beads overnight at 4 °C (*see* **Note 5**).
12. Wash the beads with MLB wash buffer (MLB lysis buffer without Complete Protease inhibitor) three times with rotation at 4 °C with 10 min intervals in between washes.
13. Add 4× LDS sample (with 0.1× volume of BME added) buffer and heat samples at 95 °C for 5 min. Quickly put samples on ice (*see* **Note 6**).
14. Centrifuge at max speed for 1 min using a tabletop centrifuge to collect residual sample after the heating.
15. Run SDS-PAGE gel. When using EZ-Link-NHS-SS-Biotin for cell surface protein biotinylation, ensure that the sample buffer contains 100 mM BME or 50 mM DTT to cleave and remove the biotin from the labeled proteins.
16. Following SDS-PAGE, transfer the separated proteins to nitrocellulose membrane using Western transfer buffer and detect the TGF-β receptors by immunoblotting using appropriate antibodies.

3.2 Affinity Crosslinking of Cell Surface TGF-β Receptors Using Radiolabeled TGF-β

Cell surface TGF-β receptors can be visualized by binding with radiolabeled TGF-β, followed by chemical crosslinking of radiolabeled TGF-β to the receptors, SDS-PAGE analysis of the ligand-bound receptors, and autoradiography. Typically, ^{125}I-labeled TGF-β is used in these experiments [101, 102]. This method has increasingly lost appeal because of concerns related to the use of ^{125}I, yet is still the only method to visualize receptors available for ligand binding. Consequently, ^{125}I-TGF-β is no longer commercially available but can still be generated by specialized contractors, or by the investigator.

3.2.1 Preparation of ^{125}I-TGF-β1

The method to radio-iodinate TGF-β is based on a modified chloramine-T method [101, 102].

1. Rehydrate 1 μg of purified lyophilized-TGF-β1 in 10 μL of solution containing 30 % acetonitrile and 0.1 % trifluoroacetic acid, and vortex vigorously for several hours before use. Adjust to pH 7.5 by the addition of 5 μL of 0.15 M $NaPO_4$ pH 7.5.
2. Add 0.5 mCi Na^{125}I to the solubilized TGF-β1, follow by sequential addition of 5 μL of 0.01 mg/mL chloramine-T for 2, 1.5, and 1 min (*see* **Note 7**).

3. Stop the reaction 1 min after the final chloramine-T addition with 20 μL of 20 mM *N*-acetyl-tyrosine, followed by 200 μL of 60 mM NaI and 200 μL of 8 M urea-1 M acetic acid. Incubate the reaction mixture for 3–5 min at room temperature.
4. Separate the radiolabeled ^{125}I-TGF-β1 from the unbound ^{125}I on a pre-packed Sephadex G25 column (0.7 cm × 20 cm) that has been prepared by washing with 5 column volumes elution buffer (4 mM HCl, 100 mM NaCl, 0.1 mg/mL BSA). Collect 500 μL fractions and measure radioactivity. Radiolabeled ^{125}I-TGF-β elutes in the void volume, and pool 3–4 peak fractions. ^{125}I-TGF-β1 is diluted to a final concentration of 0.25 ng/10 μL, assuming a 25 % recovery.
5. Aliquots can be stored at −20 °C and safely used for 30 days.

3.2.2 Crosslinking of ^{125}I-TGF-β to Cell Surface Receptors and Analysis

1. Cells are grown to confluence (10^6 cells/well) in 3 cm diameter wells of six-well plates, and subjected to the desired treatment.
2. Remove media and wash with serum-free medium, incubate cells for 12 h at 4 °C with serum-free medium containing 1 % suramin, which removes ligand from the receptor with about >90 % efficiency. All subsequent steps are carried out on ice.
3. Following two washes with ice-cold PBS containing 0.1 % BSA, incubate samples for 10 min with cold binding buffer (128 mM NaCl, 5 mM KCl, 1.2 mM $CaCl_2$, 5 mM $MgSO_4$, 50 mM HEPES pH 7.4), containing 2 mg/mL BSA.
4. Remove binding buffer and replace with 1 mL of binding buffer containing 1 ng ^{125}I-TGF-β to each well. Allow cells to bind the radiolabelled TGF-β for 4 h with gentle rocking at 4 °C. In control experiments, excess unlabelled TGF-β is added to compete out the radiolabeled TGF-β binding.
5. Aspirate binding buffer with remaining unbound TGF-β, and wash cells once with ice-cold BSA-free binding buffer.
6. To each well, add 1 mL of 150 mM disuccinimidyl suberate (DSS) in ice-cold BSA-free binding buffer. Prepare a fresh stock solution of DSS in DMSO, and dilute to 150 mM final concentration immediately before use in BSA-free binding buffer.
7. Incubate with the DSS solution for 15 min on ice with gentle rocking.
8. Wash cells twice in 2 mL of ice-cold detachment buffer (0.25 M sucrose, 10 mM Tris–Cl pH 7.4, 1 mM EDTA, 1 mM phenylmethylsulfonyl fluoride).
9. Harvest cells by scraping them in detachment buffer, transfer to microfuge tubes, and spin at 10,000 × *g* for 30 s.

10. Suspend cells in reducing electrophoresis sample buffer, homogenize and lyse cells by repeatedly passing them through a syringe to decrease viscosity.
11. Run samples in SDS-PAGE then fix, dry gel/s without staining.
12. Analyze by autoradiography.

3.3 Fluorescence-Based Detection of Cell Surface TGF-β Receptors

3.3.1 Detection of Cell Surface TGF-β Receptors by Indirect Immunofluorescence Staining

Immunofluorescence can be used to visualize TGF-β receptors or other proteins of interest. Combined with confocal or other types of high-resolution microscopy, immunofluorescence techniques have been proven to be very powerful, enabling scientists to examine and detect changes in subcellular protein localization, and to assess protein colocalization. Standard immunofluorescence combines the interaction of an unconjugated primary antibody to the protein of interest, with the use of a secondary fluorophore-conjugated antibody that recognizes the primary antibody.

Fixation

1. Treat cells with TGF-β in DMEM containing 1 % FBS for 30 min or any soluble ligand or regulator for the desired amount of time (*see* **Note 8**).
2. Remove media from cells and wash cells twice in ice-cold PBS-IHC.
3. Fix cells in 3.7 % PFA for 15 min at room temperature with gentle rocking.
4. Wash cells three times for 5 min each in PBS-IHC.

   Permeabilization

   If the primary antibody recognizes an extracellular epitope of your protein then permeabilization may not be necessary to detect cell surface protein. Check your antibody data sheet before starting the experiment. Permeabilization is required to detect target protein distributed intracellularly. The process provides access to intracellular or organellar epitopes that might be otherwise inaccessible. The TβRI antibody from Abcam recognizes peptides that reside in the cytoplasmic domain, thus permeabilization is necessary.

   (a) Permeabilize the cell membrane by incubating the cells for 10 min in PBT (PBS-IHC with 0.2 % Triton X) at RT with gentle rocking, then fix for an additional 5 min in 3.7 % PFA at room temperature.

   (b) Wash cells three times for 5 min each in PBS-IHC. Proceed to **step 5**.

Immunostaining

5. Add IHC blocking solution (3 % BSA in PBS-IHC) and incubate for 2 h at room temperature, or overnight at 4 °C.

6. Add primary antibody (TβRI, Abcam 1:50 in IHC blocking solution), and incubate overnight at 4 °C (*see* **Note 9**).
7. Rinse in PBS-IHC five times for 10 min intervals with gently rocking.
8. Incubate with fluorophore-conjugated secondary antibody (Alexa 488/568 Fluor, Invitrogen 1:400 in PBS-IHC) for 2 h at room temperature or overnight at 4 °C.
9. Rinse in PBS-IHC three to five times with 10 min intervals with gentle rocking.
10. Mount the fixed cells in Prolong Gold Antifade reagent with DAPI (Invitrogen) and seal with nail polish.
11. Analyze or store slides in dark at 4 °C.

3.3.2 Detection of Cell Surface TGF-β Receptors by Flow Cytometry

Flow cytometry allows relative quantification of receptor levels. This method is also used to isolate cell populations with a definable phenotype. Like for immunofluorescence or other immunostaining methods, the quality of the antibody is the most important determinant of the signal specificity and outcome of the experiment. Since the cells are kept intact for analyses, the antibody needs to recognize an extracellular epitope of the transmembrane receptor. Alternatively, a mutant or modified receptor expressing an extracellular tagged protein or fluorophore can be used.

Here we describe a fixed cell protocol using single-color analysis to quantify the levels of TGF-β receptors at the plasma membrane in cells expressing an N-terminally HA-tagged TβRI, using an analyzer.

Preparation of Cells and Sample Examination

1. Treat cells by adding fresh media containing TGF-β or any soluble ligand or treatment for the desired amount of time. After treatment, samples are washed in ice-cold FC staining buffer, which contains sodium azide to prevent internalization and any further changes in the levels of cell surface receptors. All subsequent steps prior to fixation are done at 4 °C.
2. Detach cells in PBS containing 0.5 mM EDTA, and collect cells by centrifugation at 350 × *g* for 5 min at 4 °C.
3. Resuspend cell pellet in FC staining buffer at a concentration of 10×10^6 cells/mL.
4. Add 100 μL of cell suspension into a 5 mL (15 × 75 mm) polystyrene round bottom tube.
5. Add the primary antibody at a 1:200 dilution (HA-mouse) or at an appropriately titrated concentration. To detect HA-tagged TβRI receptor, we use a mouse HA antibody or control mouse IgG at 0.5 mg/mL. Incubate for 30 min at 4 °C.

6. Wash off the unbound primary antibody by adding FC staining buffer to fill the tube, and centrifuge at 350 × *g* for 5 min. Repeat twice and resuspend in FC staining buffer.
7. Add the secondary antibody (Alexa 488/568 at 1:400 dilution from Invitrogen) (*see* **Note 10**).
8. Remove the unbound secondary antibody by adding FC staining buffer to fill the tube, centrifuge at 350 × *g* for 5 min. Repeat washing twice using FC-staining buffer.
9. After the last centrifugation, resuspend cells for fixation in 1 % paraformaldehyde/PBS-IHC (PFA) for 15 min on ice. To prevent cell aggregation, briefly vortex cells just after adding the fixation buffer.
10. After fixation, wash the cells by centrifugation twice in FC buffer staining buffer.
11. Filter cells through a 40–50 μm mesh. Filtering the cells assures for a homogenous single cell population.
12. The cells are ready for analysis (*see* **Notes 11** and **12**).

3.4 Immunoprecipitation

This method can be used to isolate or concentrate proteins of interest and analyze protein–protein interactions. Co-immunoprecipitated proteins, endogenously or ectopically expressed, can be identified by western blotting or by sequencing a purified protein extracted following PAGE. Transient or weak protein interactions can be identified using chemical crosslinkers (DSP, biotin, etc.), which are described in the next section.

The protocols described are used for adherent cells in monolayer that are grown in 10 cm polystyrene dish (BD Product No. 353003). If you are scaling up or down the reaction, adjust the volume of reagents accordingly.

3.4.1 Preparation of Cell Lysates

1. Treat cells by adding fresh media containing TGF-β or any other molecule of interest for the desired amount of time. Then place cell culture plates on ice. Use ice-cold reagents for rest of the experiment, unless otherwise stated.
2. To harvest cells, aspirate media and rinse cells once or twice with ice-cold PBS.
3. Remove as much as PBS using the tilting method and add 1 mL of ice-cold IP lysis buffer to each 10-cm diameter dish sample. Incubate the plates on ice for 5 min. For assays using transfected cells, add 0.3 mL of IP cell lysis buffer to each well of a 6-well plate (*see* **Note 13**).
4. Scrape cells off the plates and transfer to microcentrifuge tubes. Keep on ice.
5. Spin samples for 10 min at 10,000 × *g* at 4 °C using a microcentrifuge, and transfer the supernatants to new tubes.

3.4.2 Pre-clearing (Optional)

Pre-clearing the cell lysate helps to reduce nonspecific protein binding to irrelevant protein antigen, agarose, or Sepharose beads. Adding a pre-clearing step may reduce background. It is usually performed if an unwanted protein contaminant interferes with visualization of the protein of interested.

6. To the cell lysate, add either a control antibody or IgG from the same species as the primary antibody, and either Protein A or Protein G agarose beads (20 μL of 50 % bead slurry), depending on whether protein A or protein G will be used for immunoprecipitation (*see* **Note 14**).
7. Incubate samples at 4 °C for 30–60 min on ice.
8. Spin samples using a microcentrifuge for 10 min at 2500 ×*g*, 4 °C. Transfer the supernatants to new tubes.

3.4.3 Immuno-precipitation and Sample Examination

9. Add primary antibody to 500 μL of cell lysate and incubate with gentle rocking overnight at 4 °C. For assays using transfected cells that generally express more of the protein of interest, the incubation time can be decreased to 2–4 h at 4 °C.
10. Add pre-washed protein A or G Sepharose beads (20 μL of 50 % bead slurry). Incubate with gentle rocking for 2–4 h at 4 °C. Use wide bore pipette to prevent damage to the beads.
11. Spin the tubes in a microcentrifuge for 3 min at 2500 ×*g* at 4 °C. Wash pellet three to five times in wash IP lysis buffer at low speed 2500 ×*g* at 4 °C. Keep on ice during washes.
12. Resuspend the pellet with 20 μL of 4× LDS sample buffer. Stir, then spin the tubes in a microcentrifuge for 30 s.
13. Heat the samples to 95 °C for 5–10 min and spin the tubes in a microcentrifuge for 1 min at 10,000 ×*g*.
14. Load the samples (5–10 μL) on SDS-PAGE gel (12–15 %).
15. Analyze samples by Western blotting using antibodies of interest (*see* **Note 15**).

3.5 Immuno-precipitation Using Chemical Crosslinking

3.5.1 Immuno-precipitation Using Reversible Crosslinker DSP

Many physiological protein interactions are transient, with off-rates on the order of seconds, or are disrupted under conditions of ionic strength and detergent used in immunoprecipitations. To overcome this problem, direct protein associations or proteins in complexes can be covalently stabilized using chemical crosslinkers, prior to immunoprecipitation. Various crosslinkers with different spacer arms are available to stabilize protein interactions. We suggest the Crosslinker Application Guide on the Thermo Scientific website to find the best suited chemical crosslinker for your needs. In the protocol below, we use dithiobis [succinimidylpropionate] (DSP) also known as Lomant's reagent or DTSP, available from Thermo Scientific. DSP is an amine-reactive crosslinker that is cell-permeable, crosslinks Lys residues separated by 11.2 Å, and is cleavable, e.g., prior to SDS-PAGE, using reducing agents.

Chemical Crosslinking

1. Treat cells with TGF-β or any other molecule of interest for the desired amount of time.
2. Aspirate media and wash cells with ice-cold PBS three times.
3. Prepare 10 mM DSP in fresh DMSO, and dilute to 1 mM final concentration with ice-cold PBS immediately before use (*see* **Note 16**).
4. Add 1 mM DSP to each 10 cm dish sample on ice, and incubate for 10–30 min with gentle shaking on ice. For 10-cm dish, add 6–8 mL of 1 mM DSP solution (*see* **Note 17**).
5. Aspirate DSP and wash cells with Stop buffer (0.1 M Glycine in PBS) twice for 20 min each. Keep on ice.
6. Wash cells with PBS once and remove as much PBS as possible using the tilting method.

Preparation of Cell Lysates

7. Add 1 mL of ice-cold IP lysis buffer to each 10-cm diameter plate, and incubate the plates on ice for 5 min. For assays using transfected cells, add 0.3 mL IP lysis buffer to each well of a 6-well plate.
8. Scrape cells from the plates and transfer to 1.5 mL microcentrifuge tubes. Keep on ice.
9. Spin samples at 10,000 × *g* for 10 min at 4 °C to remove cell debris.
10. Proceed to immunoprecipitation and sample examination as detailed in Subheading 4. Immunoprecipitation.

3.5.2 Immunoprecipitation Following Biotinylation

Biotinylation can also be used in co-immunoprecipitations to detect proteins that are in complex with cell surface TGF-β receptor [57].

Cell Surface Protein Biotinylation

Follow **steps 1–6** in Subheading 3.1. Cell surface protein biotinylation for adherent monolayer cells, with changes as follows:

1. Incubate cells in EZ-link Sulfo-NHS-LC-Biotin (0.5 mg/mL in PBS) solution on ice or at 4 °C for 60 min instead of 30 min (*see* **Note 18**).

Preparation of Cell Lysates, Immunoprecipitation and Sample Examination

2. Add 1 mL of ice-cold IP lysis buffer to each 10-cm diameter plate, and incubate the plates on ice for 5 min. For assays using transfected cells, add 0.6 mL of IP lysis buffer to each well of a 6-well plate.
3. Scrape cells off the plates and transfer to 1.5 mL microcentrifuge tubes. Keep on ice.
4. Spin samples at 10,000 × *g* for 10 min at 4 °C to remove cell debris.

5. Proceed to immunoprecipitation and sample examination as detailed in Subheading 3.4. Immunoprecipitation.
6. Analyze samples by Western blotting using Streptavidin-HRP (Sigma) antibody to detect cell surface TGF-β receptors or other antibodies of interest.

4 Notes

1. Ice-cold reagents are used unless otherwise stated.
2. Biotin reagents, according to the manufacturer's instruction, need to be kept in a moisture-free container and equilibrated to room temperature before use. We usually take it from the freezer 30–60 min before use. Dissolve EZ-Link-biotin immediately before use. DO NOT prepare a stock solution of Biotin, as the NHS-ester moiety is readily hydrolyzed and will be non-reactive. AVOID buffers that contain primary amines (Tris or Glycine) as they will compete with your protein samples.
3. Pipet reagents carefully down the side of the dish/well, i.e., not directly onto the cells, to prevent cells from detaching.
4. Pre-wash [(n+1) samples×20 μL] of volume of the Neutravidin beads. For 8 samples, draw 180 μL of stock solution of NeutrAvidin beads using wide bore pipette to prevent rupturing.
5. The expression levels of TGF-β receptors differ among different cell lines. Adjust the total protein input accordingly. The amount above applies to NMuMG cells, NRK-52E cells, and MEFs.
6. As an alternative to boiling, samples biotinylated using cleavable/reversible biotin with a disulfide bond in the spacer arm, can be incubated in 4× LDS with addition of 50 mM DTT or 100 mM BME for 2 h at room temperature or 30 min at 60 °C.
7. When radiolabeling TGF-β2 or TGF-β3, use 0.1 mg/mL chloramine-T instead of 0.01 mg/mL chloramine-T recommended for TGF-β1.
8. Cells can be labeled in vivo for immunofluorescence, by incubating intact, live cells with primary antibody conjugated with fluorophore.
9. For double labeling, a second primary antibody can be added at the same time with the first antibody, assuming that each antibody is raised in different animal species. Also, it is important to test whether the two primary antibodies can be treated the same. The use of corresponding secondary antibodies coupled to different fluorophores will then distinguish the proteins of interest.
10. Different fluorophores can be used and adjusted to match the cytometer optics.

11. It is recommended that fixed samples be run on the cytometer within 1 week, preferably the same day or next day after fixation. Autofluorescence tends to increase and sample quality generally declines with long-term storage of fixed samples.
12. To isolate cells expressing the tagged protein for further experiments, use ice-cold FC staining buffer without sodium azide. The above protocol also can be applied for cell sorting with the exclusion of cell fixation step.
13. Adjusting the salt and detergent concentrations in the cell lysis buffer may be necessary. Increasing the salt or detergent concentration may decrease background signal, but may also disrupt weak interactions.
14. Use Protein A for rabbit IgG pull down and Protein G for mouse IgG pull down.
15. If the protein of interest has a molecular weight close to 50 kD, we recommend using light-chain specific secondary antibodies for western blotting to minimize masking of signal by the heavy chain IgG.
16. When diluting with PBS, the 1 mM DSP solution turns cloudy with occasional crystal formation.
17. The DSP concentration and time of incubation with DSP may be empirically adjusted, depending on the nature of the proteins and protein interactions involved.
18. It is possible to combine chemical crosslinking using DSP with cell surface protein biotinylation although this approach reduces the efficiency of either protein modification, since both treatments target lysine residues.

Acknowledgments

We thank Michael Kissner for discussion on the flow cytometry experiments. Research by the authors is sponsored by a Juvenile Diabetes Research Fund postdoctoral fellowship to E.B., an American Heart Association scientist development award to J.X., and NIH RO1 grants CA63101 and CA136690 to R.D.

References

1. Miyazono K, Derynck R (2008) The TGF-[beta] family. Cold Spring Harbor Laboratory Press, Cold Spring Harbor, NY
2. Massague J (2012) TGFbeta signalling in context. Nat Rev Mol Cell Biol 13(10): 616–630
3. Feng XH, Derynck R (2005) Specificity and versatility in TGF-beta signaling through Smads. Annu Rev Cell Dev Biol 21:659–693
4. Shi Y, Massagué J (2003) Mechanisms of TGF-beta signaling from cell membrane to the nucleus. Cell 113(6):685–700
5. Lawler S, Feng X-H, Chen R-H, Maruoka EM, Turck CW, Griswold-Prenner I et al (1997) The type II transforming growth factor-b receptor autophosphorylates not only on serine and threonine but also on tyrosine residues. J Biol Chem 272(23):14850–14859

6. Lee MK, Pardoux C, Hall MC, Lee PS, Warburton D, Qing J et al (2007) TGF-beta activates Erk MAP kinase signalling through direct phosphorylation of ShcA. EMBO J 26(17):3957–3967, PMCID: 1994119
7. Derynck R, Jarrett JA, Chen EY, Eaton DH, Bell JR, Assoian RK et al (1985) Human transforming growth factor-beta complementary DNA sequence and expression in normal and transformed cells. Nature 316(6030): 701–705
8. Rebbapragada A, Benchabane H, Wrana JL, Celeste AJ, Attisano L (2003) Myostatin signals through a transforming growth factor -like signaling pathway to block adipogenesis. Mol Cell Biol 23(20):7230–7242
9. Lee SJ, Reed LA, Davies MV, Girgenrath S, Goad ME, Tomkinson KN et al (2005) Regulation of muscle growth by multiple ligands signaling through activin type II receptors. Proc Natl Acad Sci U S A 102(50):18117–18122, PMCID: 1306793
10. Ehrlich M, Horbelt D, Marom B, Knaus P, Henis YI (2011) Homomeric and heteromeric complexes among TGF-beta and BMP receptors and their roles in signaling. Cell Signal 23(9):1424–1432
11. Luo K, Lodish HF (1997) Positive and negative regulation of type II TGF-b receptor signal transduction by autophosphorylation on multiple serine residues. EMBO J 16(8): 1970–1981
12. Derynck R, Feng XH (1997) TGF-beta receptor signaling. Biochim Biophys Acta 1333(2):105–150
13. Derynck R, Zhang YE (2003) Smad-dependent and Smad-independent pathways in TGF-beta family signalling. Nature 425(6958):577–584
14. Moustakas A, Heldin CH (2005) Non-Smad TGF-beta signals. J Cell Sci 118(Pt 16): 3573–3584
15. Galliher AJ, Schiemann WP (2007) Src phosphorylates Tyr284 in TGF-beta type II receptor and regulates TGF-beta stimulation of p38 MAPK during breast cancer cell proliferation and invasion. Cancer Res 67(8):3752–3758
16. Souchelnytskyi S, Dijke P, Miyazono K, Heldin CH (1996) Phosphorylation of Ser165 in TGF-13 type I receptor modulates TGF-fl-induced cellular responses. EMBO J 15(22):6231–6240
17. Bennett D, Alphey L (2002) PP1 binds Sara and negatively regulates Dpp signaling in Drosophila melanogaster. Nat Genet 31(4): 419–423
18. Shi W, Sun C, He B, Xiong W, Shi X, Yao D et al (2004) GADD34-PP1c recruited by Smad7 dephosphorylates TGFbeta type I receptor. J Cell Biol 164(2):291–300, PMCID: 2172339
19. Batut J, Schmierer B, Cao J, Raftery LA, Hill CS, Howell M (2008) Two highly related regulatory subunits of PP2A exert opposite effects on TGF-beta/Activin/Nodal signalling. Dev 135(17):2927–2937
20. Griswold-Prenner I, Kamibayashi C, Maruoka EM, Mumby MC, Derynck R (1998) Physical and functional interactions between type I transforming growth factor b receptors and Ba, a WD-40 repeat subunit of phosphatase 2A. Mol Cell Biol 18(11):6595–6606
21. Levy L, Hill CS (2006) Alterations in components of the TGF-beta superfamily signaling pathways in human cancer. Cytokine Growth Factor Rev 17(1-2):41–58
22. Grady WM, Markowitz SD (2002) Genetic and epigenetic alterations in colon cancer. Annu Rev Genomics Hum Genet 3:101–128
23. Galliher-Beckley AJ, Schiemann WP (2008) Grb2 binding to Tyr284 in TbetaR-II is essential for mammary tumor growth and metastasis stimulated by TGF-beta. Carcinogenesis 29(2):244–251, PMCID: 2615477
24. Weissman AM, Shabek N, Ciechanover A (2011) The predator becomes the prey: regulating the ubiquitin system by ubiquitylation and degradation. Nat Rev Mol Cell Biol 12(9):605–620, PMCID: 3545438
25. Laney JD, Hochstrasser H (1999) Substrate targeting in the ubiquitin system. Cell 97(4):427–430
26. Ebisawa T, Fukuchi M, Murakami G, Chiba T, Tanaka K, Imamura T et al (2001) Smurfl interacts with transforming growth factor-beta type I receptor through Smad7 and induces receptor degradation. J Biol Chem 276(16):12477–12480
27. Kavsak P, Rasmussen RK, Causing CG, Bonni S, Zhu H, Thomsen GH et al (2000) Smad7 binds to Smurf2 to form an E3 ubiquitin ligase that targets the TGFb receptor for degradation. Mol Cell 6:1365–1375
28. Nakao A, Imamura T, Souchelnytskyi S, Kawabata M, Ishisaki A, Oeda E et al (1997) TGF-b receptor-mediated signalling through Smad2, Smad3 and Smad4. EMBO J 16(17): 5353–5362
29. Miyazono K (2000) TGF-beta signaling by Smad proteins. Cytokine Growth Factor Rev 11(1-2):15–22
30. Hayashi H, Abdollah S, Qiu Y, Cai J, Xu Y-Y, Topper JN et al (1997) The MAD-related protein Smad7 associates with the TGF-beta receptor and functions as an antagonist of TGF-beta signaling. Cell 89(June):1165–1173

31. Ogunjimi AA, Briant DJ, Pece-Barbara N, Le Roy C, Di Guglielmo GM, Kavsak P et al (2005) Regulation of Smurf2 ubiquitin ligase activity by anchoring the E2 to the HECT domain. Mol Cell 19(3):297–308
32. Kuratomi G, Komuro A, Goto K, Shinozaki M, Miyazawa K, Miyazono K et al (2005) NEDD4-2 (neural precursor cell expressed, developmentally down-regulated 4-2) negatively regulates TGF-β (transforming growth factor-β) signalling by inducing ubiquitin-mediated degradation of Smad2 and TGF-β type I receptor. Biochem J 386:461–470
33. Komuro A, Imamura T, Saitoh M, Yoshida Y, Yamori T, Miyazono K et al (2004) Negative regulation of transforming growth factor-beta (TGF-beta) signaling by WW domain-containing protein 1 (WWP1). Oncogene 23(41):6914–6923
34. Pickart C (2001) Mechanisms underlying ubiquitination. Annu Rev Biochem 70:503–533
35. Sorrentino A, Thakur N, Grimsby S, Marcusson A, von Bulow V, Schuster N et al (2008) The type I TGF-beta receptor engages TRAF6 to activate TAK1 in a receptor kinase-independent manner. Nat Cell Biol 10(10):1199–1207
36. Mu Y, Sundar R, Thakur N, Ekman M, Gudey SK, Yakymovych M et al (2011) TRAF6 ubiquitinates TGFbeta type I receptor to promote its cleavage and nuclear translocation in cancer. Nat Commun 2:330, PMCID: 3113296
37. Inui M, Manfrin A, Mamidi A, Martello G, Morsut L, Soligo S et al (2011) USP15 is a deubiquitylating enzyme for receptor-activated SMADs. Nat Cell Biol 13(11):1368–1375
38. Eichhorn PJ, Rodon L, Gonzalez-Junca A, Dirac A, Gili M, Martinez-Saez E et al (2012) USP15 stabilizes TGF-beta receptor I and promotes oncogenesis through the activation of TGF-beta signaling in glioblastoma. Nat Med 18(3):429–435
39. Zhang L, Zhou F, Drabsch Y, Gao R, Snaar-Jagalska BE, Mickanin C et al (2012) USP4 is regulated by AKT phosphorylation and directly deubiquitylates TGF-β type I receptor. Nat Cell Biol 14(7):717–726
40. Zhang X, Berger FG, Yang J, Lu X (2011) USP4 inhibits p53 through deubiquitinating and stabilizing ARF-BP1. EMBO J 30(11):2177–2189, PMCID: 3117646
41. Kerscher O, Felberbaum R, Hochstrasser M (2006) Modification of proteins by ubiquitin and ubiquitin-like proteins. Annu Rev Cell Dev Biol 22:159–180
42. Gareau JR, Lima CD (2010) The SUMO pathway: emerging mechanisms that shape specificity, conjugation and recognition. Nat Rev Mol Cell Biol 11(12):861–871, PMCID: 3079294
43. Seeler JS, Dejean A (2003) Nuclear and unclear functions of SUMO. Nat Rev Mol Cell Biol 4(9):690–699
44. Hay RT (2005) SUMO: a history of modification. Mol Cell 18(1):1–12
45. Johnson ES (2004) Protein modification by SUMO. Annu Rev Biochem 73:355–382
46. Kang JS, Saunier EF, Akhurst RJ, Derynck R (2008) The type I TGF-beta receptor is covalently modified and regulated by sumoylation. Nat Cell Biol 10(6):654–664, PMCID: 2649123
47. Zuo W, Huang F, Chiang YJ, Li M, Du J, Ding Y et al (2013) c-Cbl-mediated neddylation antagonizes ubiquitination and degradation of the TGF-beta type II receptor. Mol Cell 49(3):499–510
48. Wells RG, Yankelev H, Lin HY, Lodish HF (1997) Biosynthesis of the type I and type II TGF-b receptors. J Biol Chem 272(17):11444–11451
49. Luga V, McLean S, Le Roy C, O'Connor-McCourt M, Wrana JL, Di Guglielmo GM (2009) The extracellular domain of the TGFbeta type II receptor regulates membrane raft partitioning. Biochem J 421(1):119–131
50. Ebner R, Chen RH, Shum L, Lawler S, Zioncheck TF, Lee A et al (1993) Cloning of a type I TGF-beta receptor and its effect on TGF-beta binding to the type II receptor. Science 260(5112):1344–1348
51. Varki A (1993) Biological roles of oligosaccharides: all of the theories are correct. Glycobiology 3(2):97–130
52. Spiro RG (2002) Protein glycosylation:nature, distribution, enzymatic formation, and disease implications of glucopeptide bonds. Glycobiology 12(4):43–56
53. Kim YW, Park J, Lee HJ, Lee SY, Kim SJ (2012) TGF-beta sensitivity is determined by N-linked glycosylation of the type II TGF-beta receptor. Biochem J 445(3):403–411, PMCID: 3462611
54. Huang S, Holzel M, Knijnenburg T, Schlicker A, Roepman P, McDermott U et al (2012) MED12 controls the response to multiple cancer drugs through regulation of TGF-beta receptor signaling. Cell 151(5):937–950
55. Overall CM, Blobel CP (2007) In search of partners: linking extracellular proteases to substrates. Nat Rev Mol Cell Biol 8(3):245–257

56. Liu C, Xu P, Lamouille S, Xu J, Derynck R (2009) TACE-mediated ectodomain shedding of the type I TGF-beta receptor downregulates TGF-beta signaling. Mol Cell 35(1):26–36, PMCID: 2740991
57. Xu P, Derynck R (2010) Direct activation of TACE-mediated ectodomain shedding by p38 MAP kinase regulates EGF receptor-dependent cell proliferation. Mol Cell 37(4):551–566
58. Atfi A, Dumont E, Colland F, Bonnier D, L'Helgoualc'h A, Prunier C et al (2007) The disintegrin and metalloproteinase ADAM12 contributes to TGF-beta signaling through interaction with the type II receptor. J Cell Biol 178(2):201–208, PMCID: 2064440
59. Xu P, Liu J, Derynck R (2012) Post-translational regulation of TGF-β receptor and Smad signaling. FEBS Lett 586(14):1871–1884
60. Lopez-Casillas F, Payne HM, Andres JL, Massague J (1994) Betaglycan can act as a dual modulator of TGF-beta access to signaling receptors: mapping of ligand binding and GAG attachment sites. J Cell Biol 124(4): 557–568
61. Guerrero-Esteo M, Sanchez-Elsner T, Letamendia A, Bernabeu C (2002) Extracellular and cytoplasmic domains of endoglin interact with the transforming growth factor-beta receptors I and II. J Biol Chem 277(32):29197–29209
62. Barbara NP, Wrana JL, Letarte M (1999) Endoglin is an accessory protein that interacts with the signaling receptor complex of multiple members of the transforming growth factor-beta superfamily. J Biol Chem 274(2): 584–594
63. Scaffidi AK, Petrovic N, Moodley YP, Fogel-Petrovic M, Kroeger KM, Seeber RM et al (2004) Alpha(v)beta(3) integrin interacts with the transforming growth factor beta (TGFbeta) type II receptor to potentiate the proliferative effects of TGFbeta1 in living human lung fibroblasts. J Biol Chem 279(36): 37726–37733
64. Kim Y, Kugler MC, Wei Y, Kim KK, Li X, Brumwell AN et al (2009) Integrin alpha3beta1-dependent beta-catenin phosphorylation links epithelial Smad signaling to cell contacts. J Cell Biol 184(2):309–322, PMCID: 2654298
65. Rudini N, Felici A, Giampietro C, Lampugnani M, Corada M, Swirsding K et al (2008) VE-cadherin is a critical endothelial regulator of TGF-beta signalling. EMBO J 27:993–1004
66. Andl CD, Fargnoli BB, Okawa T, Bowser M, Takaoka M, Nakagawa H et al (2006) Coordinated functions of E-cadherin and transforming growth factor beta receptor II in vitro and in vivo. Cancer Res 66(20):9878–9885, PMCID: 2996096
67. Jin W, Yun C, Kwak MK, Kim TA, Kim SJ (2007) TrkC binds to the type II TGF-beta receptor to suppress TGF-beta signaling. Oncogene 26(55):7684–7691
68. Jin W, Kim BC, Tognon C, Lee HJ, Patel S, Lannon CL et al (2005) The ETV6-NTRK3 chimeric tyrosine kinase suppresses TGF-beta signaling by inactivating the TGF-beta type II receptor. Proc Natl Acad Sci U S A 102(45):16239–16244, PMCID: 1283417
69. Finnson KW, Tam BY, Liu K, Marcoux A, Lepage P, Roy S et al (2006) Identification of CD109 as part of the TGF-beta receptor system in human keratinocytes. FASEB J 20(9): 1525–1527
70. Bizet AA, Liu K, Tran-Khanh N, Saksena A, Vorstenbosch J, Finnson KW et al (2011) The TGF-beta co-receptor, CD109, promotes internalization and degradation of TGF-beta receptors. Biochim Biophys Acta 1813(5):742–753
71. Bizet AA, Tran-Khanh N, Saksena A, Liu K, Buschmann MD, Philip A (2012) CD109-mediated degradation of TGF-beta receptors and inhibition of TGF-beta responses involve regulation of SMAD7 and Smurf2 localization and function. J Cell Biochem 113(1):238–246
72. Chen Y, Liu F, Massague J (1997) Mechanism of TGF-beta receptor inhibition by FKBP12. EMBO J 16(13):3866–3876
73. Huse M, Muir TW, Xu L, Chen YG, Kuriyan J, Massague J (2001) The TGF beta receptor activation process: an inhibitor- to substrate-binding switch. Mol Cell 8(3):671–682
74. Meng Q, Lux A, Holloschi A, Li J, Hughes JM, Foerg T et al (2006) Identification of Tctex2beta, a novel dynein light chain family member that interacts with different transforming growth factor-beta receptors. J Biol Chem 281(48):37069–37080
75. Datta PK, Moses HL (2000) STRAP and Smad7 synergize in the inhibition oftransforming growth factor-beta signaling. Mol Cell Biol 20(9):3157–3167
76. Tsukazaki T, Chiang TA, Davison AF, Attisano L, Wrana JL (1998) SARA, a FYVE domain protein that recruits Smad2 to the TGF-b receptor. Cell 95(6):779–791
77. Miura S, Takeshita T, Asao H, Kimura Y, Murata K, Sasaki Y et al (2000) Hgs (Hrs), a FYVE domain protein, is involved in Smad

signaling through cooperation with SARA. Mol Cell Biol 20(24):9346–9355

78. Wrighton KH, Lin X, Feng XH (2008) Critical regulation of TGFbeta signaling by Hsp90. Proc Natl Acad Sci U S A 105(27):9244–9249, PMCID: 2453700
79. Kang JS, Liu C, Derynck R (2009) New regulatory mechanisms of TGF-beta receptor function. Trends Cell Biol 19(8):385–394
80. Yamashita M, Fatyol K, Jin C, Wang X, Liu Z, Zhang YE (2008) TRAF6 mediates Smad-independent activation of JNK and p38 by TGF-beta. Mol Cell 31(6):918–924, PMCID: 2621323
81. Neil JR, Tian M, Schiemann WP (2009) X-linked inhibitor of apoptosis protein and its E3 ligase activity promote transforming growth factor-{beta}-mediated nuclear factor-{kappa}B activation during breast cancer progression. J Biol Chem 284(32):21209–21217, PMCID: 2755844
82. Schlessinger J (2002) Ligand-induced, receptor-mediated dimerization and activation of EGF receptor. Cell 110(6):669–672
83. Sigismund S, Argenzio E, Tosoni D, Cavallaro E, Polo S, Di Fiore PP (2008) Clathrin-mediated internalization is essential for sustained EGFR signaling but dispensable for degradation. Dev Cell 15(2):209–219
84. Mitchell H, Choudhury A, Pagano RE, Leof EB (2004) Ligand-dependent and -independent transforming growth factor-beta receptor recycling regulated by clathrin-mediated endocytosis and Rab11. Mol Biol Cell 15(9):4166–4178, PMCID: 515349
85. Di Guglielmo GM, Le Roy C, Goodfellow AF, Wrana JL (2003) Distinct endocytic pathways regulate TGF-beta receptor signalling and turnover. Nat Cell Biol 5(5):410–421
86. Ehrlich M, Shmuely A, Henis Y (2001) A single internalization signal from the di-leucine family is critical for constitutive endocytosis of the type II TGF-beta receptor. J Cell Sci 114(May):1777–1786
87. Penheiter SG, Mitchell H, Garamszegi N, Edens M, Dore JJE, Leof EB (2002) Internalization-dependent and -independent requirements for transforming growth factor receptor signaling via the Smad pathway. Mol Cell Biol 22(13):4750–4759
88. Yao D, Ehrlich M, Henis YI, Leof EB (2002) Transforming growth factor-beta receptors interact with AP2 by direct binding to beta2 subunit. Mol Biol Cell 13(11):4001–4012, PMCID: 133610
89. Hayes S, Chawla A, Corvera S (2002) TGFbeta receptor internalization into EEA1-enriched early endosomes: role in signaling to Smad2. J Cell Biol 158(7):1239–1249
90. Razani B, Zhang XL, Bitzer M, von Gersdorff G, Bottinger EP, Lisanti MP (2001) Caveolin-1 regulates transforming growth factor (TGF)-beta/SMAD signaling through an interaction with the TGF-beta type I receptor. J Biol Chem 276(9):6727–6738
91. Zhang XL, Topley N, Ito T, Phillips A (2005) Interleukin-6 regulation of transforming growth factor (TGF)-beta receptor compartmentalization and turnover enhances TGF-beta1 signaling. J Biol Chem 280(13): 12239–12245
92. Zuo W, Chen YG (2009) Specific activation of mitogen-activated protein kinase by transforming growth factor-beta receptors in lipid rafts is required for epithelial cell plasticity. Mol Biol Cell 20(3):1020–1029, PMCID: 2633387
93. Hu H, Milstein M, Bliss JM, Thai M, Malhotra G, Huynh LC et al (2008) Integration of transforming growth factor beta and RAS signaling silences a RAB5 guanine nucleotide exchange factor and enhances growth factor-directed cell migration. Mol Cell Biol 28(5):1573–1583, PMCID: 2258770
94. Hocevar BA, Smine A, Xu X-X, Howe PH (2001) The adaptor molecule Disabled-2 links the transforming growth factor b receptors to the Smad pathway. EMBO J 20(11):2789–2801
95. Penheiter SG, Singh RD, Repellin CE, Wilkes MC, Edens M, Howe PH et al (2010) Type II transforming growth factor-beta receptor recycling is dependent upon the clathrin adaptor protein Dab2. Mol Biol Cell 21(22):4009–4019, PMCID: 2982134
96. Choi SC, Kim GH, Lee SJ, Park E, Yeo CY, Han JK (2008) Regulation of activin/nodal signaling by Rap2-directed receptor trafficking. Dev Cell 15(1):49–61
97. Zhang L, Zhou H, Su Y, Sun Z, Zhang H, Zhang Y et al (2004) Zebrafish Dpr2 inhibits mesoderm induction by promoting degradation of nodal receptors. Science 306(5693):114–117
98. Su Y, Zhang L, Gao X, Meng F, Wen J, Zhou H et al (2007) The evolutionally conserved activity of Dapper2 in antagonizing TGF-beta signaling. FASEB J 21(March):682–690
99. Wu L, Derynck R (2009) Essential role of TGF-β signaling in glucose-induced cell hypertrophy. Dev Cell 17(1):35–48
100. Gross DN, Farmer SR, Pilch PF (2004) Glut4 storage vesicles without Glut4: transcriptional regulation of insulin-dependent vesicular traffic. Mol Cell Biol 24(16):7151–7162

101. Lyons RM, Millerf DA, Graycar JL, Moses HL, Derynck R (1991) Differential binding of transforming growth factor-beta1, -2, and -3 by fibroblasts and epithelial cells measured by affinity cross-linking of cell surface receptors. Mol Endocrinol 5(12):1887–1896
102. Frolik CA, Wakefield LM, Smith DM, Sporn MB (1984) Characterization of a membrane receptor for TGF-beta in normal rat kidney fibroblasts. J Biol Chem 259(17): 10995–11000
103. Furuhashi M, Yagi K, Yamamoto H, Furukawa Y, Shimada S, Nakamura Y et al (2001) Axin facilitates Smad3 activation in the transforming growth factor beta signaling pathway. Mol Cell Biol 21(15):5132–5141, PMCID: 87238
104. Liu W, Rui H, Wang J, Lin S, He Y, Chen M et al (2006) Axin is a scaffold protein in TGF-beta signaling thatpromotes degradation of Smad7 by Arkadia. EMBO J 25:1646–1658
105. Guo X, Ramirez A, Waddell DS, Li Z, Liu X, Wang XF (2008) Axin and GSK3-control Smad3 protein stability and modulate TGF-signaling. Genes Dev 22(1):106–120, PMCID: 2151009
106. Kwak JH, Kim SI, Kim JK, Choi ME (2008) BAT3 interacts with transforming growth factor-beta (TGF-beta) receptors and enhances TGF-beta1-induced type I collagen expression in mesangial cells. J Biol Chem 283(28):19816–19825, PMCID: 2443666
107. Ferrand N, Atfi A, Prunier C (2010) The oncoprotein c-ski functions as a direct antagonist of the transforming growth factor-{beta} type I receptor. Cancer Res 70(21):8457–8466
108. Bourguignon LY, Singleton PA, Zhu H, Zhou B (2002) Hyaluronan promotes signaling interaction between CD44 and the transforming growth factor beta receptor I in metastatic breast tumor cells. J Biol Chem 277(42):39703–39712
109. Seo SR, Ferrand N, Faresse N, Prunier C, Abecassis L, Pessah M et al (2006) Nuclear retention of the tumor suppressor cPML by the homeodomain protein TGIF restricts TGF-beta signaling. Mol Cell 23(4):547–559
110. Yamaguchi T, Kurisaki A, Yamakawa N, Minakuchi K, Sugino H (2006) FKBP12 functions as an adaptor of the Smad7-Smurf1 complex on activin type I receptor. J Mol Endocrinol 36(3):569–579
111. Huse M, Chen YG, Massague J, Kuriyan J (1999) Crystal structure of the cytoplasmic domain of the type I TGF beta receptor in complex with FKBP12. Cell 96(3):425–436
112. Kunzmann S, Wohlfahrt JG, Itoh S, Asao H, Komada M, Akdis CA et al (2003) SARA and Hgs attenuate susceptibility to TGF-beta1-mediated T cell suppression. FASEB J 17(2):194–202
113. Wrighton KH, Lin X, Feng XH (2009) Phospho-control of TGF-beta superfamily signaling. Cell Res 19(1):8–20, PMCID: 2929013
114. Lallemand F, Seo SR, Ferrand N, Pessah M, L'Hoste S, Rawadi G et al (2005) AIP4 restricts transforming growth factor-beta signaling through a ubiquitination-independent mechanism. J Biol Chem 280(30):27645–27653
115. Bai Y, Yang C, Hu K, Elly C, Liu YC (2004) Itch E3 ligase-mediated regulation of TGF-beta signaling by modulating smad2 phosphorylation. Mol Cell 15(5):825–831
116. Jin Q, Ding W, Mulder KM (2007) Requirement for the dynein light chain km23-1 in a Smad2-dependent transforming growth factor-beta signaling pathway. J Biol Chem 282(26):19122–19132
117. Partridge EA, Le Roy C, Di Guglielmo GM, Pawling J, Cheung P, Granovsky M et al (2004) Regulation of cytokine receptors by Golgi N-glycan processing and endocytosis. Science 306(5693):120–124
118. Barrios-Rodiles M, Brown KR, Ozdamar B, Bose R, Liu Z, Donovan RS et al (2005) High-throughput mapping of a dynamic signaling network in mammalian cells. Science 307(5715):1621–1625
119. Seong HA, Jung H, Choi HS, Kim KT, Ha H (2005) Regulation of transforming growth factor-beta signaling and PDK1 kinase activity by physical interaction between PDK1 and serine-threonine kinase receptor-associated protein. J Biol Chem 280(52):42897–42908
120. Zhang Y, Li X, Qi J, Wang J, Liu X, Zhang H et al (2009) Rock2 controls TGF-beta signaling and inhibits mesoderm induction in zebrafish embryos. J Cell Sci 122:2197–2207
121. Wu G, Chen YG, Ozdamar B, Gyuricza CA, Chong PA, Wrana JL et al (2000) Structural basis of Smad2 recognition by the Smad anchor for receptor activation. Science 287(5450):92–97
122. Kowanetz M, Lonn P, Vanlandewijck M, Kowanetz K, Heldin CH, Moustakas A (2008) TGFbeta induces SIK to negatively regulate type I receptor kinase signaling. J Cell Biol 182(4):655–662, PMCID: 2518705
123. Lonn P, Vanlandewijck M, Raja E, Kowanetz M, Watanabe Y, Kowanetz K et al (2012) Transcriptional induction of salt-inducible

kinase 1 by transforming growth factor beta leads to negative regulation of type I receptor signaling in cooperation with the Smurf2 ubiquitin ligase. J Biol Chem 287(16):12867–12878, PMCID: 3339966

124. Felici A, Wurthner JU, Parks WT, Giam LR, Reiss M, Karpova TS et al (2003) TLP, a novel modulator of TGF-beta signaling, has opposite effects on Smad2- and Smad3-dependent signaling. EMBO J 22(17):4465–4477

125. Watanabe Y, Itoh S, Goto T, Ohnishi E, Inamitsu M, Itoh F et al (2010) TMEPAI, a transmembrane TGF-beta-inducible protein, sequesters Smad proteins from active participation in TGF-beta signaling. Mol Cell 37(1):123–134

126. Zhu L, Wang L, Luo X, Zhang Y, Ding Q, Jiang X et al (2012) Tollip, an intracellular trafficking protein, is a novel modulator of the transforming growth factor-beta signaling pathway. J Biol Chem 287(47):39653–39663, PMCID: 3501082

127. Ferrigno O, Lallemand F, Verrecchia F, L'Hoste S, Camonis J, Atfi A et al (2002) Yes-associated protein (YAP65) interacts with Smad7 and potentiates its inhibitory activity against TGF-beta/Smad signaling. Oncogene 21:4879–4884

Chapter 2

Determining TGF-β Receptor Levels in the Cell Membrane

Long Zhang, Fangfang Zhou, Maarten van Dinther, and Peter ten Dijke

Abstract

Transforming growth factor-β (TGF-β) is a pleiotropic cytokine that signals via transmembrane TGF-β type I and type II serine/threonine kinases receptors, i.e., TβRI and TβRII. Upon TGF-β-induced receptor complex formation, the TβRII kinase phosphorylates TβRI. Subsequently, the activated TβRI induces the phosphorylation of receptor regulated SMAD2 and SMAD3, which can form heteromeric complexes with Smad4. These heteromeric SMAD complexes accumulate in the nucleus, where they regulate target gene expression. The stability and membrane localization of TβRI is an important determinant to control the intensity and duration of TGF-β signaling. TβRI is targeted for poly-ubiquitylation-mediated proteasomal degradation by the SMAD7-SMURF E3 ligase complex. We recently identified another important regulatory factor that controls TβRI levels in the cell membrane. As a strong inducer of TGF-β signaling, ubiquitin-specific protease (USP) 4 was found to directly interact with TβRI and act as a deubiquitylating enzyme, thereby stabilizing TβRI levels at the plasma membrane. This chapter introduces methods for examining cell membrane receptor (TβRI) levels.

Key words TGF-β, TβRI, Ubiquitination, E3 ubiquitin ligase, Deubiquitylating enzyme

1 Introduction

The TGF-β superfamily comprises TGF-βs, bone morphogenetic proteins (BMPs), activins, and structurally related proteins, which are multifunctional cytokines with pivotal roles in embryonic development and maintaining tissue homeostasis in adults [1, 2]. TGF-β family ligands bind to specific transmembrane type I and type II serine/threonine kinase receptor complexes. Five type II receptors, including TGF-β type II receptor (TβRII), activin type II receptor (ActII) and BMP type II receptor (BMPRII) and seven type I receptors, also termed activin receptor-like kinases (ALKs) have been identified in mammals. Upon ligand-induced heteromeric complex formation of type I and type II receptors, the constitutively active type II receptor kinase transphosphorylates specific serine and threonine residues in the GS segments in the type I receptor. Thereupon, the activated type I receptors phosphorylate selected SMADs at their two carboxy-terminal serine residues, and

Xin-Hua Feng et al. (eds.), *TGF-β Signaling: Methods and Protocols*, Methods in Molecular Biology, vol. 1344,
DOI 10.1007/978-1-4939-2966-5_2, © Springer Science+Business Media New York 2016

these phosphorylated receptor-activated SMADs (R-SMADs) can then form heteromeric complexes with a common SMAD4. Whereas TGF-β and activin signal via R-SMAD2 and -3, BMPs induce the phosphorylation of R-SMAD1, -5, and -8. Activated SMAD complexes translocate into the nucleus, where they regulate, in collaboration with other DNA binding transcription factors, co-activators and co-repressors, the transcription of target genes [1, 3, 4]. In this chapter we focus on canonical TGF-β signaling pathway via TβRII and TβRI (also termed ALK5). Conceptually, the mechanisms that underlie signaling via all TGF-β family members are very similar. Thus, remarks on TGF-β signaling likely apply to other family members.

In the receptor-activation model, TβRI acts downstream of TβRII for most, if not all, TGF-β-mediated responses, and the type I receptor thus determines the specificity of the intracellular signals. TβRI can be internalized via both clathrin-mediated and caveolin-mediated endocytosis, and caveolin mediated endocytosis turns off TGF-β signaling by promoting TβRI degradation [5, 6]. SMAD7 functions as an inhibitory SMAD by recruiting E3 ligase SMURF1/2 to the type I receptor and subsequently degrading TβRI and mitigating TGF-β signaling [7, 8]. Accordingly, the proteins that directly interact with TβRI and affect its post-transcriptional modification or the proteins that modulate the SMAD7/SMURF complex may regulate TGF-β-induced receptor internalization and SMAD activation (*see* Fig. 1).

Ubiquitin modification of TGF-β signaling components is emerging as a key mechanism of TGF-β pathway control [9, 10]. HECT-domain containing E3 ubiquitin ligases SMURFs regulate the intensity and duration of TGF-β signaling response by regulating TGF-β components both before and after signal initiation [11–14]. RING domain-containing E3 ligase ARKADIA (RNF 111) and RNF12 promotes ubiquitination and degradation of the inhibitory SMAD7, and thereby potentiates TGF-β/SMAD signaling [15–18]. The conjugating function of E3 ligases is opposed by deubiquitinating enzymes (DUBs) [19–22]. Ubiquitin-specific peptidase 15 (USP15) was identified as a deubiquitinating enzyme directly for R-SMADs and indirectly for TβRI [23, 24]. In this study, we illustrate the methods to determine TGF-β receptors in the membrane, describing our findings on the identification of USP4 as a potent enhancer of TGF-β/SMAD signaling. Deubiquitinating enzyme USP4 directly removes ubiquitin conjugation from TβRI thereby sustains TβRI levels at the plasma membrane [25] (*see* Fig. 1).

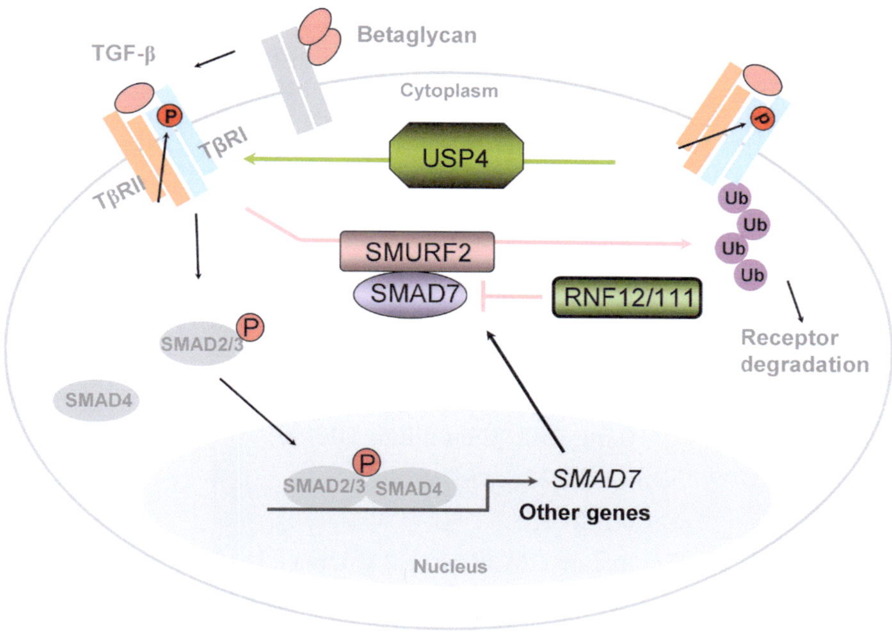

Fig. 1 Schematic representation of TGF-β signaling pathway. The extracellular dimeric TGF-β binds to type I and type II transmembrane receptors (TβRI and TβRII, respectively) that are endowed with intrinsic serine/threonine kinase domain. The type III receptor (TβRIII) is a co-receptor that facilitates binding to the signaling type I and type II receptors. While abundance of TβRIII is frequently much higher, its apparent affinity is lower than that of TβRI and TβRII. While TβRIII and TβRII can bind TGF-β by themselves, TβRI cannot bind TGF-β by itself. Thus, initially when TGF-β binds to the cell surface, it usually first interacts with TβRIII, which then presents the ligand to TβRII. The TGF-β/TβRII complex can be recognized by TβRI, and this receptor is then recruited, upon which a heteromeric receptor complex is formed, consisting of two type II receptors and two type I receptors. Subsequently, the type II receptor kinase transphosphorylates the type I receptor. Intracellular signaling is thereafter initiated by the phosphorylation of SMAD2 and SMAD3 by activated TβRI. Activated SMAD2 an Smad3 can make a complex with SMAD4, which together with other transcriptional modulators can regulate the expression of specific genes. One of these genes is the inhibitory SMAD7. SMAD7 antagonizes the activation of R-SMAD by recruiting the E3 ubiquitin ligase SMURF2 to the activated receptor at the plasma membrane and targeting this receptor for proteasomal degradation. This process is reversible through the action of deubiquitinating enzyme USP4 (and also USP15), which will stabilize TβRI at cell membrane. Smad7 itself is targeted by polyubiquitination and proteasomal degradation by E3 ubiquitin ligase RNF12 (and ARKADIA). Thus, RNF12 targets the inhibitor and potentiates TGF-β/Smad signaling

2 Materials

Cell culture reagents: Minimum Essential Medium (MEM), Dulbecco's Modified Eagle's Medium (DMEM), Phosphate buffered saline, calcium and magnesium free (PBS), fetal bovine serum (FBS), penicillin/streptomycin sulfate, nonessential amino acids (NEAA), and 0.25 % trypsin–EDTA are purchased from Invitrogen.

Antibodies: The following antibodies are used in our experiments. Anti-USP4 antibody (Sigma), anti-Flag affinity gel (Sigma), anti-Actin antibody (Sigma), anti-SMAD2 antibody (Zymed), anti-SMAD2/3 antibodies (BD), anti-phospho-SMAD2 antibody (Cell Signaling Technology), anti-Myc (9E10, Santa Cruz Biotechnology), anti-TβRI (V-22, Santa Cruz Biotechnology), anti-USP15 (Bethyl Laboratories).

2.1 Crosslinking Assay

1. PD-10 desalting columns (Sephadex G-25).
2. Column buffer: 75 mM NaCl, 4 mM HCl, 1 mg/mL BSA.
3. 2 M NaP buffer pH 7.5.
4. Chloramine T (100 μg/mL, must be made fresh. Make first at 10 mg/mL, then dilute 100×).
5. 50 mM *N*-acetyl tyrosine.
6. 60 mM KI.
7. Urea in 1 M HAc (1.2 g Urea + 1 mL of 1 M HAc).
8. Iodine-125 Radionuclide (Reductant Free) 3.7 Gbq/mL.
9. 1 μg ligand (TGF-β) in 2 μL of water.
10. PBS-B (PBS with 0.90 mM $CaCl_2$ and 0.49 mM $MgCL_2$).
11. PBS-B/BSA (PBS-B with 1 mg/mL BSA).
12. Detachment buffer: 10 mM Tris–HCl, pH 7.4, 1 mM EDTA pH 7.4, 10 % glycerol and add freshly, PMSF.
13. Solubilization buffer: 125 mM NaCl, 10 mM Tris–HCl pH 7.4, 1 mM EDTA pH 7.4, 1 % Triton X-100, and add freshly: PMSF, aprotinin, and leupeptin.
14. Fixation buffer: 25 % methanol, 7.5 % acetic acid.
15. Drying buffer: 25 % methanol, 10 % glycerol.
16. Crosslinkers (make fresh before use):
    - 54 mM disuccinimidyl suberate (DSS) in DMSO.
    - 13 mM bis[sulfosuccinimidyl]suberate (BS^3) in water.
17. An antibody recognizing your protein of interest that is suitable for immunoprecipitation, here we use USP4 antibody for the target protein, TβRI antibody as positive control and nonspecific (ns) antibody, and USP15 antibody as negative control.
18. Protein A beads.
19. 2× SDS sample buffer: 125 mM TRIS, 10 % SDS, 20 % glycerol, 10 % 2-mercaptoethanol, and bromophenol blue (20 μg/100 mL).

2.2 Cell Surface Biotinylation Assay

1. PBS-CM: PBS + 0.90 mM $CaCl_2$ + 0.33 mM $MgCl_2$.
2. PBS-CM-BSA: PBS-CM + 1 mg BSA/L.
3. Quench buffer: PBS-CM added with 50 mM NH_4Cl.
4. MESNA strip buffer: 100 mM TRIS pH 8, 100 mM NaCl, 2.5 mM $CaCl_2$, 50 mM MESNA (add freshly).
5. Sulfo-NHS-SS-Biotin.
6. Neutravidin-beads.
7. Lysis buffer.
8. 2× SDS sample buffer: 125 mM TRIS, 10 % SDS, 20 % glycerol, 10 % 2-mercaptoethanol and bromophenol blue (20 μg/100 mL).

3 Methods

3.1 Crosslinking Assay

Crosslinking assay is useful for the study of ligand-receptor interactions, as well as the determination of receptor-associated proteins. ^{125}I-labeled TGF-β ligands specifically recognize endogenous TGF-β receptors, allowing the examination of receptors-associated endogenous proteins. To identify the regulation of TβRI by USP4 at the plasma membrane, we analyzed the interaction between USP4 and ligand-associated membrane TβRI using a crosslinking assay in which we covalently affinity label cell surface proteins with ^{125}I-labeled TGF-β.

3.1.1 Experimental Procedures

Iodination of TGF-β (*See* **Note 1**)

1. Pre-equilibrate a PD-10 column with 5 mL of column buffer.
2. Transfer 2 μL of ligand (1 μg) into an Eppendorf tube. Try to perform this as quickly as possible.
3. Neutralize by adding 23 μL of 2 M NaP buffer, pH 7.5 (*see* **Note 2**).
4. Add 5 μL of ^{125}I (18.5 MBq).
5. Add 5 μL of Chloramine T.
6. Mix and leave for 2 min.
7. Add another 5 μL of Chloramine T.
8. Mix and leave for 1.5 min.
9. Add another 5 μL of Chloramine T (*see* **Note 3**).
10. Mix and leave for 1 min.
11. Stop the reaction by adding (in this particular order!) (*see* **Note 4**).
    - 20 μL of *N*-acetyl tyrosine.
    - 200 μL of KI.
    - 200 μL of Urea.

12. Mix and leave for 2–5 min.
13. Apply onto the equilibrated PD-10 column.
14. Elute the labeled ligand with column buffer. Add 500 μL of column buffer, and collect the flow through. Make ten fractions. Highest activity will be in fraction 7 or 8 (as measured by Geiger counter). Store the labeled ligand at 4 °C in a lead vial (*see* **Notes 5** and **6**).

Crosslinking of Labeled Ligands to Transmembrane or Membrane Associated Proteins (*See* **Note 7**)

1. Grow the cells that you want to use for the crosslinking experiment on a plate or flask, and make sure that they are confluent (*see* **Note 8**).
2. Wash the cells three times with 10 mL of ice-cold PBS-B/BSA.
3. Add 2 mL of ice-cold PBS-B/BSA to the plate.
4. Add 20 μL of the labeled ligand.
5. Incubate the plate for 3 h on a shaker.
6. Wash the cells two times with ice-cold PBS-B/BSA.
7. Wash the cells one time with ice-cold PBS-B.
8. Add 2 mL of ice-cold PBS-B.
9. Add 10 μL of both of the freshly prepared crosslinkers (BS3 and DSS).
10. Incubate for 15 min on a shaker (*see* **Note 9**).
11. Wash cells one time with ice-cold detachment buffer with freshly added phenylmethanesulfonylfluoride (PMSF).
12. Scrape the cells with a rubber policeman in 1 mL of detachment buffer.
13. Transfer the cells to an eppendorf tube and spin down the cells at low speed.
14. Remove supernatant and lyse the cells in 500 μL of solubilization buffer with freshly added PMSF, aprotinin, and leupeptin.
15. Incubate for 30 min on ice.
16. Spin down at maximum speed for 10 min in a cooled centrifuge.
17. Transfer lysate to a fresh tube.
18. Save 50 μL of the lysate by transferring this to a fresh tube and add an equal volume of 2× SDS sample buffer. This will be your input control.
19. Add the antibody to you protein of interest to the lysate and incubate for 3 h (rotate).
20. Add 50 μL of protein A or G beads (50 % slurry, washed with solubilization buffer) and incubate for 45 min (rotate).
21. Wash the beads four to five times with solubilization buffer.

22. Remove as much as possible of the supernatant and add 50 μL of 2× SDS sample buffer.
23. Boil samples for 5 min and run the samples on a SDS-PAGE gel (*see* **Note 10**).
24. After running the gel fix the gel for 30 min in fixation buffer at RT (shaking).
25. Incubate the gel for 45 min in drying buffer at RT (shaking).
26. Refresh drying buffer and incubate for another 45 min at RT (shaking).
27. Dry the gel on a gel-dryer (*see* **Note 11**).
28. Analyze the gel with a phosphorimager. An example of the result is shown in Fig. 2 (*see* **Notes 12** and **13**).

3.2 Cell Surface Biotinylation Assay

The proteins present on the surface of cells can be selectively modified by the water-soluble and cell-impermeable biotin analog sulfo-NHS-SS-biotin. The addition of sulfo-NHS-SS-biotin to only surface of the cells provides nonradioactive and selective labeling method of cell surface proteins and transmembrane proteins. To test the role of USP4 on TβRI levels at the plasma membrane, we compared membrane associated TβRI levels and their internalization/degradation speed in control and USP4-depleted cells by using a cell surface biotinylation assay.

3.2.1 Experimental Procedures

1. Grow COS7 cells or MDA-MB-231 cells on p94 (confluent) (*see* **Note 14**).
2. Wash cells three times with PBS-CM on ice.
3. Label cell surface proteins with 0.5 mg/mL sulfo-NHS-SS-Biotin in PBS-CM (make fresh!) for 30 min on ice (3 mL/plate).
4. Quench the reaction by washing two times with quenching buffer (5 min each, on the shaker on ice).
5. Wash two times with PBS-CM-BSA.
6. Stimulate cells in PBS-CM-BSA (total volume 2–3 mL).
7. Transfer plates to 37 °C incubator (depending on the experimental set up, time course are needed in many cases).
8. At the indicated time points transfer plates to ice.
9. Wash two times with PBS-CM (*see* **Note 15**).
10. Strip the remaining biotin from the cell surface with MESNA (5 mL, 30 min 4 °C).
11. Wash three times with PBS-CM.
12. Lyse cells in 750 μL lysis buffer with inhibitors (30 min on ice).
13. Spin 10 min at high speed with a microfuge.

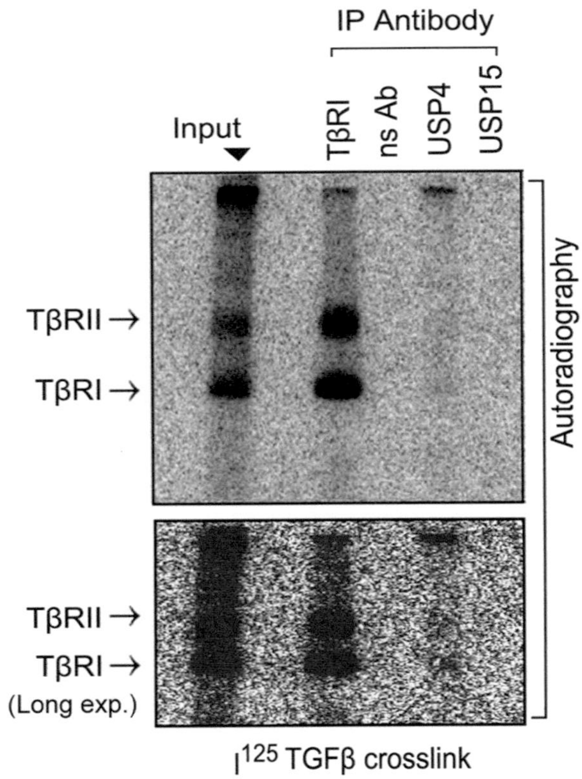

Fig. 2 USP4 interacts with endogenous TGF-β receptors. Non-transfected MDA-MB-231 cells were covalently affinity labeled with ^{125}I-labeled TGF-β ligand, cell lysates were then harvested for immunoprecipitation (IP) with TβRI, nonspecific (ns), USP4 or USP15 antibodies as indicated. Endogenous TβRI and TβRII that were immunoprecipitated by antibodies were detected by autoradiography. Crosslinked complexes of TβRI-TGFβ monomers and TβRII-TGF-β monomers are indicated by arrows. The band with high molecular weight likely represents TβRIII-TGF-β monomer crosslinked complex. The TβRI antibody not only immunoprecipitates TβRI-TGF-β monomer, but also TβRII-TGF-β monomer complex; this indicates that both receptors are part of the same heteromeric complex. The fact that USP4 (but not USP15) antibodies immunoprecipitate the TGF-β receptors indicates that intracellular USP4 protein is associated with these receptors. This Figure is reproduced with modification from Figure 2b in Zhang et al. [25]

14. Harvest supernatant.
15. Measure protein concentration.
16. Add 25 μL of neutravidin-beads (washed 4 times with 1× lysis buffer) use same amount of protein per sample (±750 μg), save 50 μL lysate for input control.
17. Incubate for 30 min (4 °C rotating).
18. Wash the beads four times with lysis buffer.
19. Add 50 μL of 2× SDS sample buffer.

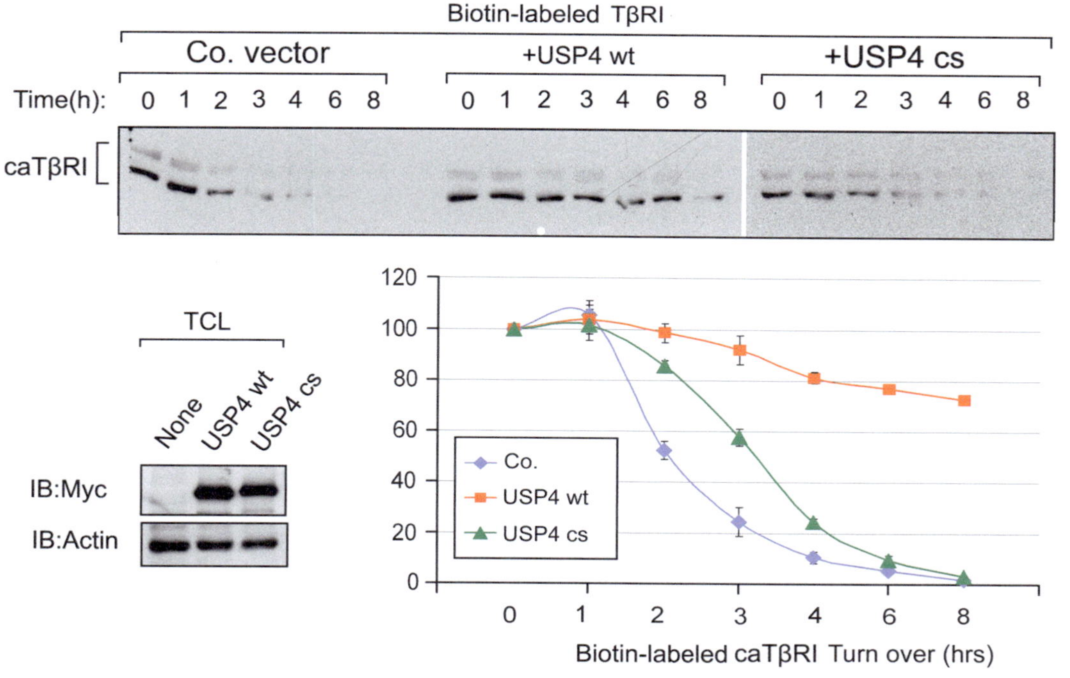

Fig. 3 USP4 increases the level of membrane-associated TβRI. COS7 cells transfected with caTβRI-Flag along with or without USP4 wild-type (wt) or a catalytically impaired/inactive mutant c311s (cs). Thirty-six hours after transfection, cells were biotinylated for 40 min at 4 °C and then incubated at 37 °C for indicated time points. The biotinylated cell surface receptors were precipitated with neutravidin beads and analyzed by anti-Flag immunoblotting. *Lower right panel*: quantification of the band intensities in *upper panel*. Quantitation of image intensity was done using Image J software. Band intensity was normalized to the $t=0$ controls. Results are shown as mean ± SD of three independent sets of experiments. Wild-type USP4 deubiquitinates TβRI, therefore inhibits the turnover of membrane-associated TβRI. Compared with USP4 WT, inactive USP4 CS mutant (carrying a point mutation in one of the key cysteines of the catalytic domain) has little effect. This Figure is reproduced with modification from Figure 3a in Zhang et al. [25]

20. Determine the biotin-labeled receptor level as well as the whole cell lysate by immunoblotting. An example of the result is shown in Figs. 3 and 4, in which cell surface biotinylated cells were treated with TGF-β for different time intervals (to induce receptor internalization and endocytosis) and membrane associated active TβRI were detected by western blotting following neutravidin beads pull-down.

4 Notes

1. The iodination (radiolabeling) of TGF-β is usually done in a specialized laboratory. When entering this laboratory wear a laboratory coat and, if required, overshoes. The labeling itself needs to be performed in a fully functional (please check) chemical hood. The radioactive Iodine that is used for labeling

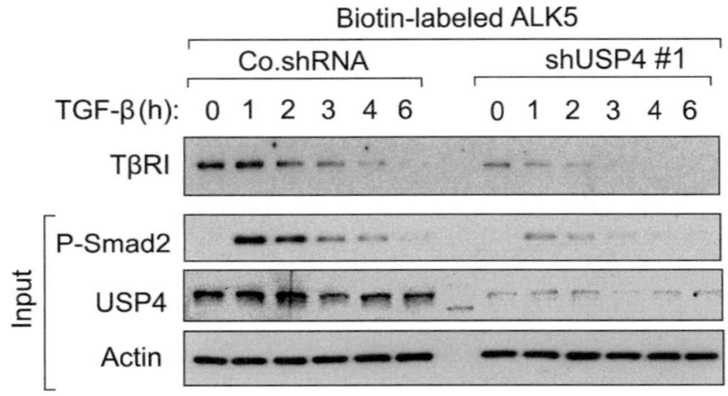

Fig. 4 USP4 depletion accelerates the turnover of TβRI and mitigates TGFβ/Smad signaling. MDA-MB-231 cells infected with control (Co.shRNA) or USP4 shRNA lentivirus (#1) were biotinylated for 40 min at 4 °C and then incubated at 37 °C with TGF-β (5 ng/mL) for indicated time to stimulate receptor internalization and treated with TGF-β (5 ng/mL) for different time points as indicated. The biotinylated cell surface receptors were precipitated with neutravidin beads and analyzed by anti-TβRI immunoblotting. Five percent whole cell lysates were loaded as input showing P-Smad2 and USP4 levels. Actin was included as a loading control for the total cell lysate. This Figure is reproduced with modification from Figure 3b in Zhang et al. [25]

is very volatile. Safety glasses and (double) gloves should be worn all the time. Hands need to be regularly checked for contamination, in particular each time after the addition of radiolabeled iodine when eppendorf tubes are opened. All additions need to be done behind a lead-enforced Perspex shield. A personal body dosimeter should be worn.

2. The 2 M NaP buffer may have some precipitate. One can remove this by warming the tube with buffer with stream of warm water. If some crystals remain, this will not cause a problem.
3. In **step 9** the third addition (and the incubation) of Chloramine T should NOT be done when labeling BMP2, 4, 6, or 7. This will lead to over-labeling of the protein and destruction of its activity. In general, for TGF-β and BMP9 three additions of Chloramine T will not destroy their activity.
4. The urea may precipitate. This can be avoided by warming the tube with urea with your hands or warm water.
5. Specific activity of ^{125}I-TGF-β is 3.7 MBq (0.1 mCi) per mL. ^{125}I-TGF-β can also be stored at −20° C in 4 mM HCl, 1 % BSA. Radiolabeled TGF-β can be stored for 1 month. Upon prolonged storage the specific activity will decrease and TGF-β may get slowly destroyed by radiation damage.
6. The work surface area (door knobs, light switches, and room in general), hands, and coat need to be checked for radioactive contamination. All radioactive waste needs to be properly disposed.

7. The volumes used in the method below are for one 10-cm dish. All steps are performed on ice to prevent internalization of the ligand/receptors during the crosslinking procedure. This experiment uses radiolabeled TGF-β, and all the precautions for working with radioactive material should be taken.
8. If the cells are not confluent you will get less specific and/or get more nonspecific signal. The TGF-β will stick very well to the plastic of the tissue culture plates. On the other hand, the cell density may also influence the number of receptors that are present on the cell.
9. By extending the treatment time with the crosslinkers (DSS and BS^3) the crosslinking efficiency can be increased. However, the chance that the monomers of ligand will crosslink, or that there will be crosslinks between type I and type II will also increase. As a result additional complexes may be visible on the gel including TGF-β monomer crosslinked to both TβRI and TβRII, TGF-β monomer crosslinked to another monomer and TβRI (or TβRII).
10. When analyzing the crosslinked complexes by gel electrophoresis, preferentially the front (small proteins) should be left on the gel. This will allow the analysis of specific bound, but not crosslinked TGF-β. As crosslinking efficiency is usually about 10–20 %, most of the specific bound TGF-β runs just above the front. Analysis of the amount of bound, but not crosslinked TGF-β can be informative; in particular when the signals of the covalently affinity labeled receptors or cell surface proteins is weak.
11. Be careful to only remove the gel from the dryer when the gel is completely dried. Otherwise the gel may crack.
12. The extent of nonspecific binding of TGF-β can be demonstrated by determining the cell-associated radioactivity in the presence of 200 (or higher) molar excess.
13. To obtain more detailed information about the apparent affinity and numbers of TGF-β receptors on cells, competition binding studies with radiolabeled (tracer) amounts of TGF-β and increasing amounts of unlabeled TGF-β and Scatchard analysis of the binding data can be performed [26, 27].
14. In general, cells with strong adhesion capacity should be chosen to perform this assay. For detecting transfected epitope tagged TβRI, COS7 cells could be used due to its high transfection efficiency and cell adhesion capacity. For the detecting of endogenous biotin labeled cell surface TβRI, experiments needs to be considerably scaled up (at least one 15 cm dish per sample).

15. To detect the amount of receptor that internalized into cells (but not yet degraded), cell surface biotin needs to be removed, go to **step 10**. Skip the **step 10** will allow the detecting of membrane associated receptor (both on the cell surface and internalized vesicles) therefore time course studies will enable observation of cell surface receptor degradation speed in different conditions. When not performing **step 10** it is not necessary to use sulfo-NHS-SS-Biotin, sulfo-NHS-Biotin would suffice.

Acknowledgements

This work was supported by a Netherlands Organization of Scientific Research grant (MW-NWO 918.66.606), Cancer Genomics Centre Netherlands, Centre for Biomedical Genetics, and Swedish Cancerfonden (09 0773).

References

1. Heldin CH, Miyazono K, ten Dijke P (1997) TGF-β signalling from cell membrane to nucleus through SMAD proteins. Nature 390 (6659):465–471
2. Massague J (1998) TGF-β signal transduction. Annu Rev Biochem 67:753–791
3. Shi Y, Massague J (2003) Mechanisms of TGF-β signaling from cell membrane to the nucleus. Cell 113(6):685–700
4. Kang JS, Liu C, Derynck R (2009) New regulatory mechanisms of TGF-β receptor function. Trends Cell Biol 19(8):385–394
5. Di Guglielmo GM, Le Roy C, Goodfellow AF, Wrana JL (2003) Distinct endocytic pathways regulate TGF-β receptor signalling and turnover. Nat Cell Biol 5(5):410–421
6. Chen YG (2009) Endocytic regulation of TGF-β signaling. Cell Res 19(1):58–70
7. Kavsak P, Rasmussen RK, Causing CG, Bonni S, Zhu H, Thomsen GH et al (2000) Smad7 binds to Smurf2 to form an E3 ubiquitin ligase that targets the TGFβ receptor for degradation. Mol Cell 6(6):1365–1375
8. Ebisawa T, Fukuchi M, Murakami G, Chiba T, Tanaka K, Imamura T et al (2001) Smurf1 interacts with transforming growth factor-β type I receptor through Smad7 and induces receptor degradation. J Biol Chem 276(16): 12477–12480
9. Lonn P, Moren A, Raja E, Dahl M, Moustakas A (2009) Regulating the stability of TGFβ receptors and Smads. Cell Res 19(1):21–35
10. De Boeck M, ten Dijke P (2012) Key role for ubiquitin protein modification in TGFβ signal transduction. Ups J Med Sci 117(2): 153–165
11. Zhu H, Kavsak P, Abdollah S, Wrana JL, Thomsen GH (1999) A SMAD ubiquitin ligase targets the BMP pathway and affects embryonic pattern formation. Nature 400 (6745):687–693
12. Lin X, Liang M, Feng XH (2000) Smurf2 is a ubiquitin E3 ligase mediating proteasome-dependent degradation of Smad2 in transforming growth factor-β signaling. J Biol Chem 275(47):36818–36822
13. Zhang Y, Chang CB, Gehling DJ, Hemmati-Brivanlou A, Delynck R (2001) Regulation of Smad degradation and activity by Smurf2, an E3 ubiquitin ligase. Proc Natl Acad Sci U S A 98(3):974–979
14. Lo RS, Massague J (1999) Ubiquitin-dependent degradation of TGF-β-activated Smad2. Nat Cell Biol 1(8):472–478
15. Zhang L, Huang H, Zhou F, Schimmel J, Pardo CG, Zhang T et al. (2012) RNF12 controls embryonic stem cell fate and morphogenesis in zebrafish embryos by targeting Smad7 for degradation. Mol Cell 46(5): 650–661
16. Niederlander C, Walsh JJ, Episkopou V, Jones CM (2001) Arkadia enhances nodal-related signalling to induce mesendoderm. Nature 410(6830):830–834

17. Episkopou V, Arkell R, Timmons PM, Walsh JJ, Andrew RL, Swan D (2001) Induction of the mammalian node requires Arkadia function in the extraembryonic lineages. Nature 410 (6830):825–830
18. Koinuma D, Shinozaki M, Komuro A, Goto K, Saitoh M, Hanyu A et al (2003) Arkadia amplifies TGF-β superfamily signalling through degradation of Smad7. EMBO J 22(24):6458–6470
19. Nijman SMB, Luna-Vargas MPA, Velds A, Brummelkamp TR, Dirac AMG, Sixma TK et al (2005) A genomic and functional inventory of deubiquitinating enzymes. Cell 123(5): 773–786
20. Yuan J, Luo KT, Zhang LZ, Cheville JC, Lou ZK (2010) USP10 regulates p53 localization and stability by deubiquitinating p53. Cell 140(3):384–396
21. Williams SA, Maecker HL, French DM, Liu JF, Gregg A, Silverstein LB et al (2011) USP1 deubiquitinates ID proteins to preserve a mesenchymal stem cell program in osteosarcoma. Cell 146(6):917–929
22. Nakada S, Tai I, Panier S, Al-Hakim A, Iemura S, Juang YC et al (2010) Non-canonical inhibition of DNA damage-dependent ubiquitination by OTUB1. Nature 466(7309):941–946
23. Inui M, Manfrin A, Mamidi A, Martello G, Morsut L, Soligo S et al (2011) USP15 is a deubiquitylating enzyme for receptor-activated SMADs. Nat Cell Biol 13(11): 1368–1375
24. Eichhorn PJ, Rodon L, Gonzalez-Junca A, Dirac A, Gili M, Martinez-Saez E et al. (2012) USP15 stabilizes TGF-β receptor I and promotes oncogenesis through the activation of TGF-beta signaling in glioblastoma. Nat Med 18(3):429–435
25. Zhang L, Zhou F, Drabsch Y, Gao R, Snaar-Jagalska BE, Mickanin C et al. USP4 is regulated by AKT phosphorylation and directly deubiquitylates TGF-β type I receptor. Nat Cell Biol 14(7):717–726
26. Frolik CA, Wakefield LM, Smith DM, Sporn MB (1984) Characterization of a membrane receptor for transforming growth factor-β in normal rat kidney fibroblasts. J Biol Chem 259(17):10995–11000
27. ten Dijke P, Iwata KK, Goddard C, Pieler C, Canalis E, McCarthy TL et al (1990) Recombinant transforming growth factor type β3: biological activities and receptor-binding properties in isolated bone cells. Mol Cell Biol 10(9):4473–4479

Chapter 3

Posttranslational Modifications of TGF-β Receptors

Xiaohua Yan and Ye-Guang Chen

Abstract

TGF-β is a prototype of the TGF-β cytokine superfamily and exerts multiple regulatory effects on cell activities. It signals through two types of membrane-bound serine/threonine kinase receptors. Upon TGF-β binding, the type II receptor TβRII recruits the type I receptor TβRI and form a functional hetero-complex. TβRII *trans*-phosphorylates the GS region of TβRI, thus triggering its kinase activity. Activated TβRI proceeds to activate downstream Smad2/3. Signal intensity and duration through the availability, activity and destiny of TGF-β receptors are finely controlled by multiple posttranslational modifications such as phosphorylation, ubiquitination, and neddylation. This chapter introduces methods for examination of these modifications of TGF-β receptors.

Key words TGF-β receptor, TβRI, TβRII, Modification, Phosphorylation, Ubiquitination, Neddylation

1 Introduction

TGF-β is a secreted polypeptide belonging to the TGF-β superfamily, which consists of 33 related proteins in human [1–6]. TGF-β and related factors play essential and diverse roles in many aspects of cellular actions, and are thus engaged in early development, tissue homeostasis and diseases such as cancer, fibrosis and skeletal disorders. TGF-β initiates signal transduction by bringing together two pairs of serine/threonine kinase receptors on cell membrane, the type II TGF-β receptor TβRII and the type I receptor TβRI/ALK5. TβRII *trans*-phosphorylates TβRI, which then proceeds to propagate signaling via phosphorylating downstream mediators, especially the Smad proteins [7–11]. As regulation of signal intensity and duration is one of the key issues in signal transduction, it is not surprising that the availability and activity of TGF-β receptors are finely controlled by multiple mechanisms, including posttranslational modifications (PTMs) such as phosphorylation, ubiquitination, sumoylation, neddylation, and others [1, 12–16].

Xin-Hua Feng et al. (eds.), *TGF-β Signaling: Methods and Protocols*, Methods in Molecular Biology, vol. 1344,
DOI 10.1007/978-1-4939-2966-5_3, © Springer Science+Business Media New York 2016

Both types of TGF-β receptors show intrinsic serine/threonine kinase activities; moreover, structural and biochemical studies suggest that they also have tyrosine kinase activities, thus functioning as dual-specificity kinases [7, 8, 11, 17]. Structurally, the two receptors are about 500 amino acids in length and consist of an *N*-glycosylated extracellular domain, a single transmembrane domain and a cytoplasmic domain; the cytoplasmic part contains a large kinase domain, which is flanked by a juxtamembrane region and a short carboxyl tail. Furthermore, the type I receptor is additionally characterized by a GS (glycine/serine) region (T_{185}TSGSGSGLPLLVQRTIART_{204} for human TβRI), which is conserved in all the type I receptors and immediately precedes the kinase domain. At the resting state, the majority of TGF-β receptors stay as monomers; upon TGF-β stimulation, the number of homodimers dramatically increases, and they form a functional receptor heterocomplex consisting of two molecules each of TβRII and TβRI [18–21]. The extracellular domains, the transmembrane domains together with the intracellular domains of TβRII and TβRI contribute to their interactions. Complex formation allows TβRII to phosphorylate theThr185, Thr186, Ser187, Ser189, and Ser191 residues in the GS region of TβRI. This phosphorylation would alter the conformation of TβRI, thereby triggering its kinase activity. Inactive TβRI can associate via its GS region with the 12 kDa FK506-binding protein FKBP12, which prevents the receptor from being phosphorylated and activated by TβRII in the absence of ligands [22, 23]. Besides the GS region, TβRII can also phosphorylate TβRI at other sites. For instance, Ser165 in the juxtamembrane region of TβRI is phosphorylated by TβRII, and this phosphorylation may modulate the specificity of receptor signaling [24]. Like other kinases, TGF-β receptors could also be auto-phosphorylated. TβRII is constitutively active and is auto-phosphorylated on Ser, Thr, and Tyr residues independently of TGF-β [8, 13, 14]. Likewise, TβRI is also auto-phosphorylated on these residues upon TGF-β stimulation, but the concrete phospho-sites remain to be defined [8, 17].

Protein phosphorylation is a reversible process regulated by kinases and phosphatases. Protein phosphatase 1c (PP1c) was firstly reported to be responsible for dephosphorylation of the Dpp type I receptor in *Drosophila melanogaster* (Dpp is a TGF-β family member), and SARA could serve as an adaptor to bridge the receptor and PP1c [25]. Subsequently, Smad7 was shown in mammalian cells to interact with growth arrest and DNA damage protein 34 (GADD34, a regulatory subunit of the PP1 holoenzyme), thus recruiting the catalytic subunit PP1c, dephosphorylating TβRI and finally leading to inactivation of the receptor [26]. Likewise, Smad7 and PP1α regulate the phospho-state of ALK1, another TGF-β type I receptor specifically expressed in endothelial cells, thereby inhibiting TGF-β/ALK1-induced Smad1/5 activation [27]. Moreover, two regulatory subunits of PP2A, Bα and Bδ, are capable of associating with TGF-β receptors and modulating TGF-β/Activin/Nodal signaling in different manners [28, 29].

Poly-ubiquitination of proteins is a crucial process that defines protein stability and turnover [1, 30–33]. In general, ubiquitination requires sequential actions of a ubiquitin-activating enzyme (E1), a ubiquitin-conjugating enzyme (E2), and a ubiquitin ligase (E3), covalently attaching a mono-ubiquitin molecule or a poly-ubiquitin chain on lysine residues of substrate proteins. The specificity of ubiquitination is largely reliant on the E3 ubiquitin ligases. TGF-β receptors can be ubiquitinated and degraded through the ubiquitin system [1, 32]. Intriguingly, unlike other receptors, TGF-β receptors usually need an adaptor, Smad7, to recruit different E3 ubiquitin ligases, especially the WW-HECT type E3 ligases, including Smurf1, Smurf2, WWP1, and NEDD4-2 [12, 15, 34]. Smad7 interacts via its PY motif with the WW domains of these E3 ligases, and recruits them to the activated TβRI. The E3 ligase-promoted ubiquitination can target the receptors for degradation either through the lysosomal pathway or through the proteasomal pathway, as degradation of ubiquitinated receptors is sensitive to both proteasome inhibitors and lysosome inhibitors, but mechanism underlying the proteasome-mediated degradation of receptors is undefined. As TβRI is complexed with TβRII, both of the receptors could be targeted for ubiquitination and degradation. In contrary to ubiquitin ligases, deubiquitinating enzymes (DUBs) remove poly-ubiquitin chains from proteins. With regard to TGF-β receptors, USP4, USP11, USP15, and UCH37 have been shown to associate with TβRI and/or Smad7, resulting in deubiquitination and stabilization of the receptor and enhancement of TGF-β signaling [35–38]. The Smad7-mediated receptor ubiquitination and stability is regulated by various mechanisms. For instance, we found that the TGF-β-inducible protein TSC-22 could associate with both the TGF-β receptor complex and Smad7 in response to TGF-β signaling [39, 40]. As both interactions require the short N-terminal fragment (1–44 amino acids) of TSC-22, these interactions are mutually exclusive. Therefore, TSC-22 antagonizes Smad7/Smurfs-mediated ubiquitination and degradation of TβRI and augments TGF-β signaling.

Other ubiquitin-related molecules, such as SUMO and NEDD8 (neural precursor cell-expressed, developmentally down-regulated 8), can also be attached to substrate proteins on lysine residues. Like ubiquitination, NEDD8-mediated neddylation, catalyzed by its own E1, E2, and E3 enzymes, modulates protein activity, stability or subcellular localization [41]. Recently, we identified Casitas B-lineage Lymphoma (c-Cbl) to act as an NEDD8 E3 ligase of TβRII [42]. TGF-β receptors can be internalized through either the clathrin-dependent pathway to propagate signaling or the caveolin-dependent pathway to promote receptor turnover and terminate signaling [43, 44]. Our findings suggest that neddylated TβRII prefers to undergo clathrin-mediated endocytosis and mediates signal transduction, and meanwhile inhibits caveolin-mediated endocytosis, thereby attenuating the ubiquitination and degradation of TβRII. Furthermore, c-Cbl-mediated TβRII neddylation could be reversed by the NEDD8-specific protease NEDP1, achieving a precise control of TβRII stability and activity.

In addition to the PTMs described above, Rik Derynck and coworkers reported that TβRI, but not TβRII or the other type I receptor members, could be sumoylated in response to TGF-β, and this modification facilitates recruitment and activation of Smad2/3 [16]. TβRI has also been reported to undergo TNF-α converting enzyme (TACE)-mediated cleavage upon TGF-β treatment, and this cleavage is promoted by TRAF6-interceded Lys63-linked poly-ubiquitination of TβRI [45–47]. Moreover, TGF-β receptors could be modified by N-linked glycosylation and core fucosylation, which in turn affect receptor intracellular transportation, activation, or endocytosis [48–51]. In this chapter, we focus on phosphorylation, ubiquitination, and neddylation of TGF-β receptors, and the methods for analyzing these PTMs (*see* **Note 1**).

2 Materials

Cell Culture Reagents: Minimum Essential Medium (MEM), Pi-free MEM, and Dulbecco's Modified Eagle's Medium (DMEM) are purchased from Invitrogen, penicillin/streptomycin sulfate and 0.25 % trypsin/EDTA from Amresco, and fetal bovine serum (FBS) from Hyclone.

Chemicals and Reagents: TGF-β1 is purchased from R&D systems, protease inhibitor cocktail from Roche, Protein A sepharose beads from Invitrogen, nickel-nitrilotriacetic acid (Ni-NTA) beads from Qiagen, and [^{32}P]orthophosphate from PerkinElmer NEN.

Antibodies: The following antibodies are used in our experiments: M2 anti-Flag antibody (Sigma), anti-myc and anti-HA antibodies (Santa Cruz Biotechnology).

2.1 Phosphorylation of TβRI

Cells: TβRI-deficient mink lung epithelial R-1B/L17 cells are obtained from Dr. Joan Massague's laboratory (Memorial Sloan Kettering Cancer Center).

DNA Constructs: The TβRI and TβRII cDNA are C-terminally tagged with HA and His, respectively, and subcloned into pCMV5 vector. Human FKBP12 is tagged with the Flag epitope sequence at the N-terminus and subcloned into pCMV5. Mutant derivatives of TβRII and FKBP12 are generated by PCR based site-directed mutagenesis.

Buffers

1. TNT buffer: 20 mM Tris–HCl, pH 7.4, 150 mM NaCl, 0.5 % (v/v) Triton X-100.
2. TNT buffer containing protease inhibitor cocktail and phosphatase inhibitors: TNT buffer containing 0.5 mM Na_3VO_4 in HEPES pH 7.5; 50 mM NaF; 30 mM $NO_4P_2O_7$.

2.2 Ubiquitination of TβRI

Cells: HEK293T.

DNA Constructs: Human TSC-22 cDNA is subcloned into pCMV-myc vector. The constitutively active (ca-) TβRI (GGD, i.e., combined mutations of L193G, P194G, and T204D) with a C-terminal HA tag is generated through PCR based site-directed mutagenesis. The ubiquitin (Ub) cDNA is N-terminally tagged with myc and His, and subcloned into pCMV5. Myc-Smurfl and Flag-Smurf2 constructs were gifts from Dr. Xin-Hua Feng (Baylor College of Medicine, Houston, USA).

Buffers

1. GTN Lysis Buffer: 6 M guanidine-HCl, 50 mM Tris–HCl, 250 mM NaCl, 5 mM imidazole, pH 8.0, and protease inhibitors.
2. Wash Buffer A: 3 M guanidine-HCl, 50 mM Tris–HCl, 250 mM NaCl, 5 mM imidazole, pH 8.0.
3. Wash Buffer B: 50 mM Tris–HCl, 250 mM NaCl, 5 mM imidazole, 0.2 % Triton X-100, pH 8.0.

2.3 Neddylation of TβRII

Cells: HEK293T

DNA Constructs: The TβRII cDNA is C-terminally tagged with HA and subcloned into pCMV5. The human c-Cbl cDNA is C-terminally tagged with a myc epitope and subcloned into pcDNA3.1(+) vector. The human NEDD8 cDNA is N-terminally tagged with Flag and inserted into pcDNA3.1(+).

Buffer

IP buffer: 50 mM Tris–HCl (pH 7.5), 150 mM NaCl, 0.5 % NP-40, 40 mM sodium pyrophosphate, 1 mM NaVO4 and protease inhibitors.

3 Methods

3.1 Phosphorylation of TβRI

In order to detect TβRII-induced phosphorylation of TβRI in vivo, we employed a TβRI-deficient cell line R-1B/L17, which is derived from the mink lung epithelial cell line Mv1Lu [52]. As Lys277 in the ATP-binding site of TβRII is essential for its kinase activity, mutation of this residue to Arg (K277R) abolishes the kinase activity [53]. Consequently, we co-expressed wild-type (wt) TβRI together with wild-type TβRII or K277R mutant in L17 cells. Cells were harvested after labeling with [^{32}P]orthophosphate and treatment with TGF-β1, and then the receptor complexes were isolated by a sequential precipitation protocol, and phosphorylated receptors were analyzed by SDS-PAGE followed by autoradiography (Fig. 1). FKBP12 has peptidyl-prolyl *cis-trans* isomerase activity, and mutation of two amino acids at or near the active site (F36Y and I90K) disrupts the interaction of FKBP12 with TβRI [23]. Therefore, to examine the effect of FKBP12 on TβRI phosphorylation, we compared the effects of wild-type FKBP12 and its mutant derivatives that fail to bind TβRI.

We assume the familiarity of the readers with several common techniques including plasmid construction, cell culture, cell transfection, SDS-PAGE, autoradiography, and immunoblotting analysis.

3.1.1 Mammalian Cell Culture and Transfection

R-1B/L17 cells are maintained in MEM, supplemented with 10 % fetal bovine serum (FBS), 100 μg/mL streptomycin sulfate and 100 U/mL penicillin (10 % FBS MEM) at 37 °C in a humidified, 5 % CO_2 incubator. To maintain L17 cells in the resting state, cells are cultured in MEM with 0.2 % FBS (0.2 % FBS MEM) for 24 h before TGF-β treatment.

Transient transfection of L17 cells is carried out by the DEAE-dextran method. Briefly, exponentially growing cells of 50–70 % confluence are incubated with MEM containing 1–1.5 mg/mL of DNA, 100 mM chloroquine and 125 mg/mL of DEAE-dextran at 37 °C for 3 h. Then, cells are shocked for 2 min with 10 %

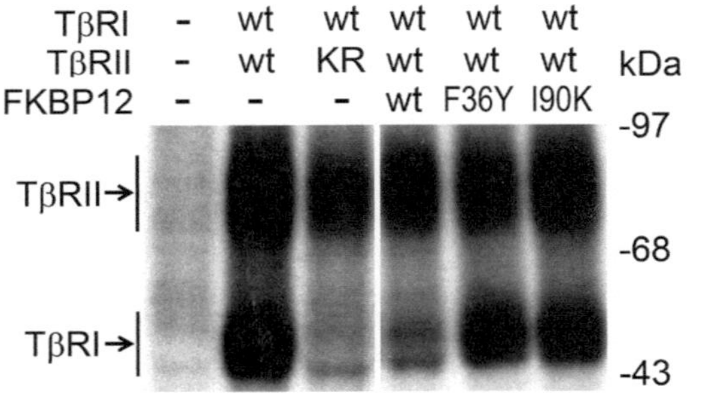

Fig. 1 FKBP12 inhibits TβRII-induced phosphorylation of TβRI. R-1B/L17 cells were co-transfected with wild-type TβRI, wild-type or kinase-defective TβRII and wild-type or binding-defective FKBP12 constructs, as indicated. Cells were labeled with [^{32}P]orthophosphate for 3 h and incubated with 1 nM TGF-β for the final 20 min. Receptor complexes were recovered by sequential precipitation via a His tag in the TβRII construct and an HA tag in the TβRI construct, analyzed by SDS-PAGE, and visualized by autoradiography

of dimethyl sulfoxide in phosphate-buffered saline (PBS) at room temperature. Subsequently, cells are washed and placed in growth medium.

Experimental Procedures:

1. Transfect L17 cells of 50–70 % confluence with constructs encoding wild-type TβRI, wild-type TβRII or its K277R mutant derivative, and wild-type or mutant FKBP12; total amount of transfected plasmids is normalized by adding empty vectors (*see* **Note 2**).
2. At 20 h post-transfection, replace the growth media with 0.2 % FBS MEM and culture cells for another day.
3. Wash cells with Pi-free MEM, three times.
4. Label cells with [^{32}P]orthophosphate (1 mCi/mL) in Pi-free MEM with 0.2 % FBS (dialyzed at 4 °C against 25 mM HEPES, pH 7.2 and 150 mM NaCl) for 3 h. For a 60-mm dish, 1 mL of labeling medium is used.
5. Treat cells with 1 nM TGF-β1 in 0.2 % FBS MEM for 20 min.
6. To harvest cells, first wash cells with ice-cold PBS once, and then lyse cells in 500 μL of TNT buffer containing protease inhibitor cocktail and phosphatase inhibitors by scraping using a cell scraper.
7. Centrifuge the samples at 18,000 g for 10 min at 4 °C with a microcentrifuge.
8. Transfer the supernatants into new 1.5 mL Eppendorf tubes.
9. Add imidazole to a final concentration of 20 mM.
10. Carry out the first round precipitation by addition of 30 μL of Ni-NTA agarose beads (Qiagen), followed by rotating for 1 h at 4 °C.
11. Wash beads by spinning at 2,500 g for 20 s at 4 °C with a microcentrifuge, displacing the supernatants, adding 750 μL of TNT buffer, and followed by rocking for 5 min at 4 °C. Repeat the wash for two more times.
12. Elute the beads-bound proteins by adding 100 μL of TNT buffer containing 250 mM imidazole and rocking for 10 min at room temperature.
13. Spin down the beads at 2,500 g for 20 s at 4 °C with a microcentrifuge.
14. Transfer the supernatants/elutes to new 1.5 mL tubes, and dilute them with 10 volumes of TNT buffer.
15. Carry out immunoprecipitation by the addition of 5 μL of anti-HA antibody (1 mg/mL) and 30 μL of Protein A sepharose, followed by rotating for 6 h (or overnight) at 4 °C.

16. Wash the sepharose with TNT buffer for four times as above.
17. After the final wash, remove TNT buffer carefully using a 1 mL syringe but keep the sepharose moist.
18. Add 20–30 μL of 2× loading buffer for each sample.
19. Boil the samples for 3–5 min, vortex for a short time period, and spin briefly.
20. Analyze the immunoprecipitates by SDS-PAGE and autoradiography.

3.2 Ubiquitination of TβRI

Modulation of the stability of TGF-β receptors is a key regulatory mechanism for TGF-β signaling. Both types of TGF-β receptors can undergo ubiquitination-mediated degradation [12, 34, 54]. To enhance the ubiquitination signal, the proteasome inhibitor MG132 is usually used in ubiquitination assay. To further facilitate detection of ubiquitinated receptors, constitutively active TβRI together with ubiquitin (Ub) and Smurf1/2 were overexpressed in cells. HEK293T is one of the most popular cell lines as it is quite easy to culture and to transfect. As our observations suggested that TSC-22 would interfere with the inhibitory activities of Smad7/Smurfs on TGF-β signaling, we examined whether ectopic TSC-22 could inhibit Smurf1/2-induced TβRI ubiquitination (Fig. 2).

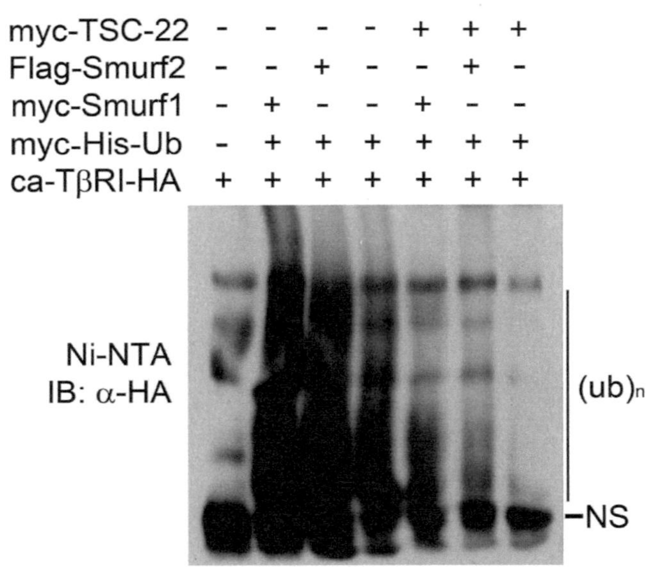

Fig. 2 TSC-22 attenuates Smurf1/2-induced ubiquitination of TβRI. HEK293T cells transfected with the indicated constructs (2 μg each, for 60-mm dishes) were treated with 50 μM MG132 for 4 h before being harvested for precipitation with nickel beads followed by anti-HA immunoblotting. *NS* nonspecific band. From Yan et al. 2012, Mol Cell Biol, 31:3700–3709

3.2.1 Mammalian Cell Culture and Transfection

HEK293T cells are maintained in DMEM, supplemented with 10 % fetal bovine serum (FBS), 100 μg/mL streptomycin sulfate, and 100 U/mL penicillin (10 % FBS DMEM) at 37 °C in a humidified, 5 % CO_2 incubator.

Cell transfection is conducted with VigoFect (Vigorous Biotechnology, Beijing, China) according to introduction.

Experimental Procedures:

1. Grow HEK293T cells in 60-mm dishes to 60–80 % confluence.
2. Transfect HEK293T cells with the constructs encoding ca-TβRI-HA, myc-His-Ub, myc-TSC-22, myc-Smurf1, and Flag-Smurf2, and the total amount of transfected plasmids is normalized with empty vectors.
3. At 40 h post-transfection, treat cells with 50 μM MG132 for 4 h in growth medium (10 % FBS DMEM).
4. Wash cells once using ice-cold PBS.
5. Place dishes on ice, add 600 μL of GTN lysis buffer containing protease inhibitors and 50 μM MG132, and detach and lyse cells by scraping using a cell scraper.
6. Transfer cell lysates into 1.5 mL Eppendorf tubes, pass the lysates through a 1 mL syringe for 10 times, sonicate, and leave on ice for 10 min.
7. Centrifuge the cell lysates at 18,000 g for 10 min at 4 °C with a microcentrifuge, and transfer the supernatants into new 1.5 mL tubes.
8. Add 30 μL of Ni-NTA beads, and rotate for 4 h to overnight at 4 °C.
9. Wash beads with buffer A twice.
10. Wash beads with buffer B twice.
11. After the final wash, add 20–30 μL of 2× loading buffer for each sample.
12. Analyze the precipitates through SDS-PAGE and anti-HA immunoblotting.

3.3 Neddylation of TβRII

Our recent findings showed that c-Cbl could promote neddylation, a ubiquitination-like posttranslational modification with NEDD8, of TβRII, which antagonizes ubiquitination and degradation of the receptor [42]. To examine c-Cbl-mediated neddylation of TβRII, we co-expressed TβRII together with NEDD8 and c-Cbl in HEK293T cells. At 40 h post-transfection, cells were harvested for neddylation assay (Fig. 3).

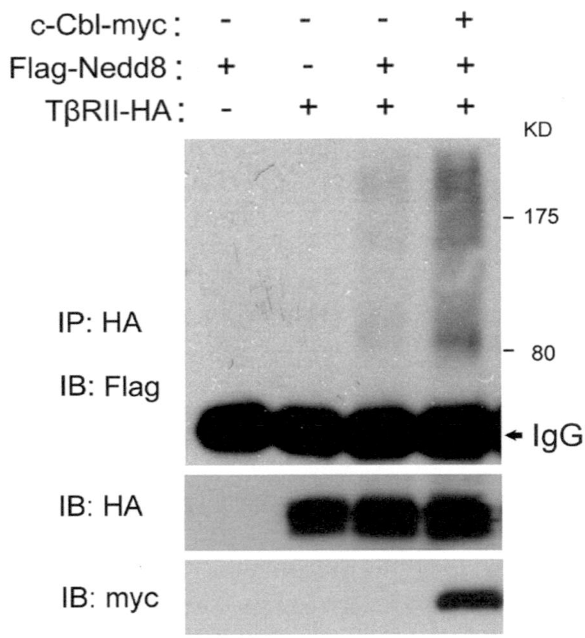

Fig. 3 c-Cbl promotes neddylation of TβRII. HEK293T cells were transfected with indicated plasmids, and the amount of TβRII-HA plasmid was adjusted to ensure comparable expression. At 40 h post-transfection, cells were harvested for anti-HA immunoprecipitation followed by anti-Flag immunoblotting analysis. From Zuo et al. 2013, Mol Cell, 49:499–510

3.3.1 Mammalian Cell Culture and Transfection

HEK293T cells are cultured and transfected as above.

Experimental Procedures:

1. Grow HEK293T cells in 60 mm dishes to 60–80 % confluence.
2. Transfect HEK293T cells with the constructs encoding TβRII-HA, Flag-NEDD8, and c-Cbl-myc, and the total amount of transfected plasmids is normalized with empty vectors.
3. At 40 h post-transfection, wash cells once using ice-cold PBS.
4. Place dishes on ice, add 600 μL of ice-clod PBS, detach cells by scraping using a cell scraper, and then transfer the cells into 1.5 mL Eppendorf tubes.
5. Collect cell pellets through centrifugation at 1,000 g for 3 min at 4 °C with a microcentrifuge.
6. Discard the supernatants, add 100 μL of 1 % SDS, and lyse cells by pipetting up and down using a 1 mL Pipetman.
7. Boil for 4 min for complete lysis.

8. Put tubes back on ice, and dilute cell lysates with 900 μL of IP buffer to reach 0.1 % of the final SDS.
9. Rotate for 30 min at 4 °C.
10. Centrifuge at 18,000 g for 10 min at 4 °C with a microcentrifuge.
11. Transfer the supernatants into new 1.5 mL tubes, and take 1/10 of lysates for total protein expression examination.
12. For the remaining lysates, perform immunoprecipitation by addition of 5 μL of anti-HA antibody (1 mg/mL) and 30 μL of Protein-A sepharose beads, followed by rotating at 4 °C overnight.
13. Wash beads with IP buffer four times.
14. Add 20–30 μL of 2× loading buffer for each sample.
15. Analyze the precipitates by SDS-PAGE followed by anti-Flag immunoblotting for neddylation detection, and TβRII expression level is examined by anti-HA immunoblotting.

4 Notes

1. Phosphorylation, ubiquitination, or neddylation of endogenous TβRI and TβRII could be detected using specific antibodies with the similar protocols described above. However, as the endogenous modification levels are much lower than those of overexpressed receptors, good antibodies, ligand stimulation, and more sensitive detection are needed.
2. The purpose of differential tagging on TβRI (HA tag) and TβRII (His tags) is to carry out sequential precipitation of the receptors. Ni-NTA beads are first used to pull down His-tagged TβRII and His-TβRII-bound HA-TβRI. [^{32}P]-labeled TβRII-TβRI complex is then precipitated by anti-HA antibody conjugated on Protein A sepharose beads.

Acknowledgement

We thank Dr. Fei Huang for critical reading of the manuscript.

References

1. Moustakas A, Heldin CH (2009) The regulation of TGFbeta signal transduction. Development 136:3699–3714
2. Massague J (2012) TGFbeta signalling in context. Nat Rev Mol Cell Biol 13:616–630
3. Ikushima H, Miyazono K (2012) TGF-beta signal transduction spreading to a wider field: a broad variety of mechanisms for context-dependent effects of TGF-beta. Cell Tissue Res 347:37–49

4. Schmierer B, Hill CS (2007) TGFbeta-SMAD signal transduction: molecular specificity and functional flexibility. Nat Rev Mol Cell Biol 8:970–982
5. Feng XH, Derynck R (2005) Specificity and versatility in TGF-beta signaling through Smads. Annu Rev Cell Dev Biol 21:659–693
6. Derynck R, Miyazono K (2008) The TGF-beta family. Cold Spring Harbor Laboratory Press, New York
7. Derynck R, Feng XH (1997) TGF-beta receptor signaling. Biochim Biophys Acta 1333:F105–F150
8. Massague J (1998) TGF-beta signal transduction. Annu Rev Biochem 67:753–791
9. Massague J, Chen YG (2000) Controlling TGF-beta signaling. Genes Dev 14:627–644
10. ten Dijke P, Hill CS (2004) New insights into TGF-beta-Smad signalling. Trends Biochem Sci 29:265–273
11. Heldin CH, Miyazono K, ten Dijke P (1997) TGF-beta signalling from cell membrane to nucleus through SMAD proteins. Nature 390:465–471
12. Huang F, Chen YG (2012) Regulation of TGF-beta receptor activity. Cell Biosci 2:9
13. Xu P, Liu J, Derynck R (2012) Post-translational regulation of TGF-beta receptor and Smad signaling. FEBS Lett 586:1871–1884
14. Wrighton KH, Lin X, Feng XH (2009) Phospho-control of TGF-beta superfamily signaling. Cell Res 19:8–20
15. Yan X, Chen YG (2011) Smad7: not only a regulator, but also a cross-talk mediator of TGF-beta signalling. Biochem J 434:1–10
16. Kang JS, Saunier EF, Akhurst RJ, Derynck R (2008) The type I TGF-beta receptor is covalently modified and regulated by sumoylation. Nat Cell Biol 10:654–664
17. Lee MK, Pardoux C, Hall MC, Lee PS, Warburton D, Qing J, Smith SM, Derynck R (2007) TGF-beta activates Erk MAP kinase signalling through direct phosphorylation of ShcA. EMBO J 26:3957–3967
18. Ehrlich M, Horbelt D, Marom B, Knaus P, Henis YI (2011) Homomeric and heteromeric complexes among TGF-beta and BMP receptors and their roles in signaling. Cell Signal 23:1424–1432
19. Zhang W, Jiang Y, Wang Q, Ma X, Xiao Z, Zuo W, Fang X, Chen YG (2009) Single-molecule imaging reveals transforming growth factor-beta-induced type II receptor dimerization. Proc Natl Acad Sci U S A 106:15679–15683
20. Zhang W, Yuan J, Yang Y, Xu L, Wang Q, Zuo W, Fang X, Chen YG (2010) Monomeric type I and type III transforming growth factor-beta receptors and their dimerization revealed by single-molecule imaging. Cell Res 20:1216–1223
21. Huang T, David L, Mendoza V, Yang Y, Villarreal M, De K, Sun L, Fang X, Lopez-Casillas F, Wrana JL, Hinck AP (2011) TGF-beta signalling is mediated by two autonomously functioning TbetaRI:TbetaRII pairs. EMBO J 30:1263–1276
22. Huse M, Chen YG, Massague J, Kuriyan J (1999) Crystal structure of the cytoplasmic domain of the type I TGF beta receptor in complex with FKBP12. Cell 96:425–436
23. Chen YG, Liu F, Massague J (1997) Mechanism of TGFbeta receptor inhibition by FKBP12. EMBO J 16:3866–3876
24. Souchelnytskyi S, ten Dijke P, Miyazono K, Heldin CH (1996) Phosphorylation of Ser165 in TGF-beta type I receptor modulates TGF-beta1-induced cellular responses. EMBO J 15:6231–6240
25. Bennett D, Alphey L (2002) PP1 binds Sara and negatively regulates Dpp signaling in Drosophila melanogaster. Nat Genet 31:419–423
26. Shi W, Sun C, He B, Xiong W, Shi X, Yao D, Cao X (2004) GADD34-PP1c recruited by Smad7 dephosphorylates TGFbeta type I receptor. J Cell Biol 164:291–300
27. Valdimarsdottir G, Goumans MJ, Itoh F, Itoh S, Heldin CH, ten Dijke P (2006) Smad7 and protein phosphatase 1alpha are critical determinants in the duration of TGF-beta/ALK1 signaling in endothelial cells. BMC Cell Biol 7:16
28. Batut J, Schmierer B, Cao J, Raftery LA, Hill CS, Howell M (2008) Two highly related regulatory subunits of PP2A exert opposite effects on TGF-beta/Activin/Nodal signalling. Development 135:2927–2937
29. Griswold-Prenner I, Kamibayashi C, Maruoka EM, Mumby MC, Derynck R (1998) Physical and functional interactions between type I transforming growth factor beta receptors and Balpha, a WD-40 repeat subunit of phosphatase 2A. Mol Cell Biol 18:6595–6604
30. Varshavsky A (2012) The ubiquitin system, an immense realm. Annu Rev Biochem 81:167–176
31. Soond SM, Chantry A (2011) How ubiquitination regulates the TGF-beta signalling pathway: new insights and new players: new isoforms of ubiquitin-activating enzymes in the E1-E3 families join the game. Bioessays 33:749–758
32. Inoue Y, Imamura T (2008) Regulation of TGF-beta family signaling by E3 ubiquitin ligases. Cancer Sci 99:2107–2112
33. Isasa M, Zuin A, Crosas B (2012) Integration of multiple ubiquitin signals in proteasome regulation. Methods Mol Biol 910:337–370
34. Kang JS, Liu C, Derynck R (2009) New regulatory mechanisms of TGF-beta receptor function. Trends Cell Biol 19:385–394

35. Zhang L, Zhou F, Drabsch Y, Gao R, Snaar-Jagalska BE, Mickanin C, Huang H, Sheppard KA, Porter JA, Lu CX, Ten Dijke P (2012) USP4 is regulated by AKT phosphorylation and directly deubiquitylates TGF-beta type I receptor. Nat Cell Biol 14:717–726
36. Al-Salihi MA, Herhaus L, Macartney T, Sapkota GP (2012) USP11 augments TGFbeta signalling by deubiquitylating ALK5. Open Biol 2:120063
37. Eichhorn PJ, Rodon L, Gonzalez-Junca A, Dirac A, Gili M, Martinez-Saez E, Aura C, Barba I, Peg V, Prat A, Cuartas I, Jimenez J, Garcia-Dorado D, Sahuquillo J, Bernards R, Baselga J, Seoane J (2012) USP15 stabilizes TGF-beta receptor I and promotes oncogenesis through the activation of TGF-beta signaling in glioblastoma. Nat Med 18:429–435
38. Wicks SJ, Haros K, Maillard M, Song L, Cohen RE, Dijke PT, Chantry A (2005) The deubiquitinating enzyme UCH37 interacts with Smads and regulates TGF-beta signalling. Oncogene 24:8080–8084
39. Yan X, Zhang J, Pan L, Wang P, Xue H, Zhang L, Gao X, Zhao X, Ning Y, Chen YG (2011) TSC-22 promotes transforming growth factor beta-mediated cardiac myofibroblast differentiation by antagonizing Smad7 activity. Mol Cell Biol 31:3700–3709
40. Xu L (2011) Averting a roadblock in transforming growth factor beta signaling. Mol Cell Biol 31:3684–3686
41. Watson IR, Irwin MS, Ohh M (2011) NEDD8 pathways in cancer. Sine Quibus Non Cancer Cell 19:168–176
42. Zuo W, Huang F, Chiang YJ, Li M, Du J, Ding Y, Zhang T, Lee HW, Jeong LS, Chen Y, Deng H, Feng XH, Luo S, Gao C, Chen YG (2013) c-Cbl-mediated neddylation antagonizes ubiquitination and degradation of the TGF-beta type II receptor. Mol Cell 49:499–510
43. Le Roy C, Wrana JL (2005) Clathrin- and non-clathrin-mediated endocytic regulation of cell signalling. Nat Rev Mol Cell Biol 6:112–126
44. Chen YG (2009) Endocytic regulation of TGF-beta signaling. Cell Res 19:58–70
45. Mu Y, Sundar R, Thakur N, Ekman M, Gudey SK, Yakymovych M, Hermansson A, Dimitriou H, Bengoechea-Alonso MT, Ericsson J, Heldin CH, Landstrom M (2011) TRAF6 ubiquitinates TGFbeta type I receptor to promote its cleavage and nuclear translocation in cancer. Nat Commun 2:330
46. Chandra M, Zang S, Li H, Zimmerman LJ, Champer J, Tsuyada A, Chow A, Zhou W, Yu Y, Gao H, Ren X, Lin RJ, Wang SE (2012) Nuclear translocation of type I transforming growth factor beta receptor confers a novel function in RNA processing. Mol Cell Biol 32:2183–2195
47. Liu C, Xu P, Lamouille S, Xu J, Derynck R (2009) TACE-mediated ectodomain shedding of the type I TGF-beta receptor downregulates TGF-beta signaling. Mol Cell 35:26–36
48. Wang X, Inoue S, Gu J, Miyoshi E, Noda K, Li W, Mizuno-Horikawa Y, Nakano M, Asahi M, Takahashi M, Uozumi N, Ihara S, Lee SH, Ikeda Y, Yamaguchi Y, Aze Y, Tomiyama Y, Fujii J, Suzuki K, Kondo A, Shapiro SD, Lopez-Otin C, Kuwaki T, Okabe M, Honke K, Taniguchi N (2005) Dysregulation of TGF-beta1 receptor activation leads to abnormal lung development and emphysema-like phenotype in core fucose-deficient mice. Proc Natl Acad Sci U S A 102:15791–15796
49. Partridge EA, Le Roy C, Di Guglielmo GM, Pawling J, Cheung P, Granovsky M, Nabi IR, Wrana JL, Dennis JW (2004) Regulation of cytokine receptors by Golgi N-glycan processing and endocytosis. Science 306:120–124
50. Kim YW, Park J, Lee HJ, Lee SY, Kim SJ (2012) TGF-beta sensitivity is determined by N-linked glycosylation of the type II TGF-beta receptor. Biochem J 445:403–411
51. Lin H, Wang D, Wu T, Dong C, Shen N, Sun Y, Sun Y, Xie H, Wang N, Shan L (2011) Blocking core fucosylation of TGF-beta1 receptors downregulates their functions and attenuates the epithelial-mesenchymal transition of renal tubular cells. Am J Physiol Renal Physiol 300:F1017–F1025
52. Boyd FT, Massague J (1989) Transforming growth factor-beta inhibition of epithelial cell proliferation linked to the expression of a 53-kDa membrane receptor. J Biol Chem 264:2272–2278
53. Wrana JL, Attisano L, Carcamo J, Zentella A, Doody J, Laiho M, Wang XF, Massague J (1992) TGF beta signals through a heteromeric protein kinase receptor complex. Cell 71:1003–1014
54. Lonn P, Moren A, Raja E, Dahl M, Moustakas A (2009) Regulating the stability of TGFbeta receptors and Smads. Cell Res 19:21–35

Chapter 4

Production, Isolation, and Structural Analysis of Ligands and Receptors of the TGF-β Superfamily

Tao Huang and Andrew P. Hinck

Abstract

The ability to understand the molecular mechanisms by which secreted signaling proteins of the TGF-β superfamily assemble their cell surface receptors into complexes to initiate downstream signaling is dependent upon the ability to determine atomic-resolution structures of the signaling proteins, the ectodomains of the receptors, and the complexes that they form. The structures determined to date have revealed major differences in the overall architecture of the signaling complexes formed by the TGF-βs and BMPs, which has provided insights as to how they have evolved to fulfill their distinct functions. Such studies, have however, only been applied to a few members of the TGF-β superfamily, which is largely due to the difficulty of obtaining milligram-scale quantities of highly homogenous preparations of the disulfide-rich signaling proteins and receptor ectodomains of the superfamily. Here we describe methods used to produce signaling proteins and receptor ectodomains of the TGF-β superfamily using bacterial and mammalian expression systems and procedures to purify them to homogeneity.

Key words TGF-β, ligand and receptor, protein expression, protein purification, structural analysis

1 Introduction

The transforming growth factor-beta (TGF-β) superfamily is comprised of a diverse family of signaling proteins, with three known family members in *C. elegans*, seven in *D. melanogaster*, and more than thirty in humans and other vertebrates [1]. The proteins of the superfamily arose as developmental factors responsible for embryonic patterning and morphogenesis in invertebrates, but have further evolved to fulfill many extra embryonic roles as organisms have diversified.

The proteins of the superfamily are produced as pre-pro proteins and are secreted either as mature disulfide-linked dimers, or as mature disulfide-linked dimers non-covalently bound to their pro-domain [2, 3]. The mature homodimers signal by binding and bringing together two transmembrane receptors, known as receptor types I and II, to form heterotetrameric complexes with

Xin-Hua Feng et al. (eds.), *TGF-β Signaling: Methods and Protocols*, Methods in Molecular Biology, vol. 1344, DOI 10.1007/978-1-4939-2966-5_4, © Springer Science+Business Media New York 2016

two type I and two type II receptors [4, 5]. The proteins of the superfamily can be divided into two phylogenetic clades based on the type I receptors they bind and Smad proteins they activate [6]. The more recently evolved members of the superfamily, including the TGF-βs, activins, nodal, and some of the GDFs and BMPs (GDF-9, -11, and -15 and BMP-15), bind and signal through type I receptors that couple to and activate R-Smads 2, 3, while the more distantly related GDFs (GDF-1, -3, -5, -7, and -10) and BMPs (BMP-2, -3, -4, -5, -6, -7, -8, -9, and -10) bind and signal through type I receptors that couple to and activate R-Smads 1, 5, and 8 [1]. The two subclasses of R-Smads, upon association with the co-mediator Smad, Smad4, assemble distinct transcriptional complexes and thus activate (or repress) distinct subsets of genes [7].

The recent structures of TGF-βs (TGF-β1 and -β3) and BMPs (BMP-2 and BMP-7) bound to the ectodomains of their receptors (TβRI and TβRII and BMPRIa/BMPRIb, ActRII/ActRIIb, respectively) show that although the TGF-βs and BMPs and their receptors share the same overall folds, they nevertheless bind and assemble their receptors in a distinct manner (Fig. 1) [8–13]. The differences are especially pronounced for the type II receptors: BMP type II receptors use the concave surface of their β-sheet to complement the convex surface of the knuckle epitope of the BMP, while the TGF-β type II receptor inserts one of its edge β-strands into the cleft between the fingertips of the TGF-β. Though the differences are less pronounced, the type I receptors also bind differently: BMP type I receptors bind to the wrist and have extensive contact with both BMP monomers, while the TGF-β type I receptor is shifted away from the wrist toward the fingertips where it contacts TβRII, the TGF-β monomer to which TβRII is bound, and to a limited extent, the adjacent monomer. These structural differences are significant as the distinct interfaces expand the range of specificity and segregate the actions of TGF-βs, which signal through receptors that activate Smads 2 and 3, away from the many BMPs and GDFs, which signal through receptors that activate Smads 1, 5, and 8 [14–17].

The determination of the structures of the TGF-βs and BMPs bound to the ectodomains of their type I and type II receptors has been highly dependent on the ability to produce and isolate highly homogenous forms of each these proteins in quantities large enough for structural studies. The objective of this review is to provide an overview of the different strategies that have been used to obtain these proteins and to characterize their binding properties and structures in vitro. The high disulfide content of both the signaling proteins and receptor ectodomains of the superfamily necessitates that procedures specifically adapted for these types of proteins be used. The primary methods that have been used for

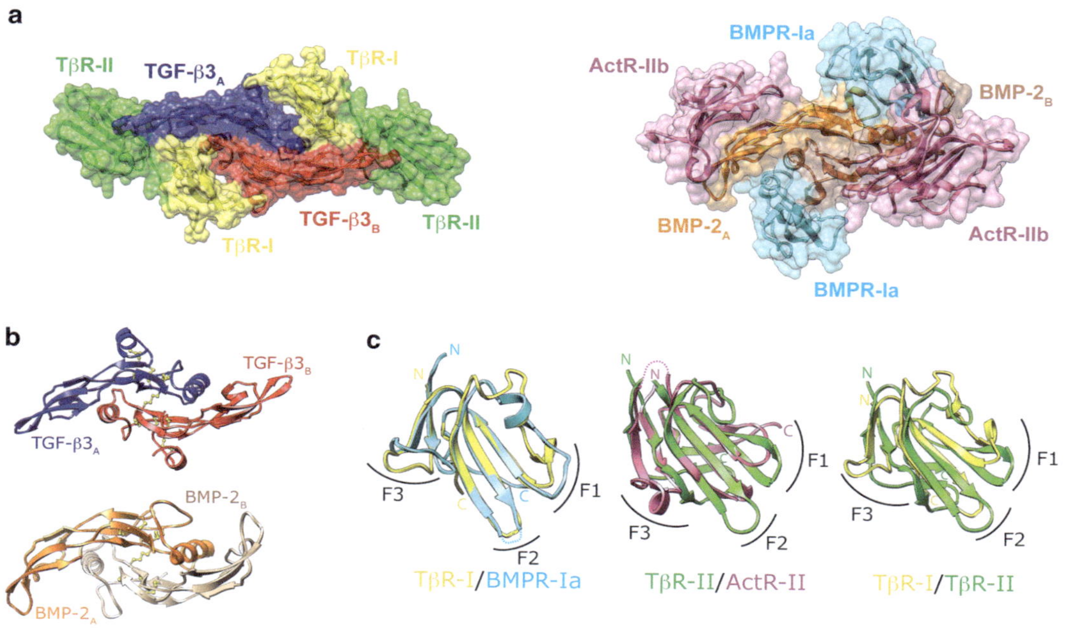

Fig. 1 Different modes of receptor complex assembly by TGF-βs and BMPs. (**a**) TGF-β (*left*) and BMP (*right*) type I receptor, type II receptor ternary complex structures. TGF-β type I and type II receptors are shaded *yellow* and *green*, respectively, and extensively contact one another. BMP type I and type II receptors are shaded cyan and magenta respectively, and do not contact one another. (**b**) Representative dimeric ligand structures, with the two monomers of TGF-β3 depicted in *blue* and *red*, and those of BMP-2 in *orange* and *brown*. (**c**) Receptor extracellular domains of the TGF-β superfamily adopt the same three finger toxin fold, as shown by an overlay of the BMP and TGF-β type I receptors on the *left* (*cyan* and *yellow*, respectively), the BMP and TGF-β type II receptors in the middle (*magenta* and *green*, respectively), and the TGF-β type I and type II receptors on the *right* (*yellow* and *green*, respectively). F1, F2, and F3 designate the three fingers of the receptor three-finger toxin fold

these purposes include overexpression as secreted proteins in cultured eukaryotic cells, and overexpression in bacteria, followed by renaturation into native signaling proteins or receptors. The methods commonly used to obtain signaling proteins of the superfamily will be described first, followed by a summary of the methods used to obtain the receptor ectodomains of the superfamily.

2 Materials

2.1 Reagents and Buffers for E. coli Expression

1. *E. coli* host strain: BL21(DE3) (Stratagene).
2. Expression constructs: The coding sequences of mature human TGF-β2 and extracellular domain of TβRII were synthesized by Genscript (Piscataway, NJ) to optimize the codon usage for *E.coli* and minimize formation of secondary structure by the mRNA. The coding sequences were inserted between the NdeI and HindIII sites in plasmid pET32a (Novagen, Madison WI).

3. Standard LB medium: Dissolve 10 g tryptone, 5 g yeast extract, and 10 g NaCl in 1 L H_2O and adjust to pH 7.4 by addition of 10 M NaOH, then sterilize by autoclaving. Add ampicillin (150 μg/mL) before use.
4. Ampicillin: Dissolve 1.5 g ampicillin in a final volume of 10 mL H_2O to yield 150 mg/mL stock solution, sterilize by filtration through a 0.22 μm filter, and store in the dark at −20 °C.
5. Isopropyl-β-D-thiogalactoside (IPTG): Dissolve 2.38 g IPTG in a final volume of 10 mL H_2O to yield 1 M stock solution, sterilize by passage through a 0.22 μm filter, and store at −20 °C.

2.2 Reagents and Buffers for Mammalian Expression

1. Cell line for transfection: CHO-lec3.2.8.1 (*see* **Note 1** in Section 4).
2. CMV-based expression construct: Construct is that initially reported by Zou et al. [18] in which the human TGF-β1 open reading frame is subcloned into pcDNA3.1(+) modified to include a glutamine synthetase gene for gene amplification. The leader peptide of TGF-β1 was replaced with that of rat serum albumin, and an eight-histidine tag was inserted immediately after the leader sequence to facilitate protein purification. In addition, Cys 33 of TGF-β1, which forms a disulfide bond with latent TGF-β1-binding protein (LTBP) was replaced by a serine residue to increase secretion.
3. Cell culture medium: DMEM/F12 medium, 10 % fetal bovine serum (FBS), antibiotics (penicillin–streptomycin), pre-warmed to 37 °C.
4. Minimal Essential Medium (MEM): Opti-MEM medium (Invitrogen).
5. Lipid transfection reagent: Lipofectamine 2000 (Invitrogen).
6. Purified plasmid DNA: Maxi Prep (Qiagen).
7. Cell culture flasks and plates.
8. Phosphate-buffered saline (PBS).
9. Glutamine-free Glasgow's Minimal Essential Medium (GMEM) (SAFC biosciences).
10. Glutamine synthase (GS) supplement (50×) (SAFC Biosciences).
11. L-methionine sulfoximine (MSX) (Sigma).
12. Serum-free medium for protein expression: CHO-S-SFM II (Gibco) or SFM4CHO-A (HyClone).
13. Dimethyl sulfoxide (DMSO).

2.3 Reagents and Buffers for Protein Purification

1. Tris(hydroxymethyl)aminomethane (Tris).
2. 2-(*N*-morpholino)ethanesulfonic acid (MES).

3. Phenylmethanesulfonyl fluoride (PMSF).
4. Dithiothreitol (DTT).
5. 3-[(3-Cholamidopropyl)dimethylammonio]-1-propanesulfonate (CHAPS).
6. Disruption buffer: 100 mM Tris, 10 mM EDTA, 1 mM PMSF, pH 8.0.
7. TGF-β solubilizing buffer: 8 M urea, 20 mM Tris, 1 % DTT (w/v), pH 8.0.
8. SP denaturing buffer: 8 M urea, 20 mM sodium acetate, 0.1 % DTT (w/v), pH 4.2.
9. TGF-β folding buffer: 100 mM Tris, 30 mM CHAPS, 1 M NaCl, 5 mM reduced glutathione, pH 9.5.
10. Source 15S buffer: 20 mM sodium acetate, 30 % isopropyl alcohol, pH 4.0.
11. Ni buffer: 50 mM Tris, 150 mM NaCl, and 10 mM imidazole at pH 8.0.
12. RII solubilizing buffer: 8 M urea, 20 mM Tris, pH 7.0
13. RII folding buffer: 200 mM Tris, 2 mM reduced glutathione (GSH), 0.5 mM oxidized-glutathione (GSSG), pH 8.0.
14. Source 15Q buffer: 20 mM MES, 1 mM PMSF, 1 mM EDTA, pH 6.0.

2.4 Equipment and Materials

1. Temperature-controlled orbital shaker.
2. Water-jacketed CO_2 incubator.
3. Refrigerated high-speed centrifuge.
4. Handheld glass/teflon homogenizer.
5. UV spectrophotometer.
6. Chromatography columns.
7. Dialysis tubing.
8. Centrifugal concentrators.
9. Amicon stirred cell concentrator.
10. Apparatus for SDS-PAGE.
11. Äkta Fast Protein Liquid Chromatography (FPLC) system (GE Healthcare) or equivalent.

3 Methods

3.1 Overview of Signaling Protein Production and Isolation

TGF-βs are disulfide-linked dimers of identical 112-residue monomers. The monomers have nine cysteines, one of which participates in the formation of the inter-chain disulfide and eight of which form four internal disulfides [19–22]. Three of the four internal disulfide bonds form a conserved structure known as a cysteine knot [23]. BMPs, GDFs, activins, and most other signaling proteins of the TGF-β superfamily share a similar structure, though the number of internal disulfide bonds varies, with some, such as activins/inhibins possessing four disulfides, but others, such as BMPs and GDFs possessing only the three that form the cystine knot (Fig. 1b).

TGF-βs and other proteins of the superfamily are produced as pre-pro proteins [2]. The pro-domains promote the maturation, including the formation of the inter-chain disulfide bond that stabilizes their dimeric structure [24, 25]. TGF-β's pro-domain, also known as latency associated peptide, or LAP, is almost three times the length of the mature signaling protein (251–280 vs. 112 residues, respectively) and contains three cysteines, two of which (C223 and C225 in TGF-β1) pair cross-wise with the same cysteines in another molecule of the pro-domain to form the pro-domain homodimer [3]. The pro-domains for proteins of the superfamily vary widely in size and sequence: all are cleaved from the mature N-terminal signaling protein prior to secretion, and while the pro-domains of some superfamily proteins, such as TGF-β, associate tightly with the mature signaling dimer and are secreted in complex with them, some, such as those for activins, do not [2]. The approaches used to produce the proteins of the superfamily include (a) overexpression of the full-length pre-pro proteins in cultured eukaryotic cells, and (b) overexpression of the mature monomers in bacteria, followed by renaturation of the overexpressed polypeptide into native dimers. The procedures commonly used to obtain proteins of the superfamily by expression in eukaryotic cells will first be described, followed by a summary of procedures used to renature bacterially expressed protein into native dimers.

3.1.1 Expression of Signaling Proteinsin Eukaryotic Cells

The majority of superfamily proteins produced in eukaryotic cells have been produced in mammalian cells, such as CHO or HEK-293 cells [18, 26–41]. The preference for mammalian cells is largely for practical reasons, as most investigators are interested in producing mammalian signaling proteins and mammalian cells have the appropriate molecular machinery in the endoplasmic reticulum (ER) for folding and proteolytic processing. Though mammalian hosts are widely used, it has been observed that the proteolytic processing of the pro domain in the ER is often incomplete. This can sometimes complicate the purification as it then becomes necessary to not only separate the mature signaling dimer from its pro-domain, but also the pro-domain-mature precursor. There are several reports of the overexpression of mammalian signaling proteins in nonmammalian cells, including human BMP-2, bovine activin A, and bovine inhibin A in insect cells [42, 43] and human activin A in yeast [44]. These expression hosts have some advantages over mammalian cells, including the absence of native binding partners that can interfere with maturation [18] and potentially higher yields [44, 45]. The higher yields have only been obtained when the protease cleavage site between the pro-domain and mature signaling dimer have been altered to more closely match the endogenous proteases in the expression host [44, 45]. Thus, successful heterologous expression has required some alterations to account for differences in molecular machinery relative to mammalian hosts, but so far these have been relegated to the enzymes responsible for proteolytic processing, not enzymes involved in catalyzing disulfide exchange or folding.

The signaling proteins are usually expressed along with their native pro-domains, although the signal peptides are often substituted to improve secretion [18, 33]. The overexpressed proteins are purified from the conditioned medium using either conventional means (ion-exchange, gel filtration, etc.) or purification tags, such as hexahistdine tags. The N-terminus of the mature signaling dimer is accessible and flexible and thus provides a suitable site for tagging, while the C-terminus is structurally ordered and buried in the interior of the structure, and thus does not provide an appropriate site for tagging. If N-terminal tags are used, they should be removed after the signaling protein has been purified as the N-terminus lies near the binding site for the type I receptor and tags attached here have been found to block receptor binding and signaling [44]. The most straightforward means of removing the tag is to engineer a protease cleavage site between the purification tag and the N-terminus and then treat with the corresponding protease after the mature signaling protein has been purified. The alternative is to place the tag downstream of the predicted signal peptide cleavage site, but before the beginning of the pro-domain. Tagging at this position has the advantage that the tag is removed along with the pro-domain after the pro-domain has been separated away from the mature signaling dimer. The disadvantage of tagging at this position is that some pro-domains, such as those for activin A, associate rather weakly with their mature domain, which may lead to significant loss of the mature signaling protein during the wash steps of the purification.

The complexes between the pro-domains and the mature signaling dimer are generally soluble and can be isolated under non-denaturing conditions. The TGF-β1:TGF-β1 pro-domain complex, for example, is isolated in 50 mM Tris, 150 mM NaCl, pH 8.0 [18, 33]. The mature signaling dimers, on the other hand, are generally poorly soluble in standard buffers at neutral pH after the pro-domain has been removed. The most common procedure is to acidify the pro-domain:signaling dimer complex after it has been purified from the medium. The acidification serves to release the mature signaling dimer from its pro-domain and to increase its charge, which generally improves solubility. Though acidification irreversibly unfolds many proteins, this is not generally observed with the disulfide-rich proteins of the TGF-β superfamily, and thus it is common for the final purification to be performed under acidic conditions. The procedures most commonly used include cation-exchange chromatography in acidic buffers (such as acetate at pH 4.0) or reverse phase chromatography with 0.05–0.1 % triflouracetic acid (and either acetonitrile or methanol for the gradient elution). The concentrated stocks of most superfamily proteins are also commonly stored under acidic conditions, such as 1 mM HCl or 100 mM acetic acid. Though it varies, most superfamily proteins are soluble when diluted 1:100 or 1:1000 from 1 mg/mL stocks in acidic conditions to phosphate buffered saline or other similar solutions near neutral pH. The procedure used to express and purify human TGF-β1 from CHO cells, as initially described by Zou, et al. (18), is described below.

3.1.2 Protocol for Expression and Purification of Human TGF-β1 Using CHO Cells

Establishment of a Stable TGF-β1-Expressing Cell Line

1. Seed CHO-lec3.2.8.1 cells at a density of 3×10^6 cells in 5 mL DMEM/F12 medium containing 10 % FBS in T-25 flask and culture them at 37 °C with 5 % CO_2.
2. After 24 h, replace the medium with 4 mL fresh DMEM/F12 medium containing 10 % FBS.
3. Dilute 10 μg of CMV-based expression plasmid (*see* Subheading 2.2 above) in 500 μL of Opti-MEM (mix A) and dilute 30 μL of lipofectamine 2000 in 500 μL Opti-MEM (mix B).
4. Incubate at room temperature for 20 min.
5. Add mix A to mix B with gentle pipetting up and down.
6. Incubate for 20 min at room temperature.
7. Add the mix A + B to cells.
8. Replace the medium (DMEM/F12 with 10 % FBS) after 24 h (*see* **Note 2**).
9. After 2 days the cells were trypsinized, seeded into ten 96-well plates, and cultured with glutamine-free GMEM-S medium supplemented with GS supplement, 10 % FBS, and 30 μM MSX (*see* **Note 3**).
10. After about 3 weeks, take the supernatant from growing wells (with visible colony) and assayed for TGF-β1 expression by ELISA (*see* **Note 4**).
11. Pick the ten highest expression clones and do the second round selection with 500 μM MSX. For each clone, seed one 96-well

plate with 100–1000 cells per well. Wait for another 3 weeks and repeat the step above.

12. After two rounds of selection and amplification, the clone with the highest expression of TGF-β1 was chosen for large-scale recombinant protein production. Make stocks of the chosen clone and store them in liquid nitrogen.

Expression of TGF-β1 from CHO Cells and Purification

1. Expand the highest expression clone of TGF-β1 into T-225 flasks or T-500 triple layer flasks and culture the cells in glutamine-free GMEM-S medium supplemented with GS supplement, 10 % FBS, and 500 μM MSX.
2. When the cultured cells reach confluence, wash the cells twice with PBS and change the medium to CHO-S-SFM II serum-free medium.
3. After 3 days, harvest the medium and replace with fresh CHO-S-SFM II serum-free medium, repeat five to six times or until desired volume (*see* **Note 5**).
4. Freeze harvested medium at −20 °C for further purification.
5. Thaw 500 mL harvested medium add imidazole to final concentration of 10 mM, filter with 0.22 μM membrane (*see* **Note 6**).
6. Equilibrate 20 mL Ni-NTA column with Ni buffer containing 50 mM Tris, 150 mM NaCl and 10 mM imidazole at pH 8.0.
7. Load 500 mL filtered medium onto Ni column and wash the column with Ni buffer for two column volumes.
8. Elute the protein by a linear concentration gradient of imidazole (10–500 mM) in Ni buffer.
9. Pool fractions that contain protein and adjust the pH to below 3.0 by adding 6 M HCl (*see* **Note 7**); dialyze the solution against 100 mM acetic acid (minimum 20-fold dilution, three times).
10. Purify TGF-β1 on Source 15S column following the same protocol as TGF-β2 (**steps 4–8** in Subheading 3.1.4 Isolation of native dimer) except using the linear gradient of 0–600 mM NaCl (*see* **Note 8**).

3.1.3 Expression of Signaling Protein Monomers in Bacteria and Renaturation into Native Dimers

The other major avenue by which signaling proteins of the superfamily have been obtained is by expression of the mature monomers in bacteria, followed by renaturation of these into disulfide-linked dimers. The majority of family members are expressed using pET vectors, which are based on the T7 promoter [47]. To optimize expression, synthetic genes in which the codons have been optimized for bacteria may be used. Though N-terminal tags are sometimes added, these are not generally required, as the proteins invariably misfold and form highly refractile inclusion bodies. The formation of inclusion bodies aids in the purification since after disrupting the cells in non-denaturing buffer, most of the contaminating soluble proteins can be removed by repeatedly washing the inclusion bodies with non-denaturing buffers containing either salt (e.g., 0.5–1.0 M NaCl) or non-ionic detergents (e.g., 0.5–2.0 % Triton X-100). The isolated inclusion bodies are almost always reconstituted in high concentrations of denaturant, such as 6 M GdmCl or 8 M urea. Typically, the denaturant is buffered at pH 8.0–9.0 and includes a reductant, such as dithiothreitol, to convert any disulfides into free cysteines. Tris is also often included, especially if urea is used as the denaturant, as its primary amine group will react with isocyanic acid which is formed from the breakdown of urea. Though typically already quite pure at this stage, the proteins are nevertheless often further purified in denaturant using ion-exchange, gel filtration, or metal affinity chromatography to remove residual contaminants that might interfere with the subsequent folding.

The native signaling dimers are most often formed by diluting the reduced denatured monomers into a much larger volume of folding buffer. The folding buffers used for most of the proteins of the superfamily are based on the buffers reported by Cerletti and coworkers for refolding the TGF-βs [48]. These include, among other components, low (mM) concentrations of redox agents, such as cysteine/cystine, reduced glutathione/oxidized glutathione, or dithiothreitol/*trans*-4,5-dihydroxy-1,2-dithiane (oxidized dithiothreitol) and relatively high concentrations (20–30 mM) of the non-denaturing detergent 3-[(3-cholamidopropyl)dimethylammonio]-1-propanesulfonate, also known as CHAPS. The particular choice of redox agents is highly protein-dependent: sometimes only reductants are used, while in other cases a combination of reductants and oxidants are used. The majority of proteins folded have included CHAPS in the folding buffer, though notably, a recent study by Ejima and coworkers found that the CHAPS derivative sodium taurodeoxycholate (TDCA) was about 10–20 times more effective than CHAPS for folding human activin A [49]. The inclusion of CHAPS, or derivatives such as TDCA, is the most expensive reagent used in the folding since to be effective they must be used at concentrations comparable or higher than their critical micelle concentrations (which for CHAPS is near 10 mM). The folding mixture sometimes also include a co-solvent: TGF-βs, for example, have been folded in buffers containing 10–20 % DMSO, while activin A has been folded without any co-solvent. The requirement for co-solvents, such as DMSO, seems to be related to the solubility of the mature signaling dimer, as activins are reasonably soluble at milligram per milliliter concentrations at pH 8.0, while TGF-βs are not. The pH of the folding buffer is almost always 8.0 or higher, which is essential as cysteine thiols have a pK_a of approximately 8.5 and are not capable of forming disulfides unless their side chain thiols are deprotonated.

The concentration of protein in the folding buffer is important: concentrations higher than about 0.4 mg/mL are generally undesirable as this tends to lead to significant precipitate and lower yields of native dimers. Though less precipitate and greater yields of native dimers can generally be obtained with lower protein concentrations, there is a limit since the formation of disulfide-linked dimers is concentration-dependent and is favored at higher protein concentrations. Thus, the protein concentration must be adjusted to achieve a balance between minimizing loss due to increased formation of aggregates at higher concentration, and a slow rate of dimer formation at low protein concentrations. Though not yet in widespread use, it has been proposed that the folding be performed in two phases, one in which native monomers are formed, and another is which the native monomers are oxidized into disulfide-linked dimers [50]. To demonstrate feasibility, BMP-2 was initially folded under relatively dilute conditions (0.2 mg/mL) in the presence of redox buffers that promote disulfide exchange (0.1 mM reduced glutathione, 0.1 mM oxidized glutathione), and after a significant fraction of protein had folded into native monomers, the protein was concentrated fivefold (1 mg/mL) and transferred into buffer that only included oxidized glutathione (25 mM). The authors reasoned that this should lead mainly to the formation of native disulfide-linked dimers since all of the other cysteine residues had already formed their native disulfide-pairings (and were therefore not accessible for misfolding). This procedure was quite successful for BMP-2, with yields roughly twofold higher than that of single phase folding at 0.2 mg/mL for a period three times longer. The two-step folding method described above is predicated on the assumption that the monomers form the cystine knot and adopt their native folds independent of the other monomer. This may not be so for all family members, but it does appear to be for some, as native monomers of both GDF-5 and TGF-β3 have been isolated [10, 51–53]. Thus, this procedure may lead to refolding of additional family members that have not been previously folded, but whether this procedure will be applicable to all family members has not yet been established.

The folding of most superfamily proteins can be monitored by running samples of the folding mixture on non-reducing SDS-gels; over a period of several days, one observes an increase in dimeric forms and a concomitant decrease in monomeric forms as the cysteines are oxidized and the proteins fold into native dimers. The primary species in the folding buffer after several days of folding include, in addition to native dimers, non-dimerized monomers and high molecular weight disulfide-linked aggregates. Ion-exchange chromatography is well-suited for isolation of the native dimers from folding mixtures owing to charge differences among the different species present. In practice, the folding mixtures are usually acidified, concentrated, and loaded onto cation-exchange columns under acidic conditions and eluted with a salt gradient. In some cases, the native dimers are directly isolated from the folding mixture using reverse-phase chromatography under acidic conditions as described above; more commonly, reverse phase chromatography is used as a polishing step following initial isolation of the native dimers using ion-exchange chromatography.

The yields of mature signaling dimer obtained using the procedures described here vary greatly. The procedure shown below typically yields 8–10 mg of TGF-β2 from each liter of bacteria, while similar procedures afford 3–4 mg of TGF-β3, and almost no TGF-β1. The differences in yield are due to differences in the efficiency of the folding—all three isoforms are produced at comparable levels and are present in equal amounts (ca. 50–60 mg/L of culture medium) prior to folding. This emphasizes the point that the folding is highly dependent on the amino acid sequence and folding conditions. The most reasonable strategy that can pursued if the folding does not lead to native dimer would be to adhere to the procedures described above for isolating the purified denatured monomers, and then screen for formation of native dimers under different folding conditions using an ELISA-based approach. The types of screens that have been previously described for folding proteins of the superfamily can be tried [48, 49] as well as other screens that have been reported for folding disulfide rich proteins more generally [54]. Though the development of a folding protocol that affords native dimers with high efficiency (defined as the amount of native dimer isolated relative to the amount of purified denatured monomer) can be time-consuming, this investment can pay off as folding efficiencies as high as 40 % have been reported, affording large quantities (20–30 mg) of highly purified dimer from just a few liters of cultured bacteria. The procedure used to bacterially express, refold, and purify human TGF-β2 is described below.

3.1.4 Protocol for E. coli Expression, Refolding, and Purification of Human TGF-β2

Expression

1. Transform *E. coli* strain BL21(DE3) with the plasmid containing TGF-β2 sequence. Spread onto a LB-agar plate supplemented with ampicillin (150 μg/mL) and incubate overnight at 37 °C.
2. Inoculate 1 L LB medium, supplemented with ampicillin (150 μg/mL), with colonies from the plate and incubate at 37 °C with orbital shaking at 250 rpm in a 2.8 L flask. Add

0.8 mM IPTG at mid-log phase (0.6 OD_{600}) to induce protein expression and incubate at 37 °C for 3 h.

3. Harvest the cells by centrifuge at 6000 × *g* for 15 min and freeze the pellet.

Recover Insoluble, Monomeric TGF-β2

1. Cell pellets from 1 L of culture were resuspended in 30 mL of disruption buffer (100 mM Tris, 10 mM EDTA, 1 mM PMSF, pH 8.0) and sonicated.
2. Centrifuge at 20,000 × *g* for 20 min and discard the supernatant (*see* **Note 9**).
3. Resuspend pellet in 100 mL disruption buffer containing 1 M NaCl.
4. Centrifuge at 20,000 × *g* for 20 min and discard the supernatant.
5. Resuspend pellet in 100 mL disruption buffer containing 1 % Triton X-100 (v/v).
6. Centrifuge at 20,000 × *g* for 20 min and discard the supernatant.
7. Resuspend pellet in 100 mL TGF-β solubilizing buffer (8 M urea, 20 mM Tris, 1 % DTT (w/v), pH 8.0), stir overnight at room temperature.
8. Add solid sodium acetate to a final concentration of 20 mM and adjust the pH to 4.0 using glacial acetic acid.
9. Centrifuge at 20,000 × *g* for 20 min and collect supernatant.

Initial Purification

1. Equilibrate 20 mL SP-sepharose cation exchange column with SP denaturing buffer (8 M urea, 20 mM sodium acetate, 0.1 % DTT (w/v), pH 4.2) for 2 column volumes.
2. Load the supernatant onto column and wash the column with SP denaturing buffer for 2 column volumes.
3. Elute with a linear 0–300 mM NaCl gradient in SP denaturing buffer.
4. Identify fractions containing purified TGF-β2 using SDS-PAGE; pool purified denatured TGF-β2 and dialyze against 100 mM acetic acid at 4 °C (minimum 20-fold dilution, 3 times).

Folding

1. Dialyzed protein is slowly diluted into TGF-β folding buffer containing 100 mM Tris, 30 mM CHAPS, 1 M NaCl, 5 mM reduced-glutathione, pH 9.5 to a final concentration of 0.2 mg/mL at 4 °C (*see* **Note 10**). Stir gently for 3–4 days at 4 °C (*see* **Note 11**).

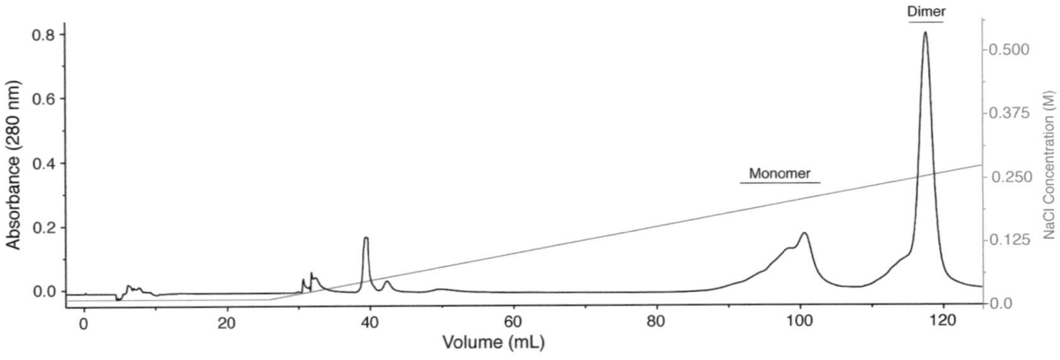

Fig. 2 Isolation of the human TGF-β2 dimer. Cation-exchange (Source 15S, GE Healthcare) elution profile of the TGF-β2 folding mixture. Peaks corresponding to monomeric and dimeric forms are labeled. Column was equilibrated in 20 mM sodium acetate, 30 % isopropanol, pH 4.0 and eluted with a linear NaCl gradient in the same buffer (as shown by the *grey line* and accompanying axis on the *right*)

Isolation of Native Dimer

1. Adjust the pH of folding mix to 3.5 using acetic acid and concentrate the mixture to roughly 50 mL using an Amicon stirred cell concentrator with a YM-10 membrane (Millipore).
2. Dialyze the concentrated refolding mix against 100 mM acetic acid (minimum 20-fold dilution, three times) (*see* **Note 12**).
3. Centrifuge at 20,000 × *g* for 20 min, collect supernatant and filter through 0.22 μM membrane (*see* **Note 13**).
4. Equilibrate a Source 15S HR 10/10 column with 20 mM sodium acetate, 30 % isopropyl alcohol, pH 4.0 (Source 15S buffer) at a flow rate of 2.0 mL/min.
5. Load sample onto Source 15S column and wash column with two column volumes Source 15S buffer.
6. Elute with a linear gradient 0–400 mM NaCl over 10 column volumes in Source 15S buffer.
7. Collect peak fractions corresponding to the dimer peak (*see* Fig. 2) and then dialyze against 100 mM acetic acid (minimum 20-fold dilution, 3 times).
8. Measure the concentration of purified TGF-β2 by OD_{280} absorbance, aliquot, and lyophilize.

3.2 Overview of Receptor Ectodomain Production and Isolation

The structurally characterized receptor ectodomains of the TGF-β superfamily, which includes the type I receptors Alk1 [55, 56], Alk3 [11, 57], Alk5 [10, 12, 58], and Alk6 [59] and the type II receptors ActRII [8, 60], ActRIIb [13, 61], BMPRII [62], and TβRII [63–65], have each been shown to adopt a three finger toxin fold, consisting of three pairs of antiparallel β-strands stabilized by a characteristic pattern of four disulfides (Fig. 1c).

The seven type I receptors, and all but one of the type II receptors of the superfamily, include one additional disulfide, while the TGF-β type II receptor, includes two. The procedures that have been used to produce the ectodomains of the superfamily are similar to those used to produce the signaling proteins and are described below.

3.2.1 Expression of Receptor Ectodomains in Eukaryotic Cells

The majority of receptor ectodomains that have been produced and studied are from mammalian species, such as humans and rodents, yet an almost equal number have been produced in mammalian and non-mamammalian hosts. The mammalian hosts used include CHO, HEK-293, and mouse myeloma cells [56, 66–70], while the non-mamammalian hosts include insect cells and the methylotrophic yeast, *Pichia pastoris* [61, 70–73]. The majority of receptors have been produced as secreted proteins, either alone as monomers, or as dimers by fusing them with the Fc region of an antibody. The Fc-dimerized receptors typically bind their cognate signaling proteins with affinities that are 1–2 orders of magnitude greater than their non Fc-dimerized counterparts [55, 66, 74], which is expected for a bivalent interaction between a covalently dimerized receptor and a covalently dimerized signaling protein. The Fc-dimerized receptor ectodomains are useful for increasing potency for use in binding studies [66] or as inhibitors in cells or animals [75–77].

The N- and C-termini are accessible and flexible in all of the type I and type II receptors of the superfamily, and thus affinity tags, such as hexahistidine tags, can be readily appended to the N- or C-termini without interfering with folding. To date, most of the ectodomains have been produced with a N-terminal tag followed by a protease cleavage site though a few have been produced with cleavable C-terminal tag. The receptors can be directly purified from the conditioned medium if they have an affinity tag. This generally yields protein free of most contaminants, though in many cases, some high molecular weight material may also be present. This high molecular weight material often disappears if the receptors are reduced prior to loading on the gel, indicating that while most of the overexpressed protein folds and adopts native disulfide pairings, some misfolds and forms disulfide-linked multimers. Thus, in many cases, receptors that have been purified with an affinity tag are further purified to eliminate multimers as well as other contaminating proteins. The multimers can be relatively easily eliminated using either size exclusion chromatography, which separates based on size, or ion-exchange chromatography, which separates based on charge (monomers have lower overall charge compared to multimers, and thus elute at lower salt concentrations than the multimers). The purification tag is often removed by treatment with appropriate protease prior to the second step of the purification, as such tags can potentially interfere with crystallization. If an affinity tag was not included in the construct, the medium is usually desalted, and the dialyzed medium is then applied to anion or cation exchange column equilibrated at a pH where the protein is expected to bind (usually determined by calculating the net charge as a function of pH using a tool such as the Scripps Protein Calculator [78]). In many cases, this affords protein that is free of most other contaminating proteins, as well as disulfide-linked multimers, though if not, an additional purification step, such as size exclusion or hydrophobic interaction chromatography, can be used. Importantly, non-reducing SDS-PAGE must be used to analyze the purity of the final preparation since reducing agents, such as β-mercaptoethanol or dithiothreitol (DTT), will convert disulfide-linked oligomers or multimers into monomers. It is important to be aware of the presence of disulfide-linked oligomers or multimers and to remove them if they are present since these will interfere in subsequent binding and structural studies.

3.2.2 Protocol for Expression of Human Fc-TβRII Using CHO Cells

Establishment of a Stable TβRII-Fc Expressing Cell Line

1. The human TβRII ectodomain is inserted in the polylinker of plasmid pFUSE-hIgG4-Fc (Invivogen) to yield an expression construct in which TβRII is in frame with a C-terminal human IgG4 Fc domain. The TβRII-Fc expression cassette is then PCR amplified and inserted downstream of the rat serum albumin signal peptide in the modified form of pcDNA3.1(+) with a glutamine synthase gene described in Subheading 2.2.
2. The CMV-based TβRII-Fc expression plasmid is then transfected into CHO-lec3.2.8.1 cells and a stable expressing clone is selected using the same procedure described in Subheading 3.1.2 above.

Expression of TβRII-Fc from CHO Cells and Purification

1. Stably transfected cells are cultured in serum free medium in T-225 flasks and the conditioned medium is collected as described in Subheading 3.1.2 above.
2. Conditioned medium is diluted threefold into 0.1 M Tris, pH 8.2 and passed over a HiTrap protein A column (GE Healthcare) equilibrated with 0.15 M NaCl, 0.01 M sodium phosphate, pH 7.2.
3. Column is washed with ten column volumes of 0.15 M NaCl, 0.01 M sodium phosphate, pH 7.2 and the protein is then eluted by passing over five column volumes of 0.1 M citrate, pH 3.0.
4. Eluted protein is adjusted to pH 7.0 by addition of 0.1 M Tris base, followed by concentration to roughly 10 mg/mL using a Amicon ultrafiltration stirred cell (Millipore).

5. One milliliter aliquots of the TβRII-Fc fusion is passed over a 1.6 × 60 cm Superdex 200 gel filtration column (GE Healthcare) equilibrated in 0.1 M Tris, pH 7.0. Peak fractions are pooled, concentrated to 10 mg/mL using an Amicon ultrafiltration stirred cell. Protein purity is checked by non-reducing SDS-PAGE and pooled stocks are stored at −20 °C until ready for use.

3.2.3 Expression of Receptor Ectodomainsin Bacteria

The other major avenue by which the receptor ectodomains have been obtained is by expression in bacteria. The ectodomains, if expressed alone, form highly refractile inclusion bodies and must be renatured [10, 52, 55, 58, 64, 73, 79]. The overall procedure parallels that used for the signaling proteins, including initial isolation of the inclusion bodies, subsequent purification of the reduced protein in denaturant, and folding by dilution into non-denaturing buffers that include redox-active species that promote disulfide exchange. The procedure used to fold and purify the human TβRII ectodomain is described in Subheading 3.2.4 below.

The receptor ectodomains, like the signaling proteins, do not require a purification tag as the overexpressed protein forms inclusion bodies, and are therefore largely pure, even before the initial purification in denaturant. Tags are however sometimes included as they aid in the isolation of the protein from the folding buffer. The buffers used to fold the receptors are generally simpler than those used to the fold the signaling proteins: most include redox-active species, such as reduced and oxidized glutathione, but other additives such as non-denaturing detergents and co-solvents are not generally required. The ectodomains generally have favorable solubility properties and can be concentrated to high concentration (10 mg/mL, or even higher) in standard buffers (Tris, Hepes, phosphate, etc.) at neutral pH. Thus, the receptors are usually concentrated following folding, and then dialyzed into a buffer suitable for subsequent ion-exchange or gel filtration chromatography. These two methods are both effective for purifying the receptors following folding as both eliminate disulfide-linked oligomers, which are the major "contaminant" present in the folding. These two purification methods do not, however, have sufficient resolution to separate native monomers from nonnative monomers, and thus in cases where the monomer pool is heterogeneous, it is recommended that the receptor be further purified using C8 or C18 reverse phase chromatography. This method of purification, which separates based on hydrophobicity, has outstanding resolution and has been found to be effective for separating native monomers away from the nonnative monomers. This type of purification denatures the receptors (since it is commonly performed in 0.1 % trifluoroacetic acid with either acetonitrile or methanol elution), but is not problematic since the disulfides remain intact and the receptors readily refold after the triflouroacetic acid and acetonitrile/methanol have been removed by either lyophilization or dialysis. The final purity of the receptor ectdomains, like the signaling proteins, should be assessed by non-reducing SDS-PAGE, since as before reductant will mask any misfolded disulfide-linked dimers and higher order oligomers.

Several receptors of the superfamily, including BMPRIa, BMPRIb, BMPRII, and TβRII have been previously produced as soluble proteins in *E. coli* by expressing them as fusion proteins with thioredoxin [62, 80–84]. The advantage of producing the receptors in this manner is that the proteins remain soluble and can be extracted from the bacteria in their native form. These types of fusion proteins are usually constructed with an intervening hexa-histidine tag and protease cleavage site between thioredoxin and the receptor. To increase the proportion of protein that remains soluble, the growth temperature can be lowered from 37 to 20 °C prior to the induction of protein expression. The protein is extracted from the cells by disrupting them in non-denaturing buffer with protease inhibitors. The proteins are usually purified by metal affinity chromatography, followed by a subsequent polishing step using gel filtration. The latter is important as the proteins produced in this manner often include some disulfide-linked dimer and higher order oligomers.

3.2.4 Protocol for Expression of the Human TβRII Ectodomain Using E. coli

Expression

1. Transform *E. coli* strain BL21(DE3) with the plasmid coding for the human TβRII ectodomain. Spread onto an LB-agar plate supplemented with ampicillin (150 μg/mL) and incubate overnight at 37 °C.
2. Inoculate 1 L LB medium, supplemented with ampicillin (150 μg/mL), with colonies from the plate and incubate at 37 °C with orbital shaking at 250 rpm in a 2.8 L flask. Add 0.8 mM IPTG at mid-log phase (0.6 OD_{600}) to induce protein expression and incubate at 37 °C for 3 h.
3. Harvest the cells by centrifugation at 6000 × *g* for 15 min and freeze the pellet.

Recover Non-soluble TβRII

1. Cell pellet from 1 L of culture is resuspended in 30 mL of disruption buffer (100 mM Tris, 10 mM EDTA, 1 mM PMSF, pH 8.0) and sonicated.
2. Centrifuge at 20,000 × *g* for 20 min and discard the supernatant (*see* **Note 14**).
3. Resuspend pellet in 100 mL disruption buffer containing 1 M NaCl.
4. Centrifuge at 20,000 × *g* for 20 min and discard the supernatant.

5. Resuspend pellet in 100 mL disruption buffer containing 1 % Triton X-100 (v/v).
6. Centrifuge at 20,000 × *g* for 20 min and discard the supernatant.
7. Resuspend pellet in 100 mL RII solubilizing buffer (8 M urea, 20 mM Tris, pH 7.0), stir overnight at room temperature.
8. Centrifuge at 20,000 × *g* for 20 min and collect supernatant.

Initial Purification

1. Equilibrate 20 mL DEAE-sepharose anion exchange column with RII solubilizing buffer (8 M urea, 20 mM Tris, pH 7.0) for two column volumes.
2. Load the supernatant onto column and wash the column with RII solubilizing buffer for two column volumes.
3. Elute with a linear 0–300 mM NaCl gradient in RII solubilizing buffer.
4. Identify fractions containing purified TβRII using reducing SDS-PAGE; pool purified denatured TβRII; add solid DTT to final concentration of 25 mM, stir for 30 min and dialyze against 100 mM acetic acid at 4 °C (minimum 20-fold dilution, 3 times).

Folding

1. Dialyzed protein is slowly diluted into RII folding buffer containing 200 mM Tris, 2 mM reduced-glutathione, 0.5 mM oxidized-glutathione, pH 8.0 to a final concentration of 0.5 mg/mL at 4 °C (*see* **Note 15**). Stir gently for 18–24 h at 4 °C in an open vessel.

Isolation of Native TβRII

1. Concentrate the folding solution to roughly 50 mL using an Amicon stirred cell concentrator with a YM-3 membrane (Millipore).
2. Dialyze the concentrated folding solution against 20 mM MES, pH 6.0 at 4 °C (minimum 20-fold dilution, 3 times).
3. Centrifuge at 20,000 × *g* for 20 min, collect supernatant and filter through 0.22 μM membrane.
4. Equilibrate a Source 15Q HR 10/10 column with 20 mM MES, 1 mM PMSF, 1 mM EDTA, pH 6.0 (Source 15Q buffer) at a flow rate of 2.0 mL/min.
5. Load sample onto Source 15Q column and wash column with 2 column volumes Source 15Q buffer.
6. Elute with a linear gradient 0–300 mM NaCl in Source 15Q buffer over 10 column volumes.

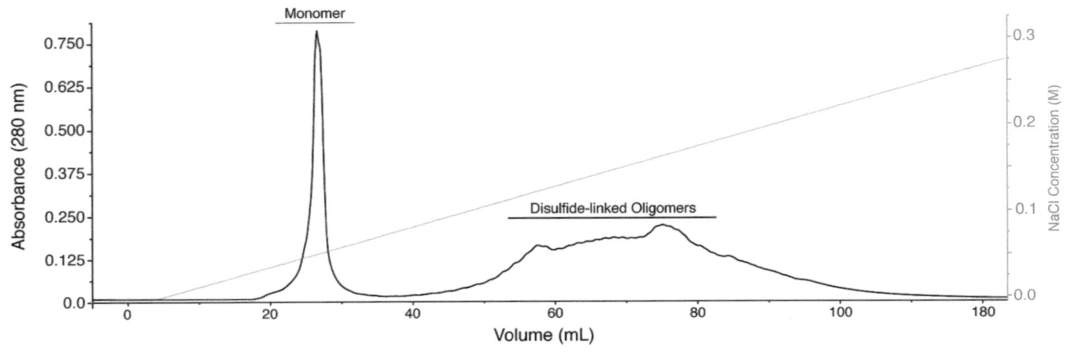

Fig. 3 Isolation of the human TβRII ectodomain. Anion-exchange (Source 15Q, GE Healthcare) elution profile of the TβRII folding mixture. Peaks corresponding to native monomer and nonnative disulfide-linked oligomeric forms are labeled. Column was equilibrated in 20 mM MES, pH 6.0 and eluted with a linear NaCl gradient in the same buffer (as shown by the *grey line* and accompanying axis on the *right*)

7. Pool the fractions corresponding to purified native receptor (*see* Fig. 3) and dialyze against Source 15Q buffer to remove salt (minimum 20-fold dilution, 3 times).
8. Measure the concentration of purified TβRII by OD_{280} absorbance, concentrate to 5–10 mg/mL, aliquot and store at −20 °C.

3.3 Methods to Assess Ligand–Receptor Binding

The binding properties of signaling proteins and receptors of the superfamily have been predominantly studied using surface plasmon resonance (SPR) [8–10, 12, 13, 55, 56, 58, 59, 66, 67, 77, 85–102]. The feature of this method that has led to its widespread use is that the signaling proteins can be readily coupled to SPR sensor surfaces at low pH where the proteins are soluble, and then the conditions can be changed to study receptor binding near neutral pH. The other advantage of performing SPR experiments with immobilized signaling proteins is that the amount of protein required is minimal (0.2–2 μg). These quantities can usually be purchased at a reasonable cost, though if they are, it is important to purchase "carrier-free" protein so that the coupling of the signaling protein to the SPR sensor chip is not blocked by the carrier.

The signaling proteins are most commonly coupled to sensor chips by a carbodiimide coupling reaction in which amino groups of lysine residues on the signaling protein are coupled to carboxylates on the surface of the sensor chip. Though reliable and easy to perform, this has the disadvantage that it results in a random coupling of the signaling protein to the sensor chip. This can result in a heterogenous surface in which some of the protein is coupled such that receptor binding is unimpaired, but another portion in which receptor binding is partially of fully impaired. The heterogeneous surfaces are generally undesirable since this complicates the pattern of receptor binding. One way to minimize heterogeneity is to form the complex in solution between the signaling protein to be studied and its receptors and in turn biotinylate the complex using an amine-reactive biotin derivative. Once modified, the biotinylated signaling protein can be separated from the receptor(s) by purifying it under denaturing conditions (such as C8 or C18 reverse phase chromatography with either acetonitrile or methanol gradient elution). This separation step also removes free biotin from signaling protein. After it has been dialyzed to remove the acetonitrile or methanol, it can be captured onto a sensor chip coated with a high density of streptavidin or neutravidin.

The binding measurements are performed by using the microfluidic system of the SPR instrument to inject the receptor ectodomain for a given period of time, followed by a disassociation period in which the instrument is switched back to the running buffer. Two types of experiments are commonly performed, kinetic and equilibrium experiments. Kinetic experiments are performed by injecting at a high flow rate for a period long enough to determine the rate of binding, but not so long that the response reaches equilibrium, while equilibrium experiments are performed by injecting at a low flow rate for a period long enough so that equilibrium is reached. Kinetic experiments are analyzed by fitting the time-dependence of the association and disassociation to the simplest mathematical model describing the binding, while equilibrium experiments are analyzed by fitting the equilibrium response (R_{eq}) as a function of concentration of the injected receptor to a standard binding equation. The kinetic experiments therefore provide the association and disassociation rate constants (k_a and k_d, respectively, as well as the K_d from k_d/k_a) as well as the maximal response (R_{max}), while the equilibrium experiments provide only the K_d and R_{max}. Kinetic experiments therefore provide more information and are therefore generally desirable, though they have the disadvantage that this comes at the cost of using larger amounts of receptor (owing to the higher flow rates). The amount of receptor used in a typical kinetic SPR experiment in which the receptor concentration is varied from 0 to 4 μM is approximately 0.2 mg.

SPR experiments, in addition to providing rates and affinities for binding of a single receptor, also affords the opportunity to study binding of more than one receptor. Such measurements can provide insight into whether the receptors compete or cooperate with one another and have been instrumental in delineating the very different modes of receptor complex assembly for the TGF-βs and BMPs [9, 10, 12]. Such experiments can be performed in one of two ways: the first and simplest is to perform a co-injection experiment in which the first half of the injection loop is filled with a near-saturating concentration of the first receptor (e.g., type II receptor), while the second half is filled with a variable concentration of the second receptor (e.g., type I receptor) on top of the same concentration of the first receptor in the first half. Such experiments can therefore be used to determine whether the first receptor (e.g., type II receptor) alters the binding of the second receptor (e.g., type I receptor), by comparison to a control experiment in which the binding of the second receptor (e.g., type I receptor) is studied in the absence of the first receptor (e.g., type II receptor).

Several alternatives to SPR have been proposed in recent years, including bio-layer inferometry [103] and microscale thermophoresis [104]. Studies using these methods for studying interactions between signaling proteins and receptors of the TGF-β superfamily have not been reported in the literature, but it is expected that these will be increasingly used in the future, owing to their lower cost and ease of use. Binding studies using titration calorimetry have not been reported, presumably because most superfamily signaling proteins are poorly soluble and must therefore be stored under conditions (low pH) where they bind their cognate receptors poorly, or not at all. Such solubility properties require that the signaling protein and receptor be contained in different buffers, which has the disadvantage that the heat evolved from mixing solvents is large compared to the heat evolved by binding.

3.4 Structural Analyses of Ligand–Receptor Complexes

The relatively small size of the signaling proteins (ca. 25 kDa) and receptor ectodomains (11–16 kDa) has enabled studies of both using NMR [20, 55, 57, 58, 64, 79, 105–107] and X-ray crystallography [19, 21, 22, 96, 98, 102, 108–110]. The strategies used to obtain the NMR solution structures and X-ray crystal structures of the signaling proteins and receptor ectodomains more or less follow the standard methodologies. The larger sizes of the complexes, on the other hand, have relegated these to being studied mostly by X-ray crystallography [8–13, 61, 99, 101]. The strategies used to obtain the structures of the signaling protein–receptor ectodomain complexes have mostly been performed by first adding an excess of receptors relative to signaling protein, followed by isolation of the complex by gel filtration. The rationale for this strategy is that it allows for the isolation of stoichiometric complexes, which due to their homogeneity, are more likely to form diffracting crystals. This strategy clearly has limits in light of weaker affinities and thus is likely to be successful for studying many of the signaling protein–receptor ectodomain complexes of the superfamily, but not all. The alternative strategy, in light of weaker affinities, is to combine the proteins in the appropriate stoichiometric ratio (two equivalents of type I or type II receptor ectodomain per equivalent of signaling homodimer), concentrate, and setup the crystallization trials at the highest protein concentration feasible (this has to be empirically determined based on the inherent solubility of the complex). The strategy is based on the presumption that the high protein concentration (typically on the order of 5–10 mg/mL) will drive the equilibrium toward complex formation given that the concentration of the complex (0.05–0.1 mM, depending on the size and concentration of the complex) is typically several orders of magnitude greater than the disassociation constant (typically 10 μM or lower). The alternative approach, in light of moderate to lower affinities, is to isotopically label either the signaling protein or receptor ectodomain with NMR-active stable isotopes (^{15}N or ^{13}C) and monitor the signals of the labeled component as increasing concentrations of the unlabeled partner are added [53]. The NMR signals of residues that lie directly in the binding site will be most strongly perturbed, thus providing a "map" of the contact surface. This procedure can be repeated with the labeling pattern reversed to obtain the complementary contact surface of the partner. These maps, together with the known structures of the two proteins, can in turn be used to construct a model of the complex using a data-driven docking algorithm, such as HADDOCK [111, 112]. This strategy has not yet been used to determine the structures of any signaling protein–receptor complexes of the TGF-β superfamily, but has been widely used to study other systems [113], and may therefore be used to study signaling protein–receptor complexes of the superfamily, especially receptors, such as some of the BMP type II receptors, that bind their cognate signaling proteins weakly.

4 Notes

1. CHO-lec3.2.8.1 cells have multiple glycosylation defects for production of glycoproteins with minimal carbohydrate heterogeneity [46].
2. It is normal from some of the cells to detach and die after transfection.
3. In each well there should be 100–1000 cells and the transfection rate is approximately 0.1 %, thus 0–1 colonies are expected in each well.
4. Colonies that do form will grow at different rates and thus the medium should be removed only when the color of the medium turns yellow. Protein expression levels detected by ELISA do not generally correlate with growth rates.
5. Replace the medium when the color changes from red to yellow; can be repeated until the cells are no longer viable (as many as 10 times). Protein secretion levels generally decrease over time.
6. Add 10 mM imidazole in the sample to prevent nonspecific binding to the Ni-NTA column.
7. Acidification of the solution will dissociate the mature TGF-β1 from LAP; however, this step will not separate the mature TGF-β1 from incompletely processed pro-domain-mature precursor (usually 5–10 %).

8. Only dimeric TGF-β will be observed (but not monomeric TGF-β as in the purification of refolded bacterial TGF-β).
9. The protein expressed in inclusion bodies will remain insoluble until dissolved in 8 M urea.
10. Small aliquots of 1 M NaOH are added during the addition to maintain the pH at 9.5.
11. The folding mixture will appear slightly cloudy and the folding can be monitored by running small aliquots of the folding mixture on a non-reducing SDS-PAGE gel.
12. Leave extra room in the dialysis tubing, because the volume will increase about 50 % after dialysis.
13. The precipitate is mostly comprised of misfolded disulfide-linked multimers; the protein that remains soluble is mostly natively folded dimers and monomers.
14. The protein expressed in inclusion bodies will remain insoluble until dissolved in 8 M urea.
15. Small aliquots of 1 M NaOH are added to maintain the pH 8.0 during the addition.

5 Summary and Future Prospects

Structural studies of signaling proteins and signaling receptors of the TGF-β superfamily have shown that the TGF-βs, evolutionary latecomers to the superfamily, diverged from the BMPs and GDFs, the ancestral signaling proteins of the superfamily, to bind and assemble their type I and type II receptors in a distinct manner [14, 16, 17]. The receptors appear to have evolved these new modes of receptor binding in a stepwise manner, first by evolution of a new mode of type I receptor binding and later by evolution of a new mode of type II receptor binding. This idea is based on the observation that evolutionary intermediates of the superfamily, such as activins and nodal, bind and signal through type I receptors that are shared with the TGF-βs, evolutionary latecomers to the superfamily, and type II receptors that are shared with many of the distantly related BMPs and GDFs. There is at present only one structure of one of these intermediates bound to a receptor (activin A bound to its type II receptor, ActRIIb, or its close homolog, ActRII) [61, 99], thus there is an urgent need to determine the structures of more of the evolutionary intermediates, such as activin, nodal, and myostatin, to their cognate type I and type II receptors. There are also several receptors yet to be structurally characterized, including the type I receptors Alk2, Alk4, and Alk7, and the type II receptor MISRII. The hope is that the principles and procedures described here for producing and studying the binding properties and structures of the signaling proteins and receptor ectodomains of the superfamily will lead to a better understanding of the different possible signaling complexes that the proteins of the superfamily form, and in turn, this will inform the underlying biology.

Acknowledgements

The author would also like to acknowledge the funding agencies that have supported the TGF-β research underway in his laboratory, including the NIH (GM58670 and CA172886), the Robert A. Welch Foundation (AQ-1842), and the Cancer Prevention and Research Institute in Texas (RP120867).

References

1. Kingsley DM (1994) The TGF-beta superfamily: new members, new receptors, and new genetic tests of function in different organisms. Genes Dev 8(2):133–146
2. Sengle G, Ono RN, Sasaki T, Sakai LY (2011) Prodomains of transforming growth factor beta (TGFbeta) superfamily members specify different functions: extracellular matrix interactions and growth factor bioavailability. J Biol Chem 286(7):5087–5099, PMCID: 3037620
3. Shi M, Zhu J, Wang R, Chen X, Mi L, Walz T et al (2011) Latent TGF-beta structure and activation. Nature 474(7351):343–349
4. Wrana JL, Attisano L, Carcamo J, Zentella A, Doody J, Laiho M et al (1992) TGF beta signals through a heteromeric protein kinase receptor complex. Cell 71(6):1003–1014
5. Yamashita H, ten Dijke P, Franzen P, Miyazono K, Heldin CH (1994) Formation of hetero-oligomeric complexes of type I and type II receptors for transforming growth factor-beta. J Biol Chem 269(31):20172–20178
6. Massague J (1998) TGF-beta signal transduction. Annu Rev Biochem 67:753–791
7. Massague J, Wotton D (2000) Transcriptional control by the TGF-beta/Smad signaling system. EMBO J 19(8):1745–1754, PMCID: 302010
8. Allendorph GP, Vale WW, Choe S (2006) Structure of the ternary signaling complex of a TGF-beta superfamily member. Proc Natl Acad Sci U S A 103(20):7643–7648, PMCID: 1456805
9. Greenwald J, Groppe J, Gray P, Wiater E, Kwiatkowski W, Vale W et al (2003) The BMP7/ActRII extracellular domain complex provides new insights into the cooperative nature of receptor assembly. Mol Cell 11(3):605–617
10. Groppe J, Hinck CS, Samavarchi-Tehrani P, Zubieta C, Schuermann JP, Taylor AB et al (2008) Cooperative assembly of TGF-beta superfamily signaling complexes is mediated by two disparate mechanisms and distinc modes of receptor binding. Mol Cell 29(2):157–168
11. Kirsch T, Sebald W, Dreyer MK (2000) Crystal structure of the BMP-2-BRIA ectodomain complex. Nat Struct Biol 7(6):492–496
12. Radaev S, Zou Z, Huang T, Lafer EM, Hinck AP, Sun PD (2010) Ternary complex of transforming growth factor-beta1 reveals isoform-specific ligand recognition and receptor recruitment in the superfamily. J Biol Chem 285(19):14806–14814, PMCID: 2863181
13. Weber D, Kotzsch A, Nickel J, Harth S, Seher A, Mueller U et al (2007) A silent H-bond can be mutationally activated for high-affinity interaction of BMP-2 and activin type IIB receptor. BMC Struct Biol 7:6, PMCID: 1802081
14. Massagué J (2008) A very private TGF-beta receptor embrace. Mol Cell 29(2):149–150
15. Nickel J, Sebald W, Groppe JC, Mueller TD (2009) Intricacies of BMP receptor assembly. Cytokine Growth Factor Rev 20(5–6):367–377
16. Hinck AP, O'Connor-McCourt MD (2011) Structures of TGF-beta receptor complexes: implications for function and therapeutic intervention using ligand traps. Curr Pharm Biotechnol 12(12):2081–2098
17. Hinck AP (2012) Structural studies of the TGF-betas and their receptors – insights into evolution of the TGF-beta superfamily. FEBS Lett 586(14):1860–1870
18. Zou Z, Sun PD (2004) Overexpression of human transforming growth factor-beta1 using

a recombinant CHO cell expression system. Protein Expr Purif 37(2):265–272

19. Daopin S, Piez KA, Ogawa Y, Davies DR (1992) Crystal structure of transforming growth factor-beta 2: an unusual fold for the superfamily. Science 257(5068):369–373
20. Hinck AP, Archer SJ, Qian SW, Roberts AB, Sporn MB, Weatherbee JA et al (1996) Transforming growth factor beta 1: three-dimensional structure in solution and comparison with the X-ray structure of transforming growth factor beta 2. Biochemistry 35(26): 8517–8534
21. Mittl PR, Priestle JP, Cox DA, McMaster G, Cerletti N, Grutter MG (1996) The crystal structure of TGF-beta 3 and comparison to TGF-beta 2: implications for receptor binding. Protein Sci 5(7):1261–1271, PMCID: 2143453
22. Schlunegger MP, Grutter MG (1992) An unusual feature revealed by the crystal structure at 2.2 A resolution of human transforming growth factor-beta 2. Nature 358(6385): 430–434
23. Sun PD, Davies DR (1995) The cystine-knot growth-factor superfamily. Annu Rev Biophys Biomol Struct 24:269–291
24. Gentry LE, Nash BW (1990) The pro domain of pre-pro-transforming growth factor beta 1 when independently expressed is a functional binding protein for the mature growth factor. Biochemistry 29(29):6851–6857
25. Gray AM, Mason AJ (1990) Requirement for activin A and transforming growth factor – beta 1 pro-regions in homodimer assembly. Science 247(4948):1328–1330
26. Bourdrel L, Lin CH, Lauren SL, Elmore RH, Sugarman BJ, Hu S et al (1993) Recombinant human transforming growth factor-beta 1: expression by Chinese hamster ovary cells, isolation, and characterization. Protein Expr Purif 4(2):130–140
27. Bustos-Valenzuela JC, Halcsik E, Bassi EJ, Demasi MA, Granjeiro JM, Sogayar MC (2010) Expression, purification, bioactivity, and partial characterization of a recombinant human bone morphogenetic protein-7 produced in human 293T cells. Mol Biotechnol 46(2):118–126
28. Chitty DW, Tremblay RG, Ribecco-Lutkiewicz M, Haukenfrers J, Zurakowski B, Massie B et al (2012) Development of BMP7-producing human cells, using a third generation lentiviral gene delivery system. J Neurosci Methods 205(1):17–27
29. Israel DI, Nove J, Kerns KM, Kaufman RJ, Rosen V, Cox KA et al (1996) Heterodimeric bone morphogenetic proteins show enhanced activity in vitro and in vivo. Growth Factors 13(3–4):291–300
30. Jones WK, Richmond EA, White K, Sasak H, Kusmik W, Smart J et al (1994) Osteogenic protein-1 (OP-1) expression and processing in Chinese hamster ovary cells: isolation of a soluble complex containing the mature and pro-domains of OP-1. Growth Factors 11(3):215–225
31. Sampath TK, Maliakal JC, Hauschka PV, Jones WK, Sasak H, Tucker RF et al (1992) Recombinant human osteogenic protein-1 (hOP-1) induces new bone formation in vivo with a specific activity comparable with natural bovine osteogenic protein and stimulates osteoblast proliferation and differentiation in vitro. J Biol Chem 267(28):20352–20362
32. Sathyamurthy M, Lee JS, Park JH, Kim YJ, Jeong JY, Jang JW et al (2012) Overexpression of PACEsol improves BMP-7 processing in recombinant CHO cells. J Biotechnol 164(2): 336–339
33. Zou Z, Sun PD (2006) An improved recombinant mammalian cell expression system for human transforming growth factor-beta2 and -beta3 preparations. Protein Expr Purif 50(1): 9–17
34. Lee SJ, McPherron AC (2001) Regulation of myostatin activity and muscle growth. Proc Natl Acad Sci U S A 98(16):9306–9311, PMCID: 55416
35. Li Q, Liu X, Wu Y, An J, Hexige S, Ling Y et al (2012) The conditioned medium from a stable human GDF3-expressing CHO cell line, induces the differentiation of PC12 cells. Mol Cell Biochem 359(1–2):115–123
36. Peng J, Li Q, Wigglesworth K, Rangarajan A, Kattamuri C, Peterson RT et al (2013) Growth differentiation factor 9:bone morphogenetic protein 15 heterodimers are potent regulators of ovarian functions. Proc Natl Acad Sci U S A 110(8):E776–E785, PMCID: 3581982
37. Takaoka K, Yoshikawa H, Hasimoto J, Masuhara K, Miyamoto S, Suzuki S et al (1993) Gene cloning and expression of a bone morphogenetic protein derived from a murine osteosarcoma. Clin Orthop Relat Res 294: 344–352
38. Ulloa L, Creemers JW, Roy S, Liu S, Mason J, Tabibzadeh S (2001) Lefty proteins exhibit unique processing and activate the MAPK pathway. J Biol Chem 276(24):21387–21396
39. Ushiro Y, Hashimoto O, Seki M, Hachiya A, Shoji H, Hasegawa Y (2006) Analysis of the function of activin betaC subunit using recombinant protein. J Reprod Dev 52(4):487–495
40. Walton KL, Makanji Y, Wilce MC, Chan KL, Robertson DM, Harrison CA (2009) A common biosynthetic pathway governs the dimer-

ization and secretion of inhibin and related transforming growth factor beta (TGFbeta) ligands. J Biol Chem 284(14):9311–9320, PMCID: 2666583

41. Papakostas TD, Pieretti-Vanmarcke R, Nicolaou F, Thanos A, Trichonas G, Koufomichali X et al (2010) Development of an efficiently cleaved, bioactive, highly pure FLAG-tagged recombinant human Mullerian Inhibiting Substance. Protein Expr Purif 70(1):32–38, PMCID: 3581853
42. Cronin CN, Thompson DA, Martin F (1998) Expression of bovine activin-A and inhibin-A in recombinant baculovirus-infected Spodoptera frugiperda Sf21 insect cells. Int J Biochem Cell Biol 30(10):1129–1145
43. Maruoka Y, Oida S, Iimura T, Takeda K, Asahina I, Enomoto S et al (1995) Production of functional human bone morphogenetic protein-2 using a baculovirus/Sf-9 insect cell system. Biochem Mol Biol Int 35(5):957–963
44. Papakonstantinou T, Harris SJ, Fredericks D, Harrison C, Wallace EM, Hearn MT (2009) Synthesis, purification and bioactivity of recombinant human activin A expressed in the yeast Pichia pastoris. Protein Expr Purif 64(2): 131–138
45. Fredericks D, Clay R, Warner T, O'Connor A, de Kretser DM, Hearn MT (2010) Optimization of the expression of recombinant human activin A in the yeast Pichia pastoris. Biotechnol Prog 26(2):372–383
46. Rosenwald AG, Stanley P, Krag SS (1989) Control of carbohydrate processing: increased beta-1,6 branching in N-linked carbohydrates of Lec9 CHO mutants appears to arise from a defect in oligosaccharide-dolichol biosynthesis. Mol Cell Biol 9(3):914–924, PMCID: 362680
47. Studier FW, Moffatt BA (1986) Use of bacteriophage T7 RNA polymerase to direct selective high-level expression of cloned genes. J Mol Biol 189(1):113–130
48. Cerletti N (2000) Process for the production of biologically active dimeric protein. US Patent 6057430
49. Ejima D, Ono K, Tsumoto K, Arakawa T, Eto Y (2006) A novel "reverse screening" to identify refolding additives for activin-A. Protein Expr Purif 47(1):45–51
50. von Einem S, Schwarz E, Rudolph R (2010) A novel TWO-STEP renaturation procedure for efficient production of recombinant BMP-2. Protein Expr Purif 73(1):65–69
51. Sieber C, Ploger F, Schwappacher R, Bechtold R, Hanke M, Kawai S et al (2006) Monomeric and dimeric GDF-5 show equal type I receptor binding and oligomerization capability and have the same biological activity. Biol Chem 387(4):451–460
52. Zuniga JE, Groppe JC, Cui Y, Hinck CS, Contreras-Shannon V, Pakhomova ON et al (2005) Assembly of TbetaRI:TbetaRII:TGFbeta ternary complex in vitro with receptor extracellular domains is cooperative and isoform-dependent. J Mol Biol 354(5):1052–1068
53. Ilangovan U, Deep S, Hinck CS, Hinck AP (2004) Sequential resonance assignments of the extracellular domain of the human TGFbeta type II receptor in complex with monomeric TGFbeta3. J Biomol NMR 29(1):103–104
54. Dechavanne V, Barrillat N, Borlat F, Hermant A, Magnenat L, Paquet M et al (2011) A high-throughput protein refolding screen in 96-well format combined with design of experiments to optimize the refolding conditions. Protein Expr Purif 75(2):192–203
55. Mahlawat P, Ilangovan U, Biswas T, Sun LZ, Hinck AP (2012) Structure of the Alk1 extracellular domain and characterization of its bone morphogenetic protein (BMP) binding properties. Biochemistry 51(32):6328–6341, PMCID: 3448977
56. Townson SA, Martinez-Hackert E, Greppi C, Lowden P, Sako D, Liu J et al (2012) Specificity and structure of a high affinity activin receptor-like kinase 1 (ALK1) signaling complex. J Biol Chem 287(33):27313–27325, PMCID: 3431715
57. Klages J, Kotzsch A, Coles M, Sebald W, Nickel J, Muller T et al (2008) The solution structure of BMPR-IA reveals a local disorder-to-order transition upon BMP-2 binding. Biochemistry 47(46):11930–11939
58. Zuniga JE, Ilangovan U, Mahlawat P, Hinck CS, Huang T, Groppe JC et al (2011) The TbetaR-I pre-helix extension is structurally ordered in the unbound form and its flanking prolines are essential for binding. J Mol Biol 412(4):601–618
59. Nickel J, Kotzsch A, Sebald W, Mueller TD (2005) A single residue of GDF-5 defines binding specificity to BMP receptor IB. J Mol Biol 349(5):933–947
60. Greenwald J, Fischer WH, Vale WW, Choe S (1999) Three-finger toxin fold for the extracellular ligand-binding domain of the type II activin receptor serine kinase. Nat Struct Biol 6(1):18–22
61. Thompson TB, Woodruff TK, Jardetzky TS (2003) Structures of an ActRIIB:activin A complex reveal a novel binding mode for TGF-beta ligand:receptor interactions. EMBO J 22(7): 1555–1566, PMCID: 152900
62. Mace PD, Cutfield JF, Cutfield SM (2006) High resolution structures of the bone morphogenetic protein type II receptor in two crystal forms: implications for ligand binding. Biochem Biophys Res Commun 351(4):831–838

63. Boesen CC, Radaev S, Motyka SA, Patamawenu A, Sun PD (2002) The 1.1 A crystal structure of human TGF-beta type II receptor ligand binding domain. Structure 10(7):913–919
64. Deep S, Walker KP 3rd, Shu Z, Hinck AP (2003) Solution structure and backbone dynamics of the TGFbeta type II receptor extracellular domain. Biochemistry 42(34):10126–10139
65. Hart PJ, Deep S, Taylor AB, Shu Z, Hinck CS, Hinck AP (2002) Crystal structure of the human TbetaR2 ectodomain – TGF-beta3 complex. Nat Struct Biol 9(3):203–208
66. Alt A, Miguel-Romero L, Donderis J, Aristorena M, Blanco FJ, Round A et al (2012) Structural and functional insights into endoglin ligand recognition and binding. PLoS One 7(2):e29948, PMCID: 3275592
67. Castonguay R, Werner ED, Matthews RG, Presman E, Mulivor AW, Solban N et al (2011) Soluble endoglin specifically binds bone morphogenetic proteins 9 and 10 via its orphan domain, inhibits blood vessel formation, and suppresses tumor growth. J Biol Chem 286(34):30034–30046, PMCID: 3191044
68. del Re E, Sidis Y, Fabrizio DA, Lin HY, Schneyer A (2004) Reconstitution and analysis of soluble inhibin and activin receptor complexes in a cell-free system. J Biol Chem 279(51):53126–53135
69. Lin HY, Moustakas A, Knaus P, Wells RG, Henis YI, Lodish HF (1995) The soluble exoplasmic domain of the type II transforming growth factor (TGF)-beta receptor. A heterogeneously glycosylated protein with high affinity and selectivity for TGF-beta ligands. J Biol Chem 270(6):2747–2754
70. Tsang ML, Zhou L, Zheng BL, Wenker J, Fransen G, Humphrey J et al (1995) Characterization of recombinant soluble human transforming growth factor-beta receptor type II (rhTGF-beta sRII). Cytokine 7(5):389–397
71. Goetschy JF, Letourneur O, Cerletti N, Horisberger MA (1996) The unglycosylated extracellular domain of type-II receptor for transforming growth factor-beta. A novel assay for characterizing ligand affinity and specificity. Eur J Biochem 241(2):355–362
72. Greenwald J, Le V, Corrigan A, Fischer W, Komives E, Vale W et al (1998) Characterization of the extracellular ligand-binding domain of the type II activin receptor. Biochemistry 37(47):16711–16718
73. Boesen CC, Motyka SA, Patamawenu A, Sun PD (2000) Development of a recombinant bacterial expression system for the active form of a human transforming growth factor beta type II receptor ligand binding domain. Protein Expr Purif 20(1):98–104
74. Komesli S, Vivien D, Dutartre P (1998) Chimeric extracellular domain type II transforming growth factor (TGF)-beta receptor fused to the Fc region of human immunoglobulin as a TGF-beta antagonist. Eur J Biochem 254(3):505–513
75. Muraoka RS, Dumont N, Ritter CA, Dugger TC, Brantley DM, Chen J et al (2002) Blockade of TGF-beta inhibits mammary tumor cell viability, migration, and metastases. J Clin Invest 109(12):1551–1559
76. Yang YA, Dukhanina O, Tang B, Mamura M, Letterio JJ, MacGregor J et al (2002) Lifetime exposure to a soluble TGF-beta antagonist protects mice against metastasis without adverse side effects. J Clin Invest 109(12):1607–1615
77. Mitchell D, Pobre EG, Mulivor AW, Grinberg AV, Castonguay R, Monnell TE et al (2010) ALK1-Fc inhibits multiple mediators of angiogenesis and suppresses tumor growth. Mol Cancer Ther 9(2):379–388
78. Putnam C. Scripps Protein Calculator v. 3.3. 2013 [updated 2013; cited]. http://www.scripps.edu/~cdputnam/protcalc.html
79. Hinck AP, Walker KP 3rd, Martin NR, Deep S, Hinck CS, Freedberg DI (2000) Sequential resonance assignments of the extracellular ligand binding domain of the human TGF-beta type II receptor. J Biomol NMR 18(4): 369–370
80. Gasparian ME, Elistratov PA, Yakimov SA, Dolgikh DA, Kirpichnikov MP (2010) An efficient method for expression in Escherichia coli and purification of the extracellular ligand binding domain of the human TGFbeta type II receptor. J Biotechnol 148(2–3): 113–118
81. Hatta T, Konishi H, Katoh E, Natsume T, Ueno N, Kobayashi Y et al (2000) Identification of the ligand-binding site of the BMP type IA receptor for BMP-4. Biopolymers 55(5):399–406
82. Kirsch T, Nickel J, Sebald W (2000) Isolation of recombinant BMP receptor IA ectodomain and its 2:1 complex with BMP-2. FEBS Lett 468(2-3):215–219
83. Marlow MS, Chim N, Brown CB, Barnett JV, Krezel AM (2000) 1H, 13C, and 15N backbone assignments of the ligand binding domain of TGFbeta type II receptor. J Biomol NMR 17(4):349–350
84. Yin H, Yeh LC, Hinck AP, Lee JC (2008) Characterization of ligand-binding properties of the human BMP type II receptor extracellular domain. J Mol Biol 378(1):191–203
85. Baardsnes J, Hinck CS, Hinck AP, O'Connor-McCourt MD (2009) TbetaR-II discriminates the high- and low-affinity TGF-beta isoforms via two hydrogen-bonded ion pairs.

Biochemistry 48(10):2146–2155, PMCID: 2801812
86. De Crescenzo G, Grothe S, Lortie R, Debanne MT, O'Connor-McCourt M (2000) Real-time kinetic studies on the interaction of transforming growth factor alpha with the epidermal growth factor receptor extracellular domain reveal a conformational change model. Biochemistry 39(31):9466–9476
87. De Crescenzo G, Grothe S, Zwaagstra J, Tsang M, O'Connor-McCourt MD (2001) Real-time monitoring of the interactions of transforming growth factor-beta (TGF-beta) isoforms with latency-associated protein and the ectodomains of the TGF-beta type II and III receptors reveals different kinetic models and stoichiometries of binding. J Biol Chem 276(32): 29632–29643
88. De Crescenzo G, Hinck CS, Shu Z, Zuniga J, Yang J, Tang Y et al (2006) Three key residues underlie the differential affinity of the TGFbeta isoforms for the TGFbeta type II receptor. J Mol Biol 355(1):47–62
89. De Crescenzo G, Pham PL, Durocher Y, Chao H, O'Connor-McCourt MD (2004) Enhancement of the antagonistic potency of transforming growth factor-beta receptor extracellular domains by coiled coil-induced homo- and heterodimerization. J Biol Chem 279(25):26013–26018
90. De Crescenzo G, Pham PL, Durocher Y, O'Connor-McCourt MD (2003) Transforming growth factor-beta (TGF-beta) binding to the extracellular domain of the type II TGF-beta receptor: receptor capture on a biosensor surface using a new coiled-coil capture system demonstrates that avidity contributes significantly to high affinity binding. J Mol Biol 328(5):1173–1183
91. Huang T, David L, Mendoza V, Yang Y, Villarreal M, De K et al (2011) TGF-beta signalling is mediated by two autonomously functioning TbetaRI:TbetaRII pairs. EMBO J 30(7):1263–1276, PMCID: 3094126
92. Li L, Orner BP, Huang T, Hinck AP, Kiessling LL (2010) Peptide ligands that use a novel binding site to target both TGF-beta receptors. Mol Biosyst 6(12):2392–2402, PMCID: 3064480
93. Mendoza V, Vilchis-Landeros MM, Mendoza-Hernandez G, Huang T, Villarreal MM, Hinck AP et al (2009) Betaglycan has two independent domains required for high affinity TGF-beta binding: proteolytic cleavage separates the domains and inactivates the neutralizing activity of the soluble receptor. Biochemistry 48(49):11755–11765, PMCID: 2796082
94. O'Connor-McCourt MD, Segarubu O, Grothe S, Tsang M, Weatherbee JA (1995) Analysis of the interaction between two TGF-beta-binding proteins and three TGF-beta isoforms using surface plasmon resonance. Ann N Y Acad Sci 766:300–302
95. Zwaagstra JC, Sulea T, Baardsnes J, Lenferink AE, Collins C, Cantin C et al (2012) Engineering and therapeutic application of single-chain bivalent TGF-beta family traps. Mol Cancer Ther 11(7):1477–1487
96. Allendorph GP, Isaacs MJ, Kawakami Y, Izpisua Belmonte JC, Choe S (2007) BMP-3 and BMP-6 structures illuminate the nature of binding specificity with receptors. Biochemistry 46(43):12238–12247
97. Allendorph GP, Read JD, Kawakami Y, Kelber JA, Isaacs MJ, Choe S (2011) Designer TGFbeta superfamily ligands with diversified functionality. PLoS One 6(11):e26402, PMCID: 3208551
98. Brown MA, Zhao Q, Baker KA, Naik C, Chen C, Pukac L et al (2005) Crystal structure of BMP-9 and functional interactions with pro-region and receptors. J Biol Chem 280(26):25111–25118
99. Greenwald J, Vega ME, Allendorph GP, Fischer WH, Vale W, Choe S (2004) A flexible activin explains the membrane-dependent cooperative assembly of TGF-beta family receptors. Mol Cell 15(3):485–489
100. Keller S, Nickel J, Zhang JL, Sebald W, Mueller TD (2004) Molecular recognition of BMP-2 and BMP receptor IA. Nat Struct Mol Biol 11(5):481–488
101. Kotzsch A, Nickel J, Seher A, Heinecke K, van Geersdaele L, Herrmann T et al (2008) Structure analysis of bone morphogenetic protein-2 type I receptor complexes reveals a mechanism of receptor inactivation in juvenile polyposis syndrome. J Biol Chem 283(9): 5876–5887
102. Saremba S, Nickel J, Seher A, Kotzsch A, Sebald W, Mueller TD (2008) Type I receptor binding of bone morphogenetic protein 6 is dependent on N-glycosylation of the ligand. FEBS J 275(1):172–183
103. Abdiche Y, Malashock D, Pinkerton A, Pons J (2008) Determining kinetics and affinities of protein interactions using a parallel real-time label-free biosensor, the Octet. Anal Biochem 377(2):209–217
104. Wienken CJ, Baaske P, Rothbauer U, Braun D, Duhr S (2010) Protein-binding assays in biological liquids using microscale thermophoresis. Nat Commun 1:100
105. Bocharov EV, Blommers MJ, Kuhla J, Arvinte T, Bürgi R, Arseniev AS (2000) Sequence-specific 1H and 15N assignment and secondary structure of transforming growth factor beta3. J Biomol NMR 16(2):179–180

106. Bocharov EV, Korzhnev DM, Blommers MJ, Arvinte T, Orekhov VY, Billeter M et al (2002) Dynamics-modulated biological activity of transforming growth factor beta3. J Biol Chem 277(48):46273–46279
107. Marlow MS, Brown CB, Barnett JV, Krezel AM (2003) Solution structure of the chick TGFbeta type II receptor ligand-binding domain. J Mol Biol 326(4):989–997
108. Griffith DL, Keck PC, Sampath TK, Rueger DC, Carlson WD (1996) Three-dimensional structure of recombinant human osteogenic protein 1: structural paradigm for the transforming growth factor beta superfamily. Proc Natl Acad Sci U S A 93(2):878–883, PMCID: 40151
109. Scheufler C, Sebald W, Hulsmeyer M (1999) Crystal structure of human bone morphogenetic protein-2 at 2.7 A resolution. J Mol Biol 287(1):103–115
110. Schreuder H, Liesum A, Pohl J, Kruse M, Koyama M (2005) Crystal structure of recombinant human growth and differentiation factor 5: evidence for interaction of the type I and type II receptor-binding sites. Biochem Biophys Res Commun 329(3): 1076–1086
111. de Vries SJ, van Dijk ADJ, Krzeminski M, van Dijk M, Thureau A, Hsu V et al (2007) HADDOCK versus HADDOCK: new features and performance of HADDOCK2.0 on the CAPRI targets. Proteins 69:726–733
112. Dominguez C, Boelens R, Bonvin AMJJ (2003) HADDOCK: a protein-protein docking approach based on biochemical or biophysical information. J Am Chem Soc 125(7): 1731–1737
113. Ritchie DW (2008) Recent progress and future directions in protein-protein docking. Curr Protein Pept Sci 9(1):1–15

Chapter 5

Phosphorylation of Smads by Intracellular Kinases

Fang Liu and Isao Matsuura

Abstract

Smad proteins transduce the TGF-ß family signal at the cell surface into gene regulation in the nucleus. In addition to being phosphorylated by the TGF-ß family receptors, Smads are phosphorylated by a variety of intracellular kinases. The most studied are by cyclin-dependent kinases, the MAP kinase family members, and GSK-3. Phosphorylation by these kinases regulates Smad activities, leading to various biological effects. This chapter describes the methods for analyzing Smad phosphorylation by these kinases.

Key words Smad, Phosphorylation, CDKs, MAPKs, GSK-3

1 Introduction

The Smad family can be divided into distinct groups [1, 2]. One group includes receptor regulated Smads (R-Smads). This includes Smad2 and Smad3 for TGF-ß and activin signaling and Smad1, Smad5, and Smad8 for BMP signaling. In response to TGF-ß or activin, the TGF-ß type I receptor or the activin type I receptor phosphorylate the SSXS motif in the C-tail of Smad2 and Smad3. Similarly, in response to BMP, the BMP type I receptors phosphorylate the SSXS motif in the C-tail of Smad1, Smad5, and Smad8. Smad4 forms complexes with R-Smads for TGF-ß, activin, and BMP signaling.

The R-Smads contain conserved N-terminal and C-terminal domains, connected by a divergent linker region [1, 2]. The linker region is rich in serine/threonine and proline, and contains several proline-directed kinase phosphorylation sites. In Smad3, this includes T179, S204, S208, and S213. In Smad2, this includes T220, S245, S250, and S255.

R-Smads are phosphorylated by cyclin-dependent kinases (CDKs) [3–8]. CDKs were discovered for regulating the cell cycle [9]. Thus far, thirteen CDKs have been identified. In addition to control the cell cycle, CDKs also regulate transcription, mRNA

Xin-Hua Feng et al. (eds.), *TGF-β Signaling: Methods and Protocols*, Methods in Molecular Biology, vol. 1344, DOI 10.1007/978-1-4939-2966-5_5, © Springer Science+Business Media New York 2016

processing, and the differentiation of nerve cells. CDK itself has little kinase activity. CDK binds a regulatory protein called cyclin. Only the cyclin-CDK complex is an active kinase. The G1 CDKs include CDK4, CDK6, and CDK2. CDK4 and CDK6 bind cyclin D, whereas CDK2 binds cyclin E and cyclin A [9]. Smad3 is phosphorylated by CDK4 and CDK2 in a cell cycle-dependent manner. The phosphorylation has been mapped to T8 in the N-terminal domain and T179 and S213 in the linker region. CDK4 and CDK2 phosphorylation of Smad3 inhibits its antiproliferative function [3, 10, 11].

The linker regions of R-Smads are also phosphorylated by CDKs and GSK-3 in response to BMP or TGF-ß. Smad1 is phosphorylated by CDKs, including CDK8 and CDK9 and GSK-3 in response to BMP [7, 8, 12, 13]. There are four putative GSK-3 phosphorylation sites in Smad1, and two of which, T202 and S210, have been confirmed phosphorylated by GSK-3 with the availability of phosphopeptide antibodies [12, 13]. CDK8 binds to cyclin C and CDK9 binds to cyclin T. CDK8 and CDK9 regulate transcription [14]. CDK8/9 phosphorylation of Smad1 increases Smad1 transcriptional activity and binding to Smurf1, which leads to Smad1 degradation [7, 8].

Smad3 is phosphorylated at the T179 and S208 by CDKs and at the S204 by GSK-3 in response to TGF-ß [6–8, 15]. GSK-3 also phosphorylates Smad3 at the T66 site in the N-terminal domain at the basal state, which leads to Smad3 degradation [16]. For TGF-ß-induced Smad3 linker phosphorylation, it appears that CDK8, CDK9 and additional CDKs may phosphorylate Smad3 [6–8]. TGF-ß-induced Smad3 linker phosphorylation can have different effects. Smurf2 has been shown to bind to the phosphorylated Smad3 linker region and induce multiple mono-ubiquitination at the C-terminal domain of Smad3. The mono-ubiquitination of Smad3 inhibits protein complex formation, thus attenuating Smad3 activity [17]. Smad3 and Smad1 have different configurations in their linker region, which may be the basis for their differential regulation.

In response to EGF, R-Smads are phosphorylated by ERK in the linker region [18–20]. In addition, TGF-ß signaling can activate Ras under certain conditions, which then activates ERK and leads to R-Smads phosphorylation [2, 21]. ERK phosphorylation of Smad3 occurs at the S208, S204, and T179 sites [20]. Smad1 contains four ERK consensus phosphorylation sites. Phosphopeptide antibodies have been raised against two of these sites S206 and S214. Both S206 and S214 are phosphorylated by ERK [12, 13]. The linker regions of R-Smads can also be phosphorylated by JNK or p38 MAPK in response to TGF-ß, HGF, PDGF, UV, or salt, some of which may occur only in certain cell types [4, 12, 22–26]. MAPK superfamily phosphorylation of R-Smads has different effects in a context-dependent manner [2, 21, 27, 28]. ERK also phosphorylates Smad4 at T276, a consensus ERK phosphorylation site in the linker region [29]. The phosphorylation increases TGF-ß-induced nuclear accumulation of Smad4.

In addition, several other intracellular kinases phosphorylate Smads. Calcium calmodulin-dependent kinase II (CaMKII) phosphorylates Smad2 and Smad4, and to a much less extent, Smad3. The CaMKII phosphorylation of Smad2 occurs in the linker at S240 and S260 and in the N-terminal domain at S110, and the phosphorylation results in the inhibition of TGF-ß-inducible nuclear accumulation of Smad2 [30]. PKC directly phosphorylates Smad3 and Smad2. The phosphorylation has been mapped to the S37 and S70 in the N-terminal domain of Smad3 and S47 and S110 in the N-terminal domain of Smad2. PKC phosphorylation of Smad3 inhibits its DNA binding activity [31]. Casein kinase 1 gamma 2 (CKIγ2) binds specifically to Smad3 and phosphorylates Smad3 at S418 in the C-terminal domain. The phosphorylation promotes the ubiquitination and degradation of activated Smad3 [32]. Misshapen (Msn) kinases belong to the Ste20 kinase family (known as MAP4K). Recent study has shown that Msn also phosphorylates all R-Smads (except Smad3) in the C-terminal domain at T312 in *Drosophila* Mad, S322 in Smad1 and S324 in Smad2 at the analogous position. The phosphorylation inhibits Smad interaction with the type I receptors of the TGF-ß family [33]. In addition, Mps1 phosphorylates Smad2 and Smad3 at the SSXS motif in the C-tail in response to nocodazole-induced microtubule depolymerization [34]. The methods for phosphorylation by these kinases are described in detail in the original papers. The methods described below are also useful for studying phosphorylations by these kinases.

2 Materials

Cell Culture: Mv1Lu mink lung epithelial cells are cultured in MEM with 10 % FBS and 1 % penicillin/streptomycin.

Chemicals and Reagents: Protease inhibitor cocktail (100×) contains 1 mg/mL leupeptin, 1 mg/mL antipain, 5 mg/mL aprotinin, 10 mg/mL soybean trypsin inhibitor, and 10 mg/mL benzamidine hydrochloride. These inhibitors are from Sigma. The cocktail is made in 10 mM Tris–HCl, pH 7.4, 1 mM EDTA, then aliquot and store at −70 °C. Alternatively, one can use the complete protease inhibitor cocktail tablets (Roche). ^{32}P-orthophosphate is from Perkin-Elmer. Protein A and protein G beads are from GE Healthcare Life Sciences. Ni-NTA-agarose and GSH agarose are from Sigma. Bacterially expressed and activated ERK2 is from Millipore. Recombinant GSK-3β is from New England Biolabs.

Antibodies: CDK2, CDK4, CDK7, CDK8, CDK9, and Rb antibodies are from Santa Cruz Biotechnology. Smad3 antibody is from Invitrogen. Smad2 pS245/pS250/pS255 phosphopeptide antibody is from Cell Signaling Technology. Smad3 phosphopeptide antibodies, including the pS204 antibody, are described previously [3].

2.1 In Vitro Kinase Assay Using Immunoprecipitated Endogenous CDKs

Lysis buffer: 50 mM Tris–HCl, pH 7.5, 150 mM NaCl, 0.5 % NP40, 50 mM NaF, 1 mM Na_3VO_4 (sodium orthovanadate), 5 mM β-glycerophosphate, 1 mM DTT, protease inhibitor cocktail, and 1 mM phenylmethylsulfonyl fluoride (PMSF). Protease inhibitor cocktail is described under Subheading 2.

IP kinase buffer: 50 mM HEPES, pH 7.4, 15 mM $MgCl_2$, 1 mM EGTA, 0.1 % Tween 20, 1 mM DTT, 50 μM ATP.

2.2 In Vivo Phosphorylation

Phosphatase inhibitors: 50 mM NaF, 10 mM $Na_4O_7P_2$ (sodium pyrophosphate), 1 mM Na_3VO_4, 5 mM β-glycerophosphate.

Lysis buffer: 10 mM Tris (pH 7.8), 150 mM NaCl, 1 mM EDTA, 1 % NP40, 1 μg/mL RNaseA, phosphatase inhibitors, protease inhibitor cocktail, 1 mM PMSF.

Washing buffer: 10 mM Tris (pH 7.8), 150 mM NaCl, 1 mM EDTA, 1 % NP40, phosphatase inhibitors, 1 mM PMSF.

3 Methods

To study phosphorylation, one will usually examine the amino acid sequence to determine whether it contains a consensus phosphorylation site for a particular kinase. CDKs and MAPKs are proline-directed kinases. The minimal requirement is an S/TP motif. The consensus phosphorylation site for CDK2 is S/TPXK/R [9]. The consensus phosphorylation site for CDK4, CDK8, or CDK9 seems not defined, as each of them has only a few substrates identified. The ERK consensus phosphorylation site is PXS/TP. JNK and p38 are distinct and do not require the upstream proline. The consensus phosphorylation site for GSK-3 is S/TXXXS/TP [35], where the N-terminal serine or threonine is the target for phosphorylation. Although X can be any amino acid, it is often proline. The C-terminal serine or threonine is a residue that has undergone a priming phosphorylation by a different kinase. Although not strictly required, the priming phosphorylation enhances the efficiency of phosphorylation by GSK-3 100–1000-fold for most substrates [35]. In addition to the examination of the consensus sites, one can also use computer programs, such as the Scansite software program, to determine which site is recognized as a phosphorylation site for a kinase. This program can run under high stringency as well as under low stringency. While all these are very useful information, consensus sites or Scansite program identified sites do not always predict the actual phosphorylation sites.

Phosphorylation experiments usually start with ^{32}P labeling. Phosphopeptide antibodies raised against each of the putative phosphorylation sites are important tools to study phosphorylation. They allow one to determine which of the putative phosphorylation sites is indeed phosphorylated in vivo and in vitro. Below are methods for phosphorylation by CDKs, MAPKs, and GSK-3. In addition to ^{32}P labeling, one can use phosphopeptide antibodies in nonradioactive experiments to determine which sites are phosphorylated.

3.1 In Vitro Kinase Assay Using Immunoprecipitated Endogenous CDKs

The use of immunoprecipitated endogenous CDK has the advantage of physiological relevance. On the other hand, it is possible that another kinase can associate with the immunoprecipitated CDK. To overcome this concern, one can preincubate the CDK antibody with the antigen peptide to determine whether the peptide dramatically reduces the phosphorylation. If it does, it provides strong evidence that it is the CDK, not an associated kinase, that phosphorylates the substrate.

In the assay, one usually includes a negative control and a positive control. GST can be included as a negative control if the substrate is a GST fusion protein. For positive control, one usually uses the best-characterized substrate. For CDK2 and CDK4, the best-characterized substrate is Rb [9]. For CDK8 and CDK9, the best-characterized substrate is the C-terminal domain (CTD) of RNA polymerase II [14].

Experimental Procedure:

1. Mv1Lu mink lung epithelial cells on a 100 mm plate are lysed with 1.2 mL of lysis buffer on a shaker at 4 °C for 1 h (*see* **Notes 1** and **2**).
2. Collect the lysate in an eppendorf tube and spin in microcentrifuge at 12,000 x g at 4 °C for 20 min.
3. Transfer the supernatant to a new eppendorf tube.
4. Measure the protein concentration of the lysates.
5. Remove appropriate volume of lysates containing 240 μg protein and make the volume 1 mL with the lysis buffer. Add 1.2 μg of an affinity-purified CDK antibody (*see* **Note 3**) or IgG control antibody and incubate at 4 °C for 2 h with constant rocking. After the initial experiment yields positive result, repeat the experiment with the CDK antibody with or without preincubation with the antigen peptide.

6. Add 10 μL of 1:1 mixture of protein A+protein G beads (30 μL of 33 % suspension), and further incubate on the rocker at 4 °C for 1 h.
7. Spin down the beads. Wash the beads three times with the lysis buffer and 3 times with IP kinase buffer (50 mM HEPES, pH 7.4, 15 mM $MgCl_2$, 1 mM EGTA, 0.1 % Tween 20, 1 mM DTT, 50 μM ATP). For the washing steps, one can omit the protease inhibitor cocktail in the lysis buffer.
8. For ^{32}P labeling, the kinase reaction is carried out in 30 μL containing 50 mM HEPES, pH 7.4, 15 mM $MgCl_2$, 1 mM EGTA, 0.1 % Tween 20, 1 mM DTT, 50 μM ATP, 5 μCi [γ-^{32}P]ATP (3000 Ci/mmol), and 1 μM GST-Smad3 or other substrate at 30 °C for 1 h (*see* **Note 4**). The volume of the protein A/G beads is 10 μL. Add the components above in 20 μL, so the final volume is 30 μL.
9. The kinase reaction is terminated by addition of 10 μL of 4× SDS gel loading buffer.
10. Heat the samples at 100 °C for 5 min. Half of the supernatant (15 μL) is subjected to SDS PAGE. Phosphorylated proteins are visualized by autoradiography. An example of CDK4 phosphorylation of Smad3 and Smad2 in this assay is shown in Fig. 1. An example of CDK8 and CDK9 phosphorylation of Smad3 in this assay is shown in Fig. 2.
11. For quantitative analysis, phosphorylated protein bands are excised from the gel and counted for radioactivity.
12. For nonradioactive phosphorylation, the kinase reaction is carried out in 30 μL containing the immunoprecipitated kinase, 50 mM HEPES, pH 7.4, 15 mM $MgCl_2$, 1 mM EGTA, 0.1 % Tween 20, 1 mM DTT, 0.5 mM ATP, and 0.4 μM GST-Smad3 or other substrate at 30 °C for 1 h. The kinase reaction is terminated by addition of 10 μL of 4× SDS protein gel sample buffer. The reaction products are then analyzed by immunoblotting using Smad3 phosphopeptide antibodies or other phosphopeptide antibodies.

3.2 In Vitro Kinase Assay Using Reconstituted Cyclin-CDK

In this method, His tagged cyclin and GST-CDK are expressed and purified from bacteria. Cyclin and CDK are then mixed and activated by CAK (CDK activating kinase) in the presence of ATP and Mg^{2+}. One can use HeLa extracts as a source for CAK. The activated cyclin-CDK complexes are then purified.

Regarding the molar ratio to mix cyclin with CDK, our titration experiments indicate that depending on different preps, the ratio can vary from 1.5:1 to 5:1 for optimal activity. Beyond the ratio of 5:1, increase of the amount of cyclin usually does not further increase the activity. It is well known that CDK4 purified from

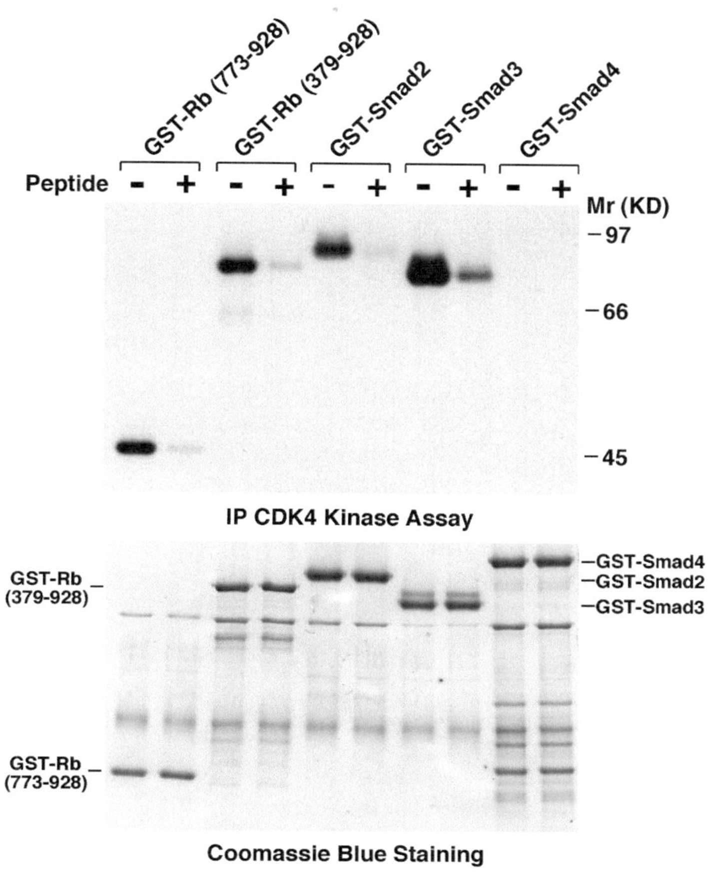

Fig. 1 CDK4 can phosphorylate Smad3 and Smad2 in vitro. CDK4 immunoprecipitated from 240 μg Mv1Lu cell lysates with 1.2 μg CDK4 antibody in the absence or presence of 9 μg of the antigen peptide (mouse CDK4 aa 282–303) was used in a kinase assay with 1 μM substrates. GST-Rb was included as a positive control. GST-Rb (773–928) contains the proline-rich region, and GST-Rb (379–928) contains in addition the Rb pocket domain. The *upper panel* shows the ^{32}P autoradiogram. The *lower panel* shows the Coomassie blue staining of the gel. Reproduced from *Nature* 2004;430:226–231 [3]

bacteria is not as active as CDK2, and thus requires a larger amount in a kinase reaction. An alternative method is to purify CDK4 from baculovirus.

The protocols below are the kinase assays for in vitro reconstituted cyclin D-CDK4, cyclin E-CDK2, and cyclin A-CDK2. Kinase assays for in vitro reconstituted other cyclin-CDK complexes can be similarly performed.

3.2.1 In Vitro Kinase Assay Using Reconstituted Cyclin D-CDK4

Cyclin D1-CDK4 is reconstituted as described previously [36] with slight modifications, such as no $MnCl_2$ in the reaction. The reconstituted cyclin D1-CDK4 is purified by GSH-agarose beads.

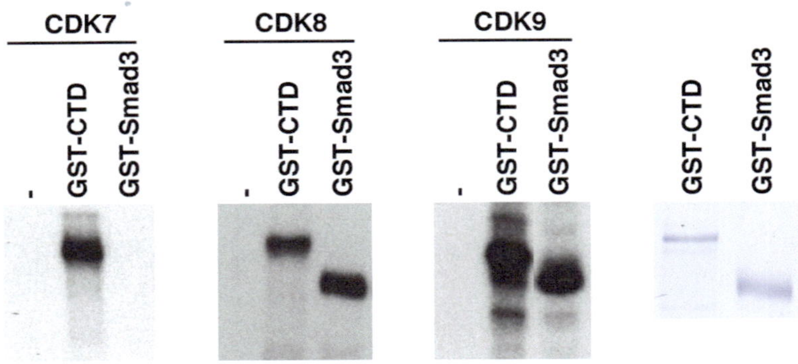

Fig. 2 CDK8 and CDK9 can phosphorylate Smad3 in vitro. CDK7, CDK8, or CDK9 was immunoprecipitated from Mv1Lu cell lysates and subjected to an in vitro kinase assay using GST-Smad3 or GST-CTD as a control. The – lanes were from immunoprecipitation by IgG and using GST-Smad3 as a substrate. The *right panel* shows Coomassie blue staining of GST-CTD and GST-Smad3 used in the kinase reaction

Experimental Procedure

1. Express and purify each of GST-CDK4 and 6His-cyclin D1 from *E. coli*.
2. Mix GST-CDK4 (13 μg), 6His-cyclin D1 (45 μg), and HeLa S100 extracts (600 μg) in a 500 μL reaction containing 50 mM HEPES, pH 7.5, 15 mM $MgCl_2$, 1 mM EGTA, 1 mM Na_3VO_4, 5 mM β-glycerophosphate, 1 mM DTT, 1 mM ATP, and 1 mM PMSF. Incubate at room temperature for 1 h.
3. Add 50 μL of GSH-agarose beads. Add Tween 20 to final concentration of 0.1 %. Rock the mixture at 4 °C for 1 h.
4. Wash the beads six times with a buffer containing 80 mM HEPES, pH 7.5, 15 mM $MgCl_2$, 1 mM EGTA, 1 mM Na_3VO_4, 5 mM β-glycerophosphate, 0.1 % Tween 20, 1 mM DTT, and 1 mM PMSF.
5. Elute proteins on the beads with 130 μL of elution buffer containing 10 mM GSH, 50 mM HEPES, pH 7.5, 10 mM $MgCl_2$, 1 mM EGTA, 0.1 % Tween 20, and 1 mM DTT. Measure protein concentration (*see* **Note 5**).
6. 500 ng of reconstituted 6His-cyclin D1/GST-CDK4 is used in a 20 μL kinase reaction containing 35 mM HEPES, pH 7.5, 10 mM $MgCl_2$, 1 mM EGTA, 0.1 % Tween 20, 1 mM DTT, 15 μM ATP, 5 μCi [γ-^{32}P]ATP, and 0.4 μM GST-Smad3 or other substrate (*see* **Note 6**). Incubate the reaction at 30 °C for 40 min.
7. The kinase reaction is terminated by addition of 4× SDS gel loading buffer.

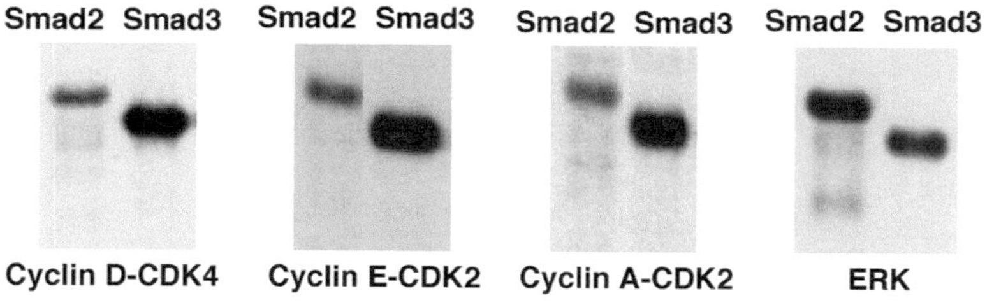

Fig. 3 Reconstituted G1 cyclin-CDKs and recombinant ERK can phosphorylate Smad3 and Smad2 in vitro. Bacterially expressed and in vitro reconstituted G1 cyclin-CDKs and recombinant activated ERK were used in a kinase reaction containing 0.4 μM GST-Smad2 or GST-Smad3. The results also suggest that Smad3 is a better substrate than Smad2 for CDK4 and CDK2, whereas Smad2 is a better substrate than Smad3 for ERK

8. Load half of the sample on SDS PAGE followed by autoradiography. An example of cyclin D1-CDK4 phosphorylation of Smad3 and Smad2 in this assay is shown in Fig. 3.

3.2.2 In Vitro Kinase Assay Using Reconstituted Cyclin E-CDK2 and Cyclin A-CDK2

Cyclin E-CDK2 and cyclin A-CDK2 are reconstituted similarly. The activated cyclin E-CDK2 or cyclin A-CDK2 is extensively purified by successive Ni-NTA-agarose and GSH-agarose chromatography.

Experimental Procedure:

1. Express and purify each of GST-CDK2, 6His-cyclin E, and 6His-cyclin A from *E. coli*.
2. Mix GST-CDK2 (60 μg) with 6His-cyclin E (300 μg) or with 6His-cyclin A2 (300 μg) in a 2 mL eppendorf tube. Incubate at 30 °C for 15 min.
3. Add HeLa S100 extracts (192 μg) and 20 mM MOPS, pH 7.4, 10 mM $MgCl_2$, 1 mM EGTA, 1 mM Na_3VO_4, 5 mM β-glycerophosphate, 1 mM DTT, 1 mM ATP, and 1 mM PMSF in 1.2 mL. Incubate at 30 °C for 30 min.
4. Add 0.3 mL of Ni-NTA-agarose beads (50 % suspension).
5. Rock at 4 °C for 1 h.
6. Pack into a column.
7. Wash the column with 2 mL buffer 1 containing 50 mM sodium phosphate, pH 8.0, 1 M NaCl, 1 % NP40, 1 mM DTT, 1 mM benzamidine, 1 mM PMSF.
8. Wash the column with 2 mL buffer 2. Buffer 2 is the same as buffer 1 except with 0.5 M NaCl and no NP40.
9. Elute the column with buffer 2 containing 0.25 M imidazole.

10. Measure protein concentration of the fractions, and combine the appropriate fractions.
11. Add 100 μL of GSH agarose beads to the combined fractions.
12. Rock at 4 °C for 1 h.
13. Pack into a column. Reload the flow through into the column.
14. Wash the column with 1.5 mL of MTPBS (15 mM Na_2HPO_4, 4 mM NaH_2PO_4, 150 mM NaCl), pH 7.4, 1 mM benzamidine, 1 mM PMSF.
15. Elute the column with a buffer containing 50 mM Tris–HCl, pH 8.0, 5 mM GSH, 1 mM DTT, 1 mM benzamidine, 1 mM PMSF.
16. Measure protein concentration of the fractions, and combine the appropriate fractions (*see* **Note 7**).
17. The reconstituted cyclin E-CDK2 and cyclin A-CDK2 can be frozen down in aliquots for future use. Add BSA to 0.5 μg/mL. Add glycerol to 10 %. Store at −70 °C.
18. 13 ng of 6His-cyclin E/GST-CDK2 or 96 ng of 6His-cyclin A/GST-CDK2 is used in a 20 μL kinase reaction containing 35 mM HEPES, pH 7.5, 10 mM $MgCl_2$, 1 mM EGTA, 0.1 % Tween 20, 1 mM DTT, 100 μM ATP, 2 μCi [γ-^{32}P]ATP, and 0.4 μM GST-Smad3, GST-Smad2, or other substrate (*see* **Note 8**). Incubate the reaction at 30 °C for 40 min.
19. The kinase reaction is terminated by addition of 4× SDS gel loading buffer.
20. Half of the sample was analyzed for phosphorylation by SDS PAGE and autoradiography. An example of cyclin E-CDK2 and cyclin A-CDK2 phosphorylation of Smad3 and Smad2 in this assay is shown in Fig. 3.

3.3 In Vitro Kinase Assay of MAP Kinases

In vitro kinase assay for MAP kinase family members, including ERK, JNK, and p38, can be performed using immunoprecipitated kinase from ligand-stimulated mammalian cells. The activated ERK, JNK, or p38 from bacteria or baculovirus is also available commercially and can be used in the in vitro kinase assays. Below is a protocol using a commercial recombinant activated ERK2. Kinase assays using a commercial activated JNK or p38 can be performed using a similar protocol.

Experimental Procedure:

1. For ^{32}P labeling, the kinase reaction contains 20 mM MOPS-NaOH, pH 7.4, 10 mM $MgCl_2$, 1 mM EGTA, 100 μM ATP, 3 μCi [γ-^{32}P]ATP (3000 Ci/mmol), 0.4 μM GST-Smad2,

GST-Smad3 or other substrate, and 26 ng recombinant activated ERK2 (16 units) in 30 μL. Incubate at 30 °C for 30 min.

2. Add 10 μL of 4× SDS protein gel sample buffer to terminate the reaction.
3. Load half of the reaction onto SDS PAGE, followed by autoradiography. An example of ERK2 phosphorylation of Smad3 and Smad2 in this assay is shown in Fig. 3 (*see* **Note 9**).
4. For nonradioactive phosphorylation, the kinase reaction contains 20 mM MOPS-NaOH, pH 7.4, 10 mM $MgCl_2$, 1 mM EGTA, 0.5 mM ATP, 0.3 μM GST-Smad3 or other substrate, and 26 ng recombinant activated ERK2 (16 units) in 30 μL. The reaction is carried out at 30 °C for 30 min, and then terminated by addition of 10 μL of 4× SDS protein gel sample buffer. The reaction products are analyzed by immunoblotting using Smad3 phosphopeptide antibodies or other phosphopeptide antibodies.

3.4 In Vitro Kinase Assay of GSK-3ß

1. For ^{32}P labeling, the kinase reaction contains 20 mM Tris–HCl, pH 7.5, 10 mM $MgCl_2$, 5 mM DTT, 50 μM ATP, 5 μCi [γ-^{32}P]ATP (3000 Ci/mmol), 0.5 μM GST-Smad3 or other substrate, and 1 μL of GSK-3ß (100 units) in 30 μL. Incubate at 30 °C for 1 h (*see* **Note 10**).
2. Add 10 μL of 4× SDS protein gel sample buffer to terminate the reaction.
3. Load half of the sample onto SDS PAGE, followed by autoradiography.
4. For nonradioactive phosphorylation, the kinase reaction is carried out in 30 μL containing 20 mM Tris–HCl, pH 7.5, 10 mM $MgCl_2$, 5 mM DTT, 0.5 mM ATP, 0.5 μM GST-Smad3, GST-Smad2 or other substrate, and 1 μL of GSK-3ß (100 units) at 30 °C for 1 h. For GST-Smad3, 10 % of the reaction products are analyzed by immunoblotting using the Smad3 pS204 antibody. For GST-Smad2, 20 % of the reaction products are analyzed by Smad2 pS245/S250/S255 antibody (*see* **Note 11**). Examples of the results are shown in Fig. 4.

3.5 In Vivo Phosphorylation

The protocol below analyzes Smad3 phosphorylation by G1 CDKs in a cell cycle-dependent manner. The protocol can also be used with minor modifications for analysis of Smad phosphorylation by other kinases, such as by ERK in response to EGF or by CDK and GSK3 in response to TGF-ß. For such analysis, cells will be serum-starved prior to the addition of a ligand. To confirm that a particular kinase is responsible for the phosphorylation, one can use chemical inhibitors, knockdown of the kinase, and cells from knockout animals.

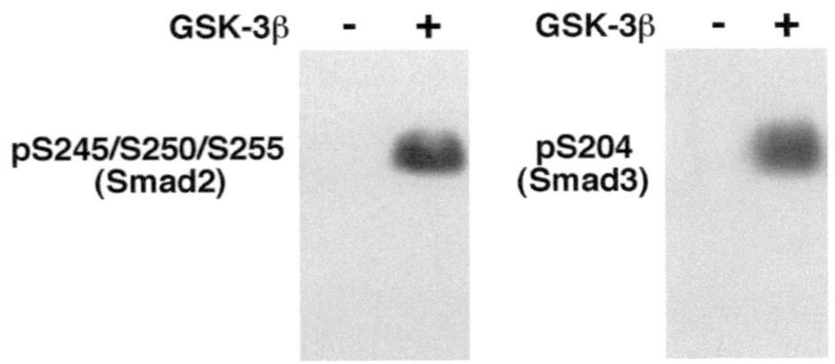

Fig. 4 GSK-3β can phosphorylate Smad3 and Smad2 in vitro. Recombinant GSK-3β was used in a kinase reaction containing 0.5 μM GST-Smad2 or GST-Smad3. Phosphorylation of Smad2 was analyzed by immunoblotting with the pS245/pS250/pS255 phosphopeptide antibody, which recognizes phosphorylation occurred within these three sites. Phosphorylation of S204 in Smad3 was analyzed by immunoblotting with the pS204 phosphopeptide antibody

Experimental Procedure:

1. Synchronize Mv1Lu cells on 100 mm plates at the G0/G1 phase by contact inhibition in complete medium (*see* **Note 12**).
2. Split the cells and plate into fresh medium. For ^{32}P labeling, plate cells on 60 mm dishes. In parallel, plate cells on two other sets of 100 mm dishes at the same density. One set of the plates will be used for FACS analysis for cell cycle population. Another set of the plates will be lysed for analyzing protein levels or activities.
3. After release from growth arrest for various time periods (*see* **Note 13**), cells are washed twice with phosphate-free medium, and then incubate in phosphate-free medium for 30 min.
4. Label with 1 mCi ^{32}P-orthophosphate in 1 mL for 1.5 h.
5. Remove ^{32}P-label, wash with cold PBS four times. Then add 1 mL of lysis buffer to the plates, shake at 4 °C for 1 h.
6. Transfer the lysates to eppendorf tubes. Spin at 12,000 x g for 30 min.
7. Transfer the supernatant to eppendorf tubes.
8. Add 45 μL of protein A+protein G beads (33 % suspension) to pre-clear. Rock at 4 °C for 30 min. Spin down briefly.
9. Transfer the supernatant to eppendorf tubes, add Smad3 polyclonal antibody to a final concentration of 2.5 μg/mL, rock at 4 °C for 8 h (minimally 5 h).
10. Add 45 μL of protein A+protein G beads (33 % suspension). Rock at 4 °C for 1.5 h. Spin down briefly.

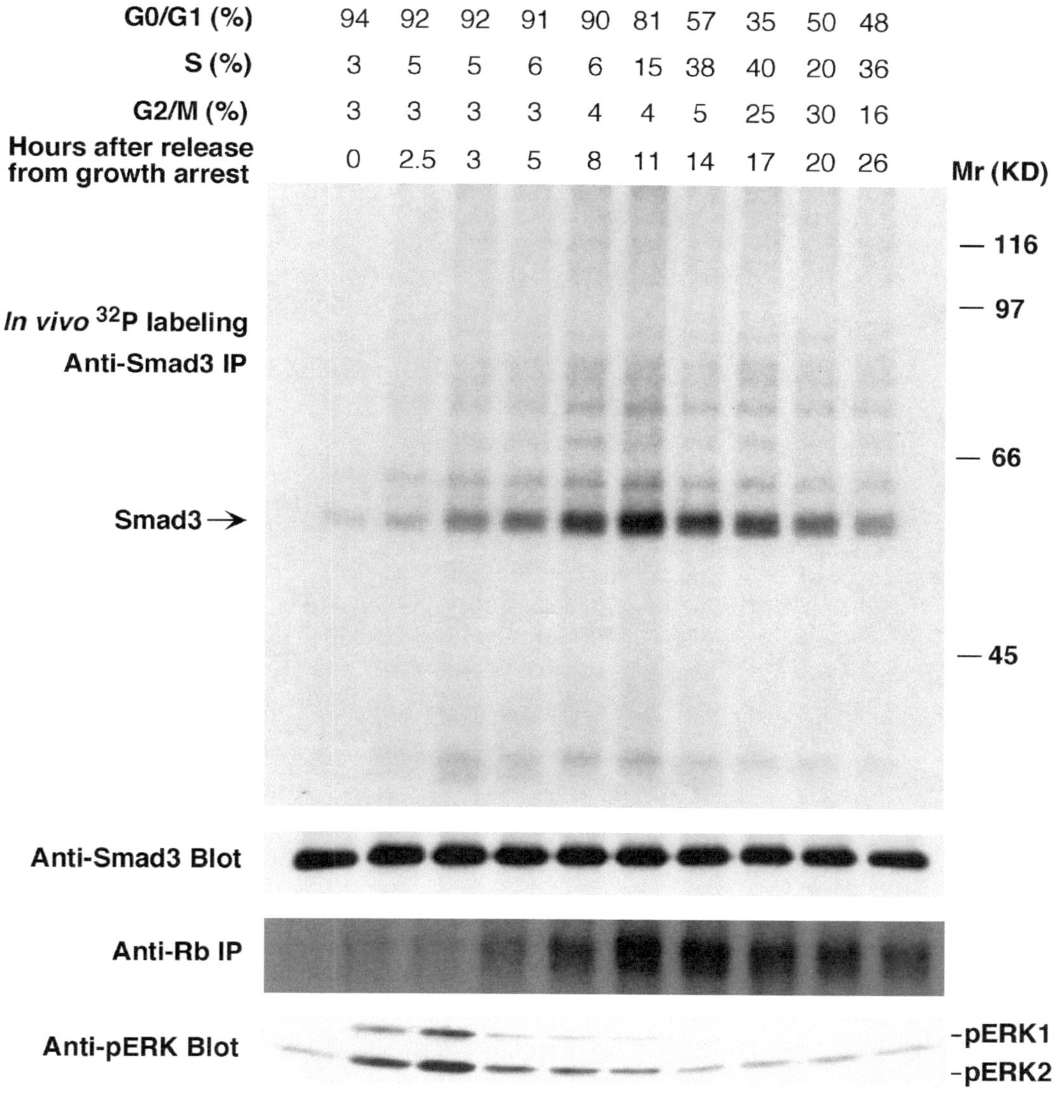

Fig. 5 Smad3 is phosphorylated by endogenous G1 CDKs in vivo. Mv1Lu cells were synchronized by contact inhibition, then plated into fresh medium, labeled with ^{32}P-orthophosphate, immunoprecipitated (IP) first with a Smad3 antibody and subsequently with a Rb antibody. The ERK activity profile, analyzed by a phosphotyrosine antibody that recognizes only activated ERK (pERK1 and pERK2) in unlabeled lysates, suggests that ERK may contribute to Smad3 phosphorylation at very early, but not at the peak, phosphorylation time points. Reproduced from *Nature* 2004;430:226–231 [3]

11. Transfer the supernatant to fresh tubes. Add Rb antibody to a final concentration of 2.5 μg/mL. Incubate at 4 °C overnight on a rocker. Then add protein A+protein G beads, and follow the same procedures as for Smad3 antibody.
12. Wash the beads five times, each time with 1 mL of washing buffer rock for 10 min. Spin down briefly.

13. Add 40 μL of 1.2× protein gel sample buffer to the beads, incubate at 37 °C for 10 min (*see* **Note 14**).
14. Spin down. Transfer the supernatant to eppendorf tubes.
15. Load half of the sample on SDS PAGE. Analyze by autoradiography. An example of the assay is shown in Fig. 5.
16. For nonradioactive phosphorylation, after release from growth arrest for various time periods, cells will be lysed and analyzed by phosphopeptide antibodies to determine which of the putative sites are phosphorylated by G1 CDKs.

4 Notes

1. We use Mv1Lu cells, because the cells grow rapidly. Another excellent cell line is the human U2OS osteosarcoma cell line. We examined G1 CDK activities in a variety of cell lines. These two cell lines have highest levels of G1 CDK activities.
2. The Mv1Lu cells are seeded for ~40 h. At the time of lysis, the cells are ~80–90 % confluence. Mv1Lu cells at 80–90 % are still rapidly growing.
3. For CDK4, one must use affinity-purified antibody.
4. Inclusion of cold ATP in a kinase reaction makes the background low.
5. The concentration is usually 50–60 ng/μL 6His-cyclin D1/GST-CDK4 in our experiments.
6. For reconstituted 6His-cyclin D1/GST-CDK4, we use fresh prep each time.
7. The concentration of the combined fractions was 26 ng/μL for 6His-cyclin E/GST-CDK2 and 32 ng/μL for 6His-cyclin A/GST-CDK2 in this experiment. The volume was ~230 μL.
8. The reconstituted cyclin E-CDK2 and cyclin A-CDK2 are much more active than the reconstituted cyclin D1-CDK4. Therefore, much less cyclin E-CDK2 or cyclin A-CDK2 is used in the kinase reaction compared to cyclin D1-CDK4.
9. The results in Fig. 3 suggest that Smad2 is a better substrate than Smad3 for ERK. For more quantitative assessment, one can perform kinase and substrate titration experiments to determine their apparent Km and Vmax. The greater the ratio of Vmax/Km, the better the substrate.
10. For Smad1, it was shown that in an in vitro kinase assay, phosphorylation by GSK-3ß requires downstream priming phosphorylation [12]. For Smad3, GSK-3ß can directly phosphorylate it without priming phosphorylation in vitro [6, 15]. One possibility

is that the Smad1 study was carried out with a small amount of GSK-3ß, whereas the studies on Smad3 used more GSK-3ß in the kinase reactions. It should be noted that in vivo, in response to TGF-ß, Smad3 phosphorylation by GSK-3ß at the S204 site requires the priming phosphorylation at the S208 site [6, 15].

11. Even an excellent phosphopeptide antibody can recognize the nonphosphorylated substrate if too much substrate is loaded. For each phosphopeptide antibody, the Western blot conditions, including how much protein to load onto a gel, need to be optimized.
12. For contact inhibition, use complete medium, not serum-free medium. After cells become confluent, maintain the cells in the culture for 4–5 days and change medium every the other day to keep cells healthy.
13. The 0 time point is the contact inhibited cells. Greater than 90 % cells should be in the G0/G1 phase at the 0 time point.
14. For in vivo ^{32}P-labeling experiments, incubation at 37 °C for 10 min, instead of boiling the sample, may reduce background.

Acknowledgement

We thank Drs. X.-H. Feng, K. Matsuzaki, X.-F. Wang, and Y. E. Zhang and our laboratory members for discussions and assistance during the course of our phosphorylation studies. The phosphorylation project was supported by NIH grant CA93771.

References

1. Feng XH, Derynck R (2005) Specificity and versatility in tgf-beta signaling through Smads. Annu Rev Cell Dev Biol 21:659–693
2. Massagué J (2012) TGFβ signalling in context. Nat Rev Mol Cell Biol 13:616–630
3. Matsuura I, Denissova NG, Wang G, He D, Long J, Liu F (2004) Cyclin-dependent kinases regulate the antiproliferative function of Smads. Nature 430:226–231
4. Matsuzaki K, Kitano C, Murata M, Sekimoto G, Yoshida K, Uemura Y, Seki T, Taketani S, Fujisawa J, Okazaki K (2009) Smad2 and Smad3 phosphorylated at both linker and COOH-terminal regions transmit malignant TGF-beta signal in later stages of human colorectal cancer. Cancer Res 69:5321–5330
5. Baughn LB, Di Liberto M, Niesvizky R, Cho HJ, Jayabalan D, Lane J, Liu F, Chen-Kiang S (2009) CDK2 phosphorylation of Smad2 disrupts TGF-beta transcriptional regulation in resistant primary bone marrow myeloma cells. J Immunol 182:1810–1817
6. Wang G, Matsuura I, He D, Liu F (2009) Transforming growth factor-beta-inducible phosphorylation of Smad3. J Biol Chem 284:9663–9673
7. Alarcón C, Zaromytidou AI, Xi Q, Gao S, Yu J, Fujisawa S, Barlas A, Miller AN, Manova-Todorova K, Macias MJ, Sapkota G, Pan D, Massagué J (2009) Nuclear CDKs drive Smad transcriptional activation and turnover in BMP and TGF-beta pathways. Cell 139:757–769
8. Gao S, Alarcón C, Sapkota G, Rahman S, Chen PY, Goerner N, Macias MJ, Erdjument-Bromage H, Tempst P, Massagué J (2009) Ubiquitin ligase Nedd4L targets activated Smad2/3 to limit TGF-beta signaling. Mol Cell 36:457–468
9. Morgan David O (2007) The cell cycle: principles of control, 1st edn. New Science Press, London

10. Zelivianski S, Cooley A, Kall R, Jeruss JS (2010) Cyclin-dependent kinase 4-mediated phosphorylation inhibits Smad3 activity in cyclin D-overexpressing breast cancer cells. Mol Cancer Res 8:1375–1387
11. Cooley A, Zelivianski S, Jeruss JS (2010) Impact of cyclin E overexpression on Smad3 activity in breast cancer cell lines. Cell Cycle 9:4900–4907
12. Sapkota G, Alarcón C, Spagnoli FM, Brivanlou AH, Massagué J (2007) Balancing BMP signaling through integrated inputs into the Smad1 linker. Mol Cell 25:441–454
13. Fuentealba LC, Eivers E, Ikeda A, Hurtado C, Kuroda H, Pera EM, De Robertis EM (2007) Integrating patterning signals: Wnt/GSK3 regulates the duration of the BMP/Smad1 signal. Cell 131:980–993
14. Hsin JP, Manley JL (2012) The RNA polymerase II CTD coordinates transcription and RNA processing. Genes Dev 26:2119–2137
15. Millet C, Yamashita M, Heller M, Yu LR, Veenstra TD, Zhang YE (2009) A negative feedback control of transforming growth factor-beta signaling by glycogen synthase kinase 3-mediated Smad3 linker phosphorylation at Ser-204. J Biol Chem 284: 19808–19816
16. Guo X, Ramirez A, Waddell DS, Li Z, Liu X, Wang XF (2008) Axin and GSK3-ß control Smad3 protein stability and modulate TGF-b signaling. Genes Dev 22:106–120
17. Tang LY, Yamashita M, Coussens NP, Tang Y, Wang X, Li C, Deng CX, Cheng SY, Zhang YE (2011) EMBO J 30:4777–4789
18. Kretzschmar M, Doody J, Massague J (1997) Opposing BMP and EGF signalling pathways converge on the TGF-ß family mediator Smad1. Nature 389:618–622
19. Kretzschmar M, Doody J, Timokhina I, Massague J (1999) A mechanism of repression of TGF-ß/Smad signaling by oncogenic Ras. Genes Dev 13:804–816
20. Matsuura I, Wang G, He D, Liu F (2005) Identification and characterization of ERK MAP kinase phosphorylation sites in Smad3. Biochemistry 44:12546–12553
21. Derynck R, Zhang YE (2003) Smad-dependent and Smad-independent pathways in TGF-beta family signalling. Nature 425:577–584
22. Mori S, Matsuzaki K, Yoshida K, Furukawa F, Tahashi Y, Yamagata H, Sekimoto G, Seki T, Matsui H, Nishizawa M, Fujisawa J, Okazaki K (2004) TGF-ß and HGF transmit the signals through JNK-dependent Smad2/3 phosphorylation at the linker regions. Oncogene 23:7416–7429
23. Yamagata H, Matsuzaki K, Mori S, Yoshida K, Tahashi Y, Furukawa F, Sekimoto G, Watanabe T, Uemura Y, Sakaida N, Yoshioka K, Kamiyama Y, Seki T, Okazaki K (2005) Acceleration of Smad2 and Smad3 phosphorylation via c-Jun NH(2)-terminal kinase during human colorectal carcinogenesis. Cancer Res 65:157–165
24. Yoshida K, Matsuzaki K, Mori S, Tahashi Y, Yamagata H, Furukawa F, Seki T, Nishizawa M, Fujisawa J, Okazaki K (2005) Transforming growth factor-beta and platelet-derived growth factor signal via c-Jun N-terminal kinase-dependent Smad2/3 phosphorylation in rat hepatic stellate cells after acute liver injury. Am J Pathol 166:1029–1039
25. Kamaraju AK, Roberts AB (2004) Role of Rho/ROCK and p38 MAP kinase pathways in TGF-ß-mediated Smad-dependent growth inhibition of human breast carcinoma cells in vivo. J Biol Chem 280:1024–1036
26. Furukawa F, Matsuzaki K, Mori S, Tahashi Y, Yoshida K, Sugano Y, Yamagata H, Matsushita M, Seki T, Inagaki Y, Nishizawa M, Fujisawa J, Inoue K (2003) p38 MAPK mediates fibrogenic signal through Smad3 phosphorylation in rat myofibroblasts. Hepatology 38: 879–889
27. Liu F (2008) Regulation of Smad activity by phosphorylation. In: TGF-ß in cancer therapy. Jakowlew S (ed) Humana Press, Totowa, USA, pp 105–123
28. Wrighton KH, Lin X, Feng XH (2009) Phospho-control of TGF-beta superfamily signaling. Cell Res 19:8–20
29. Roelen BA, Cohen OS, Raychowdhury MK, Chadee DN, Zhang Y, Kyriakis JM, Alessandrini AA, Lin HY (2003) Phosphorylation of threonine 276 in Smad4 is involved in transforming growth factor-beta-induced nuclear accumulation. Am J Physiol Cell Physiol 285:C823–C830
30. Wicks SJ, Lui S, Abdel-Wahab N, Mason RM, Chantry A (2000) Inactivation of Smad-transforming growth factor ß signaling by Ca(2+)-calmodulin-dependent protein kinase II. Mol Cell Biol 20:8103–8111
31. Yakymovych I, Ten Dijke P, Heldin CH, Souchelnytskyi S (2001) Regulation of Smad signaling by protein kinase C. FASEB J 15:553–555
32. Guo X, Waddell DS, Wang W, Wang Z, Liberati NT, Yong S, Liu X, Wang XF (2008) Ligand-dependent ubiquitination of Smad3 is regulated

by casein kinase 1 gamma 2, an inhibitor of TGF-beta signaling. Oncogene 27:7235–7247

33. Kaneko S, Chen X, Lu P, Yao X, Wright TG, Rajurkar M, Kariya K, Mao J, Ip YT, Xu L (2011) Smad inhibition by the Ste20 kinase Misshapen. Proc Natl Acad Sci U S A 108:11127–11132
34. Zhu S, Wang W, Clarke DC, Liu X (2007) Activation of Mps1 promotes transforming growth factor-beta-independent Smad signaling. J Biol Chem 282:18327–18338
35. Doble BW, Woodgett JR (2003) GSK-3: tricks of the trade for a multi-tasking kinase. J Cell Sci 116:1175–1186
36. Phelps DE, Xiong Y (1997) Assay for activity of mammalian cyclin D-dependent kinases CDK4 and CDK6. Methods Enzymol 283: 194–204

Chapter 6

Analysis of Smad Phosphatase Activity In Vitro

Tao Shen, Lan Qin, and Xia Lin

Abstract

Phosphorylation of Smad1/5/8 at the C-terminal SXS motif by BMP type I receptors is one of the most critical events in BMP signaling. Conversely, protein phosphatases that dephosphorylate phospho-Smad1/5/8 can consequently prevent or terminate BMP signaling. PPM1H is an undercharacterized phosphatase in the PPM family. We recently demonstrated that PPM1H can dephosphorylate Smad1 in the cytoplasm and block BMP signaling responses in cellular assays. Here we describe in vitro method showing that PPM1H is a *bona fide* phosphatase for Smad1/5/8. PPM1H is produced as GST fusion protein in *E. coli*, and purified against glutathione sepharose beads. Bacterially purified recombinant PPM1H possesses phosphatase activity toward artificial substrate *para*-nitrophenyl phosphate (pNPP). Recombinant PPM1H also dephosphorylates immuno-purified phosphorylated Smad1 in test tubes. These direct in vitro phosphatase assays provide convincing evidence demonstrating the role of PPM1H as a specific phosphatase for P-Smad1.

Key words In vitro, Phosphatase, pNPP, Immunoprecipitation, Western blotting

1 Introduction

In eukaryotic organisms, signal transduction pathways are often regulated by dynamic interplay between protein kinases and phosphatases. Phosphorylation of Smad1/5/8 at the C-terminal SXS motif by BMP type I receptors is one of the most critical events in BMP signaling [1]. Protein phosphatases are anticipated to dephosphorylate phospho-Smad1/5/8 and consequently prevent or terminate BMP signaling. Identification of the specific Smad phosphatase has been one of the major tasks in BMP/TGF-β biology. Until recently, a few phosphatases have been shown to dephosphorylate Smads in the BMP/TGF-β signal transduction pathways [2]. We recently undertook a genomic screen and identified PPM1A as a general R-Smads phosphatase, which can dephosphorylate Smad2 and Smad3 [3] and also Smad1 [4] at the

Xin-Hua Feng et al. (eds.), *TGF-β Signaling: Methods and Protocols*, Methods in Molecular Biology, vol. 1344,
DOI 10.1007/978-1-4939-2966-5_6, © Springer Science+Business Media New York 2016

C-terminal SXS motif. PPM1A is a metal ion dependent phosphatase belonging to PP2C protein phosphatases family. PPM1A is demonstrated to be predominantly nuclear [3]. In addition, Smad1/5/8 are also inactivated by *S*mall *C*-terminal domain *P*hosphatases SCP1/2/3 [5] in the nucleus. Intriguingly, a mitochondrial phosphatase PDP also dephosphorylates Smad1, but its mechanism remains elusive [6]. Most recently, we identified a PP2C family phosphatase called PPM1H (Protein Phosphatase 1H, Magnesium-dependent) as a novel phosphatase towards dephosphorylation of BMP-specific Smads [7].

In our study, PPM1H activity on Smad1 dephosphorylation is confirmed by using a series of biochemical assays. First, the dephosphorylation of Smad1 by PPM1H is confirmed by expression of Smad1 and PPM1H in 293T cells (Fig. 1). Phosphorylation of Smad1 is induced by co-transfection of a constitutively activated BMP type I receptor (BMPRIA), ALK3(Q233D). Dephosphorylation of Smad1 by PPM1H is then detected by Western blotting using a specific anti-P-Smad1 antibody. To rule out the possibility that PPM1H activates another phosphatase in cells that could be the direct Smad1 phosphatase, we have also carried out several in vitro cell-free phosphatase activity assays. Direct dephosphorylation of Smad1 by PPM1H is confirmed by using in vitro phosphatase assays. This protocol is focused on the method of dephosphorylation of Smad1 in vitro.

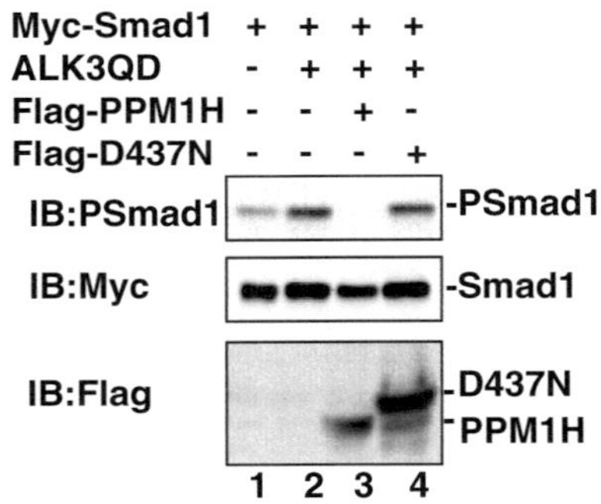

Fig. 1 *PPM1H*, but not PPM1H D437N, *dephosphorylates Smad1 in 293T cells*. Myc-tagged Smad1 and FLAG-tagged PPM1H or PPM1H D437N mutant were co-transfected into HEK293T cells for 48 h, and the cell lysates were subjected to Western Blotting with the antibody specific for the phospho-SXS motif of Smad1/5/8 to determine the effect of PPM1H on phosphorylation of Smad1. A constitutively active mutant of BMP receptor ALK3 (Q233D) was co-transfected to induce Smad1 phosphorylation (*lane 2*). PPM1H reduced the ALK3 (Q233D)-induced P-Smad1 level (*lane 3*–5). Equal level of total Myc-tagged Smad1 protein in each sample was confirmed by immune-blotting with an anti-Myc antibody

2 Materials

1. *Expression vector:* Smad1 is obtained by RT-PCR and cloned into the EcoRI (5′) and SalI (3′) of the pXF6F (a derivative of pRK5, Genentech) vector with an N-terminal FLAG tag. The constitutively active BMPRIA (ALK3(Q233D)) is in the pCDNA3 (Invitrogen) vector with an N-terminal HA tag. PPM1H is obtained by RT-PCR and cloned into the EcoRI (5′) and SalI (3′) of the pXF6F or pGTX3XF (with a fused GST tag and a PreScission protease site).
2. *Antibodies:* P-Smad1 (Cell Signaling, Cat #9511), Smad1 (Life Technologies), PPM1H (ABGENT, #AP9093c), Myc (Thermo Scientific, 9E10), FLAG (Sigma, M2), FLAG beads (Sigma, Cat # A2220).
3. *Glutathione Sepharose beads* (GE) and *Protein A/G beads* (Pierce, Cat # 20421).
4. *Cell culture:* 293T cells is cultured in Dulbecco's modified Eagle's medium (DMEM, Life Technologies Cat # 11965-092) with 10 % fetal bovine serum (FBS, Life Technologies Cat # 16000-044).
5. *Bacteria culture media and transformation reagent:* BL21 competent cells; SOC and LB; LB plate (Ampicillin 100 μg/mL); 42 °C Water Bath.
6. *Transfection Reagent:* Lipofectamine LTX (Invitrogen, Cat # 15338-030) and Plus (Invitrogen, Cat # 11514-015) Reagents; Opti-MEM I Reduced Serum Medium (Life Technologies Cat # 31985-062).
7. *IPTG* (Isopropyl β-D-1-thiogalactopyranoside, Sigma, Cat # 16758).
8. *PreScission Protease* (GE, Cat # 27-0843-01).
9. *FLAG Lysis Buffer:* 25 mM Tris–HCl, pH 7.5, 300 mM NaCl, and 1 % Triton X (EMD, Cat # 648462).
10. *GST Lysis Buffer:* 50 mM Tris–HCl, pH 7.5, 100 mM NaCl, and 1 mM EDTA.
11. *GST Washing Buffer:* 25 mM Tris–HCl, pH 7.5, 100 mM NaCl, 1 mM EDTA, and 1 % Triton X (EMD, Cat # 648462).
12. *PreScission cleavage buffer:* 50 mM Tris–HCl, pH 7.5, 150 mM NaCl, 1 mM EDTA, 1 mM dithiothreitol (DTT).
13. *pNPP* (*p*-Nitrophenyl phosphate, Sigma, A3469).
14. *Protein Phosphatase Buffer A*: 40 mM Tris–HCl, pH 8.0, 20 mM KCl and 2 mM dithiothreitol (DTT).
15. *Protein Phosphatase Buffer B:* 40 mM Tris–HCl, pH 8.0, 20 mM KCl, 2 mM dithiothreitol (DTT) and 30 mM $MgCl_2$ or 10 mM $MnCl_2$.
16. *VERSAmax tunable Micro-plate Reader.*

3 Methods

In the in vitro phosphatase assay, purified protein phosphatases directly catalyze the dephosphorylation reaction of the phosphorylated proteins in the phosphatase buffer. Purified recombinant protein phosphatase PPM1H can be prepared in bacteria (to be discussed later). As PPM1H's substrate, Smad1 is purified by immunoprecipitation of exogenously expressed Smad1 in mammalian cells. To ensure the majority of Smad1 is in its phosphorylated status, the constitutively active type I receptor kinase ALK3QD is co-expressed with Smad1 in cultured cells.

3.1 HEK293T Cell Culture and Transfection

Experimental procedure:

1. Culture HEK293T cells in DMEM containing 10 % FBS at 37 °C with 5 % CO_2.
2. The day before transfection, subculture HEK293T cells in 10-cm plate. The cell number seeded should produce 70–90 % confluence on the day of transfection (*see* **Note 1**).
3. For each plate of cells to be transfected, dilute 3 μg of pXF6F-Smad1 and 2 μg of pCDNA3-ALK3QD in 500 μL of Opti-MEM I Reduced Serum Media, and then add 7.5 μL of Plus reagent and mix them gently.
4. Dilute 7.5 μL of Lipofectamine LTX reagent in 500 μL of Opti-MEM I Reduced Serum Medium.
5. Add diluted DNAs to diluted Lipofectamine LTX reagent. Mix them gently.
6. After 15 min incubation, add DNA mixture directly to cells and mix gently by rocking the plate back and forth.
7. Incubate the cells at 37 °C in a 5 % CO_2 incubator for another 48 h before assaying for transgene expression. The medium could be changed 24 h after transfection (Optional).

3.2 HEK293T Cell Lysis and Immunoprecipitation

Experimental Procedure:

1. Aspirate the medium completely and place the plate on ice. Wash the cells with ice cold PBS twice (*see* **Note 2**). Tilt the plate to aspirate PBS completely.
2. Add 1 mL of ice cold FLAG Lysis Buffer containing freshly added protease inhibitors (1 μg/mL aprotinin, 1 μg/mL leupeptin, and 1 mM phenylmethylsulfonyl fluoride) and phosphatase inhibitors (1 mM NaF, 1 mM Na_3VO_4).
3. Scrape cells off the plate using a cold plastic cell scraper.
4. Gently transfer the cell suspension into a precooled 1.5 mL micro centrifuge tube.

5. Nutate for 30 min at 4 °C.
6. Centrifuge in microcentrifuge at 4 °C, 14,000 g for 20 min.
7. Transfer the supernatant (i.e., cell lysates) to a fresh tube on ice, and discard the pellet.
8. Mix 50 μL of the lysate with 50 μL of 2× SDS loading dye, which serves as the INPUT. The rest of cell lysates are used for subsequent immunoprecipitation.
9. Add 20–30 μL of Protein-A Beads into the lysate and nutate at 4 °C for 60 min.
10. Spin down in microcentrifuge at 4 °C, 14,000 g for 1 min. Transfer the supernatant to a fresh tube kept on ice and discard the pellet beads.
11. For antibody pull-down, add 30 μL of FLAG Beads to the supernatant and nutate at 4 °C for 60 min (If the antibody conjugated beads are not available, *see* **Note 3**).
12. Wash the pellet beads (FLAG) with 1 mL of FLAG Lysis Buffer for three times by nutating at 4 °C for 5 min per wash.
13. Dilute pellet beads (with the Smad1 IP product) in 100 μL of 20 % glycerol PBS containing freshly added protease inhibitors and kept in −80 °C for later use (*see* **Note 4**).

3.3 Analysis of Smad1 Phosphorylation by Western Blotting

In vitro dephosphorylation assays for specific substrates depend on phospho-specific antibodies which specifically recognize the phosphorylated substrates. Here we use the P-Smad1 antibody against phosphorylated Smad1 at the SXS motif. Before the in vitro phosphatase assay is performed, the phosphorylation status of the immune-purified Smad1 is examined by SDS-PAGE-Western Blotting using the specific anti-P-Smad1 antibody.

3.4 Molecular Cloning of GST-Tagged PPM1H in an *E. coli* Expression Plasmid

PPM1H is obtained by RT-PCR and cloned into the EcoRI (5′) and SalI (3′) of the pGTX3XF (with a fused GST tag on the N-terminal and a PreScission protease site).

3.5 Expression and Purification of GST-Tagged PPM1H

Recombinant Glutathione S-transferase (GST) fusion proteins are generated and purified per manufacturer's instruction (Amersham Biosciences).

Experimental Procedure:

1. Thaw BL21 competent cells on ice (50 μL in tube for one transformation).
2. Add 1 μg of pGTK3XF-PPM1H DNA directly into the competent cells and mix by flicking the tube gently.
3. Leave the tube on ice for 20 min.

4. Heat-shock the cells by placing the tube in 42 °C water bath for 1 min.
5. Place the tube on ice for another 20 min immediately after the heat shock.
6. Add 900 μL of SOC media and culture the cells at 37 °C for 1 h in a shaker.
7. Spread 100 μL of the cells to a LB agar plate containing ampicillin (100 μg/mL).
8. Incubate the plate at 37 °C for 18 h.
9. Following transformation, pick one colony and culture it in 10 mL of LB medium containing ampicillin (100 μg/mL) with shaking at 37 °C, overnight.
10. Transfer bacterial culture to 200 mL of LB medium containing ampicillin (100 μg/mL) and further adjust the volume with LB medium until OD 600 nm = 0.1–0.2.
11. Incubate at 37 °C with shaking and monitor OD 600 nm frequently until the OD 600 nm = 0.5–0.6 (*see* **Note 5**).
12. Add IPTG (200 mM) to a final concentration of 200 μM and incubate the culture at 16 °C with shaking, overnight (*see* **Note 5**).
13. Cool down the culture on ice for 30 min, spin down at 4,000×*g* for 20 min and discard the supernatant.
14. Resuspend the bacterial pellet with 10 mL of GST Lysis Buffer (plus 0.4 mg/mL lysozyme) with protease inhibitors.
15. Divide the 10 mL lysate into two 15-mL Corning tubes. For sonication, keep the lysate on ice. Sonicate with 20 pulses, 50 % duty cycle, 3 output. For each sample/tube, sonication is carried out five times, 1 min each time.
16. Add 20 % Triton X-100 to a final concentration of 1 %. Nutate at 4 °C for 30 min.
17. Centrifuge the samples at 15,000 ×*g*, 4 °C for 20 min. Collect the supernatant and add 250 μL of 50 % Glutathione Sepharose 4B beads. Nutate at 4 °C for 1 h.
18. Spin down the beads at 4,000 ×*g*, 4 °C for 5 min; carefully remove the supernatant. Wash the beads with GST Washing Buffer with protease inhibitors three times. Nutate for 5 min at 4 °C each time per wash.
19. Resuspend the beads with 2 mL of PBS with protease inhibitors; add 70 % glycerol to a final concentration of 15 %. Aliquot to 1.5 mL Eppendorf tubes and store in –80 °C freezer.
20. Run 10 % SDS-PAGE to verify the purification quality of GST-PPM1H.

3.6 Removal of GST Tag by Enzymatic Cleavage

Experimental Procedure:

1. Wash the GST-PPM1H bound Glutathione Sepharose beads with 10× bed volumes of PreScission cleavage buffer. Bed volume is equal to 0.5× the volume of the 50 % Glutathione Sepharose slurry used.
2. Add 1× bed volumes of PreScission cleavage buffer with PreScission protease (10 units per mg of GST-PPM1H). Nutate at 4 °C, overnight.
3. Spin down the beads at 600 × *g*, 4 °C for 10 min. Carefully collect the supernatant into 1.5 mL Eppendorf tube and add 70 % glycerol to a final concentration of 15 %.
4. (Bradford Assay) Measure the concentration of recombinant PPM1H. First plot a concentration standard cure using BSA (0, 5, 15, 25, 50 ng/mL); Dilute 5 μL of PPM1H protein sample or BSA standard samples in 795 μL sterile water, add 200 μL Bradford Reagent; vortex well and set them aside for 15 min; pipette 200 μL of each sample into a clear 96-well plate and scan on the VERSAmax tunable Micro-plate Reader at 595 nm.
5. Run 10 % SDS-PAGE to monitor the quality of PPM1H.

3.7 Phosphatase Activity Test (pNPP Assay) and Quantification

Experimental Procedure:

1. Prepare PPM1H Protein Phosphatase Buffer (*see* **Note 6**).
2. Dilute recombinant PPM1H in a proper Protein Phosphatase Buffer (5, 10 ng/μL).
3. Mix 10 μL of diluted PPM1H and 50 μL pNPP in 140 μL Protein Phosphatase Buffer on a clear 96-well plate with cover. Incubate at 37 °C for 30 min. All reactions are triplicated.
4. Stop the reaction by adding 50 μL of stop solution (1 M NaOH). Mix well by quickly tapping the plate.
5. Read the absorbance in the VERSAmax tunable Micro-plate Reader at 405 nm.
6. The PPM1H phosphatase activity is evaluated by the absorbance reading (*see* Fig. 2).

3.8 Perform Smad1 Dephosphorylation Reaction In Vitro

Experimental Procedure:

1. Prepare PPM1H Protein Phosphatase Buffer (*see* **Note 6**).
2. Dilute recombinant PPM1H in a proper Protein Phosphatase Buffer (0, 1, 2.5, 5, 7.5, 15 ng/μL).
3. Mix 10 μL of diluted PPM1H and 5 μL of Smad1 IP product in 10 μL of Protein Phosphatase Buffer in 1.5 mL Eppendorf tube. Incubate at 37 °C for 30 min (*see* **Note 7**). All reactions are duplicated.

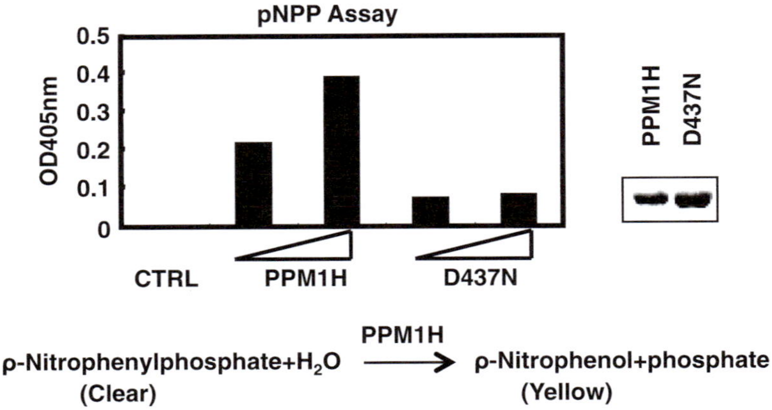

Fig. 2 Recombinant PPM1H protein possesses phosphatase activity in vitro. PPM1H and PPM1H D437N proteins, which were expressed in and purified from *E. coli* cells, were incubated with *para*-nitrophenyl phosphate (pNPP) substrate solution (Sigma) at 37 °C for 30 min in the in vitro phosphatase buffer. The reaction was stopped by adding 50 μL of 1 M NaOH. Absorption of dephosphorylated products (*yellow color*) at 405 nm was measured in VERSAmax tunable Microplate Reader

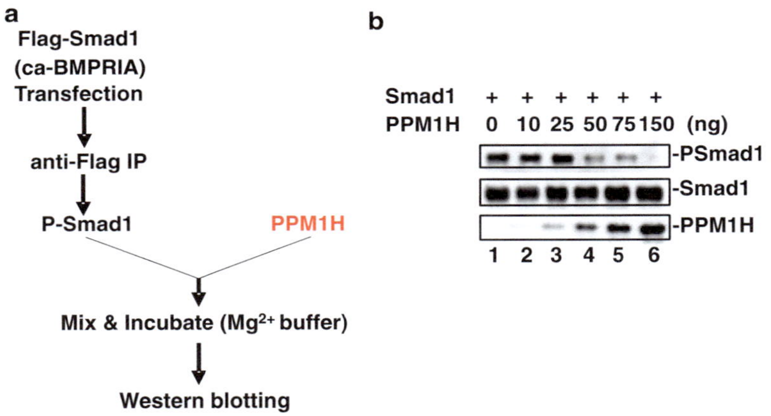

Fig. 3 Recombinant PPM1H dephosphorylates P-Smad1 in the presence of Mg^{2+} or Mn^{2+}. (a) The flowchart of PPM1H in vitro dephosphorylation assay. (b) Immunopure Flag-Smad1 was incubated with recombinant PPM1H protein at 37 °C for 60 min in the in vitro phosphatase buffer in the presence of Mg^{2+} (30 mM) or Mn^{2+} (10 mM). The reaction mixture was analyzed by Western blotting using indicated antibodies

4. Stop the reaction by adding 25 μL of 2× SDS loading buffer. Boil samples at 95 °C for 5 min.
5. Samples are subjected to Western Blotting analysis using specific antibody of P-Smad1. The result of dephosphorylated Smad1 by PPM1H is shown in Fig. 3.

4 Notes

1. Cell density should be 70–90 % confluent at time of transfection. Improper density of cells affects transfection efficiency.
2. In the following procedures, if not specifically mentioned, place all cell lysates or protein samples on ice.
3. Add 4–5 μg of the anti-Flag antibody to the cell lysate; Nutate at 4 °C for 3 h then add 30 μL of Protein A/Agarose bead slurry; Nutate at 4 °C for another 1 h.
4. It is always better to use the fresh IP products in the in vitro phosphatase assays.
5. Induction parameters including IPTG concentration, period of induction and the cell concentration when IPTG is added are optimized in order to increase the enzyme activity of soluble proteins.
6. It is always better to use fresh-prepared buffer for the reaction.
7. The reaction parameters including protein phosphatase concentration, amount of substrate and reaction time are optimized to maximize phosphatase performance.

References

1. Hoodless PA, Haerry T, Abdollah S, Stapleton M, O'Connor MB, Attisano L et al (1996) MADR1, a MAD-related protein that functions in BMP2 signaling pathways. Cell 85(4):489–500
2. Wrighton KH, Lin X, Feng XH (2009) Phospho-control of TGF-beta superfamily signaling. Cell Res 19(1):8–20
3. Lin X, Duan X, Liang YY, Su Y, Wrighton KH, Long J et al (2006) PPM1A functions as a Smad phosphatase to terminate TGFbeta signaling. Cell 125(5):915–928
4. Duan X, Liang YY, Feng XH, Lin X (2006) Protein serine/threonine phosphatase PPM1A dephosphorylates Smad1 in the bone morphogenetic protein signaling pathway. J Biol Chem 281(48):36526–36532
5. Knockaert M, Sapkota G, Alarcon C, Massague J, Brivanlou AH (2006) Unique players in the BMP pathway: small C-terminal domain phosphatases dephosphorylate Smad1 to attenuate BMP signaling. Proc Natl Acad Sci U S A 103(32):11940–11945, PMCID: 1567677
6. Chen HB, Shen J, Ip YT, Xu L (2006) Identification of phosphatases for Smad in the BMP/DPP pathway. Genes Dev 20(6):648–653, PMCID: 1413280
7. Shen T, Sun C, Zhang Z, Xu N, Duan X, Feng XH et al (2014) Specific control of BMP signaling and mesenchymal differentiation by cytoplasmic phosphatase PPM1H. Cell Res 24(6):727–741, PMCID: 4042171

Chapter 7

Three-dimensional Mammary Epithelial Cell Morphogenesis Model for Analysis of TGFß Signaling

Juliet Rashidian and Kunxin Luo

Abstract

Culturing mammary epithelial cells in laminin-rich extracellular matrices (three dimensional or 3D culture) offers significant advantages over that in the conventional two-dimensional (2D) tissue culture system in that it takes into considetation the impact of extracellular matrix (ECM) microenvironment on the proliferation, survival, and differentiation of mammary epithelial cells. When grown in the 3D culture, untransformed mammary epithelial cells undergo morphogenesis to form a multicellular and polarized acini-like structure that functionally mimics the differentiated alveoli in the pregnancy mammary gland. This process is subjected to regulation by many growth factors and cytokines. The transforming growth factor-ß (TGFß) is a multipotent cytokine that regulates multiple aspects of development and tumorigenesis. In addition to its effects on epithelial cell proliferation, survival, and differentiation, it is also a potent regulator of the cell–matrix interaction. Thus, the 3D culture model may recapitulate the complex in vivo epithelial cell microenvironment and allow us to fully evaluate the role of TGFß signaling in multiple aspects of normal and cancerous cell behavior. In this chapter we provide detailed protocols for growing mammary epithelial cells in the 3D Matrigel for analysis of signaling pathways.

Key words 3D culture, Matrigel, Acini, MCF10A, TGFß

1 Introduction

Four decades ago, Emerman and colleague showed that mammary epithelial cells could be cultured and induced to differentiate on floating collagen membranes [1]. A few years later, Bissell and Petersen developed a three-dimensional (3D) culture system using either collagen gels or laminin-rich extracellular matrices (lrECM) [2–6]. By studying differentiation and morphogenesis of mammary epithelial cells in this 3D culture model, important insights regarding the critical roles of extracellular matrix (ECM) microenvironment on the proliferation, survival, and differentiation of mammary epithelial cells have been revealed [reviewed in [7]]. Since then, various immortalized and non-transformed human

Xin-Hua Feng et al. (eds.), *TGF-β Signaling: Methods and Protocols*, Methods in Molecular Biology, vol. 1344,
DOI 10.1007/978-1-4939-2966-5_7, © Springer Science+Business Media New York 2016

mammary epithelial cell lines, inculding MCF10A and HMT-3522 [2, 8–12] as well as mouse mammary epithelial cell lines, including EpH4 and TAC-2 [13, 14] have been cultured and studied in 3D models. Primary mouse and human mammary epithelial cells have also been grown in basement membrane gels in 3D cultures [2–4].

The 3D culture system offers significant advantages over the conventional two-dimensional (2D) tissue culture system. When grown in the 3D IrECM or floating collagen, untransformed mammary epithelial cells undergo morphogenesis to form a multicellular and polarized acini-like structure that functionally mimics the differentiated epithelial alveoli [15]. Contact of cells with basement membrane in 3D initiates proliferation followed by establishment of polarity, which then stops proliferation and triggers apoptosis of inner cell layers necessary for lumen formation and acini maturation [2, 3]. Cells in the acini also deposit basement membrane (BM) components such as collagen IV [2] and laminin V [16]. This 3D culture model thus allows us to examine critical aspects of normal epithelial cell biology including polarity and morphological differentiation, features not well preserved in 2D cultures. Another important feature of the 3D model is its ability to distinguish between untransformed mammary epithelial cells and malignant breast cancer cells. In the conventional 2D culture, breast cancer cells are often morphologically indistinguishable from the untransformed mammary epithelial cells. In the 3D culture, however, malignant breast cancer cell lines fail to differentiate into the organized acini-like structure displayed by the untransformed epithelial cells, but grow into large, irregular colonies instead [15]. In addition, breast cancer cells also show differences in gene expression patterns and signaling activities that are not apparent in the 2D culture [17, 18].

1.1 3D Cultures and TGFß Signaling

TGFß is a multipotent cytokine that regulates many cellular processes in both untransformed and cancer cells, including proliferation, survival, differentiation, migration, and invasion as well as cell–matrix interaction [19–25]. The activities and functions of TGFß signaling in tumorigenesis are also context-dependent. In some cases, breast cancer cells in the 2D culture respond differently to TGFß from when they were cultured in 3D. Although most studies in the past on TGFß signaling have been performed in tissue culture cell lines in 2D cultures, more and more recent investigations have employed the cells in 3D cultures as an ex vivo model to study the impact of TGFß signaling on polarity, differentiation, cell–ECM interaction and malignant progression. For example, Park et al. showed that TGFß cooperated with low dose radiation to promote disruption of epithelial interactions and neoplastic progression of untransformed human mammary epithelial cells [26]. Using cancer cell lines in the 3D culture, the effects of TGFß and its antagonists on cancer cell growth, malignant transformation, migration, invasion, EMT, and angiogenesis have been investigated [27–31].

Because the 3D acini differentiation model mimics the alveolar differentiation process in the pregnant glands, we have recently investigated the signaling mechanism by which SnoN, a negative regulator of the Smad proteins, regulates mammary gland alveologenesis and lactation using both primary mammary epithelial cells and MCF10A cell lines in the 3D model [32]. We found that mammary epithelial cells isolated from the SnoN knockout mice that displayed severe defects in alveologenesis in vivo were also defective in forming acini-like structures in the 3D culture [32]. Using this system, we have revealed a novel pathway by which SnoN regulates STAT5 expression and activation in mammary epithelial cells. Moreover, using HMT-3522-S1 cells in the 3D cultures, we have also demonstrated that cytoplasmic SnoN antagonizes TGFß signaling through sequestering the Smad proteins [33]. Thus, the 3D culture may serve as a powerful model system in investigating the activity and regulation of TGFß signaling in mammary epithelial cell function and breast cancer progression.

1.2 The MCF10A Cell Line

The MCF10A cell line is a spontaneously immortalized, nontransformed epithelial cell line derived from the mammary tissue of a patient with a fibrocystic disease [12]. It has a stable and near-diploid karyotype, lacks anchorage-independent growth or the ability to form a tumor in nude mice and is dependent on exogenous growth factors and hormones for proliferation and survival [12]. A few modest genetic abnormalities, including deletion of the ARF locus and amplification of the *myc* gene have been reported for MCF10A, but the p53 status appears to be normal in these cells [34–36]. Muthuswamy et al. first reported studying cell signaling and cell matrix interaction using MCF10A in 3D culture [37]. In the 3D matrigel, MCF10A cells undergo growth factor-dependent proliferation followed by the induction of luminal apoptosis, organization of an apico-basal polarity, and the deposition of a basal lamina, resulting in the formation of acinar structures that resemble those formed in vivo [36–40]. Compared to other non-malignant human mammary epithelial cell lines, MCF10A is relatively easy to grow and maintain and is responsive to TGFß stimulation. The establishment of MCF10A-based breast cancer progression model also makes it possible to study malignant progression within the identical genetic background [41].

2 Materials

Cell culture reagents and chemicals: Dulbecco's Modified Eagle's Medium (DMEM)/F12 and horse serum, 0.25 % Trypsin–EDTA, Newborn Calf Serum (NCS), penicillin–streptomycin are purchased from Gibco. Bovine insulin, hydrocortisone, cholera toxin and ethidium bromide are purchased from Sigma-Aldrich. Human insulin is puechased from Santa Cruz Biotechnology. Epidermal Growth Factor (EGF) is purchased from PeproTech, TGFß1 from R&D System, rhodamine-phalloidin from Molecular Probes, paraformaldehyde from Electron Microscopy Sciences, Cultrex basement membrane extract (Matrigel) from Trevigen, 8-well glass chamber slide from Lab-Tek, and mounting medium containing DAPI from Vector Laboratories.

Antibodies: Mouse anti-E-cadherin antibody and anti-GM130 (Golgi matrix protein of 130) antibody are purchased from BD Transduction Laboratories. Rat anti-α6 integrin antibody (MAB1378) is purchased from Chemicon, Rabbit anti-Ki67 antibody from Zymed and anti-active caspase-3 antibody from Cell Signaling.

PBS: 137 mM NaCl, 2.7 mM KCl, 10 mM $Na_2HPO_4 \cdot 2H_2O$, 2 mM KH_2PO_4; pH 7.4.

PBS containing 2 % paraformaldehyde.

PBS containing 0.5 % Triton X-100.

PBS containing 0.1 M glycine.

IF buffer: PBS + 10 % Newborn Calf Serum (NCS) + 0.02 % Triton X-100.

RIPA buffer: 1 % Nonidet P-40, 0.2 % SDS, 0.5 % sodium deoxycholate, 150 mM sodium chloride, 50 mM Tris–HCl, pH 7.4, 10 mM sodium fluoride, 1 mM sodium orthovanadate, 10 mM β-glycerophosphate, 5 μg/mL aprotinin, 5 μg/mL leupeptin, 100 μM phenylmethylsulfonyl fluoride (PMSF).

3 Methods

3.1 Growing MCF10A Cells in the 2D Culture

1. Cells are grown in a 5 % CO_2 humidified incubator at 37 °C.
2. Culture medium (Table 1) is DMEM/F12 supplemented with:
   - 5 % Horse serum.
   - 20 ng/mL Epidermal growth factor (EGF).
   - 10 μg/mL Bovine or human insulin.
   - 0.5 μg/mL Hydrocortisone.
   - 100 ng/mL Cholera toxin.
   - Penicillin–streptomycin.
3. Cells should be passaged every 3 days or more frequently.

Table 1
Medium recipes for growing MCF10A cells in 2D and 3D cultures*

Components[#]	Culture medium (final concentrations)	Resuspension medium (final concentrations)	3D medium (final concentrations)
DMEM/F12[a]	470 mL	395 mL	485 mL
Horse serum	25 mL (5 %)	100 mL (20 %)	10 mL (2 %)
Insulin[b] (10 mg/mL stock in dH_2O +1 % glacial acetic acid)	500 μL (10 μg/mL)		500 μL (10 μg/mL)
EGF (100 μg/mL stock in dH_2O)	100 μL (20 ng/mL)		
Cholera toxin (1 mg/mL stock in dH_2O)	50 μL (100 ng/mL)		50 μL (100 ng/mL)
Hydrocortisone (1 mg/mL stock in ethanol)	250 μL (0.5 μg/mL)		250 μL (0.5 μg/mL)
Pen/strep (100× solution)	5 mL	5 mL	5 mL

*Sterilize each medium by filtering through a 0.2 μm filter
[#]Aliquot insulin, EGF, and hydrocortisone stock solutions and store at −20 °C. Store cholera toxin aliquots at 4 °C
[a]DMEM/F12 is commercially available or you can prepare DMEM and F12 separately and then mix them at an 1:1 ratio
[b]Insulin from both bovine or human sources could be used to make the medium

3.2 Passaging MCF10A Cells in the 2D Culture

1. Aspirate the culture medium and wash the cells with phosphate-buffered saline once.
2. Trypsinize cells with 2 mL of 0.25 % Trypsin–EDTA.
3. Immediately aspirate the trypsin but leave a trace of the solution on the plate. If excess trypsin is not removed the cells will clump and aggregate.
4. Incubate the cells at room temperature for 5 min and then in the incubator (37 °C) for another 5 min.
5. Gently tap the bottom of the dish to dislodge the cells. If cells are not completely dissociated from the dish, leave them for a few minutes more in the incubator.
6. Once all cells are detached, add 2–3 mL of resuspension medium (Table 1) and pipette to break up cell clumps. Transfer the cell suspension to a 15-mL conical tube.
7. Rinse the dish with another 2 mL of resuspension medium and transfer it to the conical tube. If you have multiple dishes, it is recommended to process two dishes at a time since cells will re-attach if they are not resuspended soon after being dislodged.

8. Spin down the cells at $150 \times g$ for 5 min.
9. Aspirate the medium and resuspend the cells in 5 mL of culture medium and count the cells. Plate 4×10^5 cells into a 10 cm^2 tissue culture dish with 10 mL culture medium. This will give you a 50 % confluent culture the next day.

The morphology of MCF10A cells should be monitored carefully. In sub-confluent monolayer cultures, MCF10A cells tend to grow in clusters. In near-confluent cultures, MCF10A cells display the cobblestone appearance that is characteristic of epithelial cells (Fig. 1a, b). If cells exhibit a more spindle-like shape and fibroblastic appearance, they most likely will not form proper acini-like structures when grown in 3D. These cells should be discarded. This usually occurs when one batch of cells has been passaged for a long time. It is important to keep track of the passage numbers.

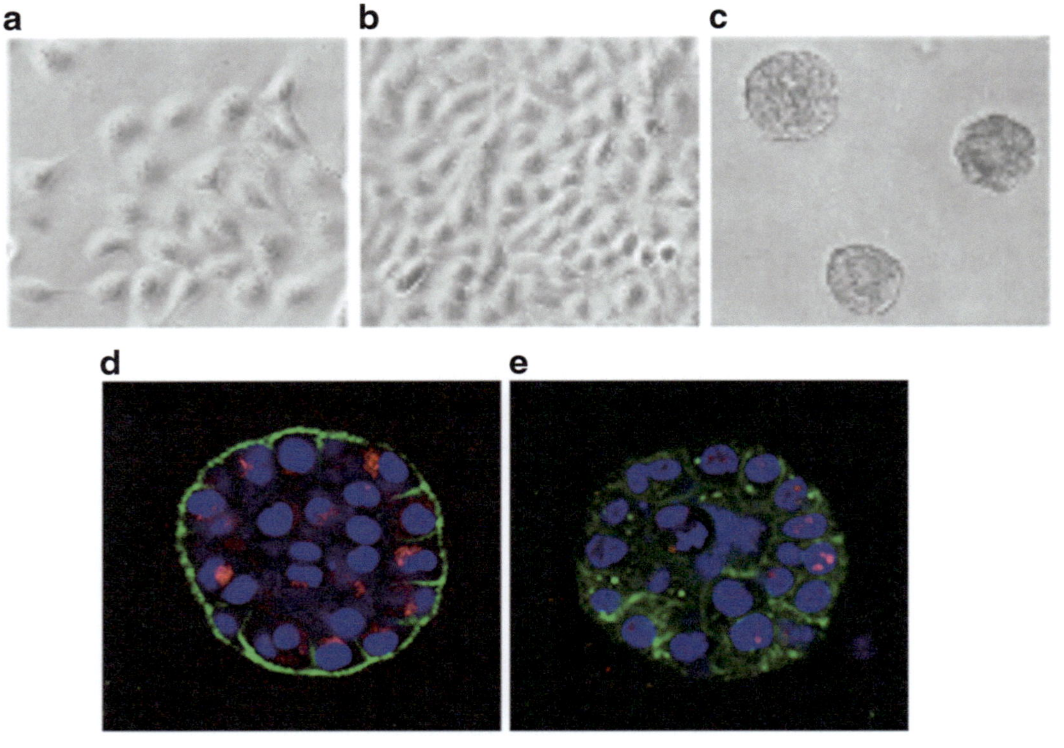

Fig. 1 MCF10A cells grown in 2D and 3D cultures. Morphology of (a) sub-confluent and (b) confluent cultures of MCF10A cells in 2D under a phase-contrast microscope. (c) Phase-contrast micrograph of MCF10A spherical acini cultured on Matrigel for 6 days. (d, e) Representative confocal microscopic imaging of MCF10A acini grown on Matrigel for 6 days. Acini are immunostained for (d) α6 integrin (*green*) and GM130 (*red*) or (e) E-cadherin (*green*) and ki67 (*red*). Nuclei are stained with DAPI (*blue*)

3.3 TGFß Growth Inhibition Assay

1. Trypsinze and resuspend the MCF10A cells in culture medium as described above and count the cells.
2. Plate 5×10^3 cells in each well of a 24-well plate.
3. Add an increasing concentrations of TGFß to the cultures. We suggest 0, 20, 50 and 100 pM of TGFß1.
4. Incubate the plate at 37 °C for 4 days.
5. Determine the relative cell growth by counting cells and calculating the percentage of TGFß1-treated cells relative to that of unstimulated cells (0 pM).

3.4 TGFß Stimulation for Analysis of Various Aspects of Epithelial to Mesenchymal Transdifferetiation

1. Prepare a suspension of MCF10A cells in culture medium and grow them on glass coverslips placed in wells in a 6-well plate. Put 1×10^5 cells per well and incubate at 37 °C overnight.
2. The next day, treat the cells with 50 pM TGFß1 and incubate for 2–3 days.
3. The cells are now ready for various EMT analysis.

3.5 3D Culture of Mammary Epithelial Cells

Mammary epithelial cells are usually grown in a matrix of reconstituted basement membrane derived from Engelbreth–Holm–Swarm (EHS) tumors, commercially available as Matrigel [42]. The matrigel contains matrix components laminin, collagen IV, and entactin. The exogenous laminin in the Matrigel is necessary and essential for the morphogenesis of mammary epithelial cells in 3D culture [16]. Mammary epithelial cells can be cultured using both total embedment method and overlay method. In the total embedment method, epithelial cells are completely embedded within the Matrigel. The gelled bed contains culture media with growth factors and hormones that are essential for proliferation and survival [39]. In the overlay method, the Matrigel is first coated into the well to form a 1 mm-thick bed. Next, epithelial cells are placed on top of this bed. In both methods cells undergo morphogenesis and form acinus-like structures. While the embedment culture may resemble the in vivo environment more, the overlay method consumes less Matrigel and more importantly, makes it easier to harvest the acini for subsequent analysis.

Although we use MCF10A as the primary cell line to describe the procedure, the following protocol can be applied to both untransformed mammary epithelial cell lines (HMT3522-S1 cells, MCF10A cells) or breast cancer cell lines (MCF7 or MDA-MB-231 cells) [2, 8, 9, 18, 43, 44].

3.6 MCF10A Cells Overlay Culture on 3D Matrigel

1. Upon receiving, thaw the Growth Factor Reduced Matrigel on ice and make 1 mL-aliquots in precooled small tubes. Once aliquoted, the Matrigel can be stored at −80 °C. Since there is lot-to-lot variability, it is advised to test individual lots before purchasing large quantities of Matrigel.

2. Thaw an aliquot of Matrigel at 4 °C overnight (*see* **Note 1**).
3. Using the tip of a P-200 pipette, spread 60–70 μL of Matrigel evenly to each well of an 8-well glass chamber slide. This has to be done rapidly before the gel solidifies. Keeping the temperature of the Matrigel mix low by precooling the slide chambers and the pipette tips may help to delay the solidification. It is also important not to generate air bubbles while coating the slide chambers with Matrigel.
4. Place the chamber in a cell culture incubator for at least 20 min to allow the basement membrane to solidify.
5. While waiting for the Matrigel to solidify, prepare the cell suspension. Trypsinize the cells and resuspend them in 2–3 mL of resuspension medium (Table 1). We usually use cells from a sub-confluent dish that has been seeded 24 h before.
6. Spin the cells at 150 × *g* for 5 min.
7. Aspirate the medium and resuspend the cells in 2 mL of 3D medium (Table 1). Gently pipette cells several times with a P-1000 pipette to break up the cell clumps and obtain a single-cell suspension. Count the cells and adjust the volume to achieve a 10,000 cells/mL suspension (*see* **Note 2**).
8. Prepare a stock of 3D medium containing 4 % Matrigel and 10 ng/mL EGF. If you want to include growth factors other than EGF, add it to this medium at twice the final concentration. Use medium that has been cooled on ice to avoid solidification of Matrigel. Make 200 μL of this stock for each well of the chamber slide. It is recommended to prepare stock solution for "*n* + 1" assays (*see* **Note 3**).
9. Mix the cells suspension (**step** 7) with the Matrigel-containing medium at a 1:1 ratio and plate 400 μL of the mixture on top of the solidified Matrigel in each pre-coated well (**step 4**). This will generate 2000 cells/well in medium containing 2 % Matrigel and 5 ng/mL EGF (*see* **Note 4**).
10. Gently transfer the chamber slide to a 5 % CO_2 humidified incubator at 37 °C (Day = 0).
11. Refeed the cells with 3D medium containing 2 % Matrigel and 5 ng/mL EGF every 4 days (Days = 4, 8, 12, etc). MCF10A cells will form acini-like structures by Day 5 or 6 and later begin to form a hollow lumen in the center of the acini.

3.7 Protocol for Treatment of TGFß in the 3D Matrigel Cutlure

1. Follow the **steps 1**–7 of the protocol for MCF10A cells overlay culture on Matrigel in Subheading 3.6.
2. Add 800 pg/mL (32 pM) TGFß1 to the 3D medium (4 % Matrigel + 10 ng/mL EGF). This will give you the final

concentration of 400 pg/mL (16 pM) of TGFß after the medium is mixed with cell suspension at the 1:1 ratio.

3. Plate 400 μL of the mixture on top of the solidified Matrigel in each pre-coated well. Follow the rest of the protocol as described in **steps 10** and **11** from Subheading 3.6.

3.8 Three-Dimensional Morphogenesis of MCF10A Acini

Mammary epithelial cells seeded on the Matrigel will undergo proliferation and morphogenesis to form acini-like structures (Figs. 1c and 2). Two populations of cells within each acinus appear on days 5–8: a well-polarized outer layer of cells that is in direct contact with ECM and a set of inner cells that are poorly polarized and lack contact with ECM. Since long-term survival of epithelial cells requires direct interaction with the ECM, the inner cells eventually express activated caspase activities and undergo apoptosis [45–48]. By day 8, the increased apoptotic signaling as well as growth control signals in the inner cells cooperate to direct the formation of a hollow lumen in the 3D model (Fig. 2) [37, 38]. The process of 3D morphogenesis can be monitored by examining the proliferation and apoptosis, polarity and cell–cell adhesion of the acini. The following protocols provide details on how to analyze these parameters.

3.9 Indirect Immunofluorescence Staining of 3D Acini

1. Fixation: Aspirate the medium from each well of the chamber slide and add 500 μL of freshly prepared 2 % paraformaldehyde (in PBS). Incubate for 20 min at room temperature with gentle rocking.
2. Permeabilization: Aspirate the paraformaldehyde solution and add 500 μL of PBS containing 0.5 % Triton X-100. Incubate for 12 min at 4 °C. Alternatively, you can also incubate the acini in PBS containing 0.4 % Triton X-100 for 40 min at 4 °C.

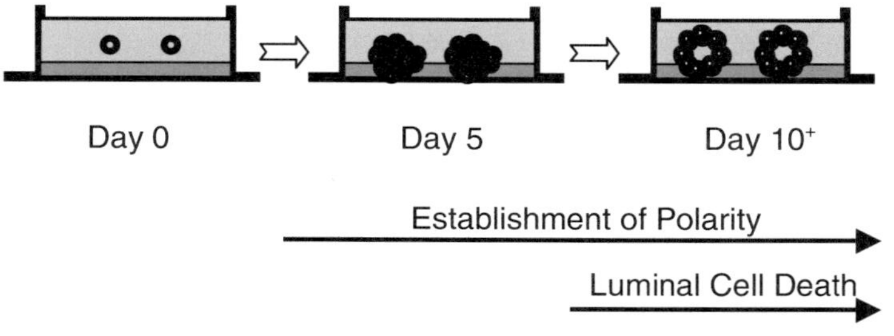

Fig. 2 Stages of MCF-10A acinar morphogenesis in 3D culture

3. Washing: Aspirate the permeabilization buffer and wash the wells with 500 μL of PBS containing 0.1 M glycine for 15–20 min at room temperature with gentle rocking. Repeat the wash two more times (*see* **Note 5**).
4. Blocking: Aspirate the washing buffer and add 200 μL of IF buffer (PBS + 10 % Newborn Calf Serum (NCS) + 0.02 % Triton X-100) containing 1 % BSA. Incubate for 1–2 h at room temperature.
5. Primary antibody binding: Aspirate the blocking buffer and add 100–200 μL of primary antibody solution prepared in IF buffer. Incubate at 4 °C overnight (*see* **Notes 6** and **7**).
6. Washing: Aspirate the primary antibody solution and wash the wells with 500 μL of IF buffer for 15–20 min with gentle rocking. Repeat the wash two more times.
7. Secondary antibody binding: Add 100–200 μL of fluorescence-conjugated secondary antibody solutions prepared in IF buffer at a concentration recommended by the manufacturer and incubate for 50 min at room temperature (*see* **Note 8**).
8. Washing: Aspirate the secondary antibody solution and wash the wells once with 500 μL of IF buffer and twice with 500 μL of PBS for 15–20 min each with gentle rocking.
9. A mounting medium containing DAPI is used to counterstain the nuclei and mount the slides. To do so, aspirate the last washing buffer and remove the gasket from the slide. Next, put a small drop of mounting medium directly on each well and place a 60 × 24 mm coverslip on the entire slide. Let the mounting medium spread evenly on the slide and let it dry at room temperature for a few hours. Slides can be stored at 4 °C for up to 1 week or at −20 °C for up to 2 months.

3.10 Analysis of Cell Death in Acini by Ethidium Bromide Staining

1. Remove medium and wash each well with 500 μL of PBS.
2. Prepare 1 μM ethidium bromide solution in PBS and add 100–200 μL to each well. Incubate for 15–30 min at 37 °C.
3. Remove ethidium bromide solution and replace with PBS to avoid a high autofluorescence background.
4. Ethidium bromide positive cells are detected by the rhodamine channel on a microscope equipped with a mercury lamp.

3.11 Imaging the Stained Acini

Although the acini in the 3D Matrigel culture could be visualized by phase contrast microscopy (Fig. 1c), confocal microscopy provides much higher quality images as it can provide cross sections through the entire acinus (Fig. 1d, e). To acquire high quality images it is important to: (1) optimize the immunostaining condition; (2) optimize confocal microscopy parameters to minimize the noises of the image. These noises are due to the three-dimensional structure and thickness of the acini in Matrigel. You can reduce the noises by increasing the acquisition time and either line-averaging or frame-averaging, as well as reducing the pinhole size [36].

3.12 Prepare Protein Lysates from the 3D Culture

Protocol 1 (for analysis of intracellular proteins) (*see* **Note 9**):

1. Aspirate the medium and wash each well with 500 μL of ice-cold PBS.
2. Add 2–3 volumes of the ice-cold 0.25 % trypsin + 0.1 % EDTA or PBS containing 2.5 mM EDTA to each well. Detach the Matrigel from the well by gently scraping the bottom with a pipette tip.
3. Shake gently for 15–30 min to allow Matrigel to dissolve.
4. Transfer the solution to a conical tube. Wash the well one more time with PBS–EDTA and transfer it to the conical tube. Gently shake the tube for 15–30 min.
5. Invert the tube gently and check to see if the Matrigel has dissolved completely without visible gel fragments. If not, wait longer and/or add more trypsin + EDTA or PBS–EDTA buffer.
6. Centrifuge the colonies at 100 × *g* (or less) for 3 min. Aspirate the supernatant, lyse the cells with an appropriate extraction buffer on ice for 20 min.
7. Transfer the cell lysate to a microcentrifuge tube and centrifuge the lysates for 15 min at maximal speed in a microcentrifuge at 4 °C. Transfer the supernatant to a new tube. The clarified lysate is ready for protein quantification or can be frozen at −80 °C for storage. Extracted proteins may be stored at −80 °C for several months.

Protocol 2 (for detection of cell surface proteins, protein complexes, and phospho proteins) (*see* **Note 10**):

1. Aspirate the medium and wash each well with 500 μL of ice-cold PBS.
2. Lyse the embeded cells in Matrigel directly by adding RIPA buffer to each well. Incubate on ice for 20 min.
3. Transfer the cell lysate to a microcentrifuge tube and centrifuge the lysates for 15 min at maximal speed in a microcentrifuge at 4 °C. Transfer the supernatant to a new tube. The clarified lysate can be frozen at −80 °C for storage.

4 Notes

1. Always handle Matrigel on ice; it will solidify at room temperature.
2. It is critical to have a single-cell suspension for the 3D culture. Otherwise, multiple acini will grow on top of each other.
3. As the stock solution of EGF is 100 μg/mL, it is advised to generate a fresh working solution of 10 μg/mL. Do not store the working solution for extended use since EGF in a solution at a concentration less than 100 μg/mL is unstable.
4. You may need to adjust the number of cells to be plated in each well. In our laboratory we usually seed 1800–2000 cells/well. This gives us single and isolated acini after 3D morphogenesis.
5. You can pause the procedure at this step by aspirating the last washing buffer and store the chamber slide at −20 °C. Otherwise, directly proceed to **step 4**.
6. The following antibodies have been used to analyze cell–cell junctions, basolateral polarity, apical polarity, proliferation, and apoptosis in 3D cultures. A more complete list of other antibodies used for analysis of MCF10A acini has been provided by Debnath et al. [36].
   - Cell–cell junctions: E-cadherin mouse antibody, 1:200 dilution.
   - Basolateral polarity: α6 integrin (MAB1378) rat antibody, 1:200 dilution.
   - Apical polarity: Golgi matrix protein of 130 (GM130) mouse antibody, 1:200 dilution.
   - Proliferation marker: Ki67 rabbit antibody, 1:500 dilution.
   - Apoptosis marker: Active caspase 3 rabbit antibody, 1:200 dilution.
7. It is possible to combine two different primary antibodies for a double immunostain as along as the two antibodies are from the different animal species and can be distinguished readily by secondary antibodies.
8. Secondary antibodies are light sensitive. Perform the incubation step at a dark place or cover the chamber slides with foil.
9. (a) This protocol is not recommended for the detection of cell surface proteins, protein complexes, and phospho proteins. (b) Perform all steps on ice.
10. (a) Quantification of protein concentration in the lysates is not possible due to the abundance of proteins in the Matrigel. (b) Perform all steps on ice.

Acknowledgement

We would like to thank members of Dr. Mina Bissell's group for their assistance. The work is supported by NIH RO1 CA101891 and RO1 DK090347 to K. Luo. J. Rashidian is supported by the Susan G. Komen For The Cure KG101263.

References

1. Emerman JT, Pitelka DR (1977) Maintenance and induction of morphological differentiation in dissociated mammary epithelium on floating collagen membranes. In Vitro 13(5):316–328
2. Petersen OW, Ronnov-Jessen L, Howlett AR, Bissell MJ (1992) Interaction with basement membrane serves to rapidly distinguish growth and differentiation pattern of normal and malignant human breast epithelial cells. Proc Natl Acad Sci U S A 89(19):9064–9068
3. Barcellos-Hoff MH, Aggeler J, Ram TG, Bissell MJ (1989) Functional differentiation and alveolar morphogenesis of primary mammary cultures on reconstituted basement membrane. Development 105(2):223–235
4. Li ML, Aggeler J, Farson DA, Hatier C, Hassell J, Bissell MJ (1987) Influence of a reconstituted basement membrane and its components on casein gene expression and secretion in mouse mammary epithelial cells. Proc Natl Acad Sci U S A 84(1):136–140
5. Emerman JT, Bartley JC, Bissell MJ (1981) Glucose metabolite patterns as markers of functional differentiation in freshly isolated and cultured mouse mammary epithelial cells. Exp Cell Res 134(1):241–250
6. Lee EY, Lee WH, Kaetzel CS, Parry G, Bissell MJ (1985) Interaction of mouse mammary epithelial cells with collagen substrata: regulation of casein gene expression and secretion. Proc Natl Acad Sci U S A 82(5):1419–1423
7. Weigelt B, Bissell MJ (2008) Unraveling the microenvironmental influences on the normal mammary gland and breast cancer. Semin Cancer Biol 18(5):311–321
8. Rizki A, Weaver VM, Lee SY, Rozenberg GI, Chin K, Myers CA et al (2008) A human breast cell model of preinvasive to invasive transition. Cancer Res 68(5):1378–1387
9. Wang F, Weaver VM, Petersen OW, Larabell CA, Dedhar S, Briand P et al (1998) Reciprocal interactions between beta1-integrin and epidermal growth factor receptor in three-dimensional basement membrane breast cultures: a different perspective in epithelial biology. Proc Natl Acad Sci U S A 95(25): 14821–14826
10. Weaver VM, Petersen OW, Wang F, Larabell CA, Briand P, Damsky C et al (1997) Reversion of the malignant phenotype of human breast cells in three-dimensional culture and in vivo by integrin blocking antibodies. J Cell Biol 137(1):231–245
11. Shay JW, Tomlinson G, Piatyszek MA, Gollahon LS (1995) Spontaneous in vitro immortalization of breast epithelial cells from a patient with Li-Fraumeni syndrome. Mol Cell Biol 15(1):425–432
12. Soule HD, Maloney TM, Wolman SR, Peterson WD Jr, Brenz R, McGrath CM et al (1990) Isolation and characterization of a spontaneously immortalized human breast epithelial cell line, MCF-10. Cancer Res 50(18):6075–6086
13. Niemann C, Brinkmann V, Spitzer E, Hartmann G, Sachs M, Naundorf H et al (1998) Reconstitution of mammary gland development in vitro: requirement of c-met and c-erbB2 signaling for branching and alveolar morphogenesis. J Cell Biol 143(2): 533–545
14. Soriano JV, Pepper MS, Nakamura T, Orci L, Montesano R (1995) Hepatocyte growth factor stimulates extensive development of branching duct-like structures by cloned mammary gland epithelial cells. J Cell Sci 108(Pt 2):413–430
15. Lee GY, Kenny PA, Lee EH, Bissell MJ (2007) Three-dimensional culture models of normal and malignant breast epithelial cells. Nat Methods 4(4):359–365
16. Gudjonsson T, Ronnov-Jessen L, Villadsen R, Rank F, Bissell MJ, Petersen OW (2002) Normal and tumor-derived myoepithelial cells differ in their ability to interact with luminal breast epithelial cells for polarity and basement membrane deposition. J Cell Sci 115(Pt 1):39–50
17. Anders M, Hansen R, Ding RX, Rauen KA, Bissell MJ, Korn WM (2003) Disruption of 3D tissue integrity facilitates adenovirus infection

by deregulating the coxsackievirus and adenovirus receptor. Proc Natl Acad Sci U S A 100(4):1943–1948

18. Kenny PA, Lee GY, Myers CA, Neve RM, Semeiks JR, Spellman PT et al (2007) The morphologies of breast cancer cell lines in three-dimensional assays correlate with their profiles of gene expression. Mol Oncol 1(1):84–96
19. Moses H, Barcellos-Hoff MH (2011) TGF-beta biology in mammary development and breast cancer. Cold Spring Harb Perspect Biol. 3(1):a003277
20. Derynck R, Akhurst RJ, Balmain A (2001) TGF-beta signaling in tumor suppression and cancer progression. Nat Genet 29(2):117–129
21. Wu MY, Hill CS (2009) Tgf-beta superfamily signaling in embryonic development and homeostasis. Dev Cell 16(3):329–343
22. Heldin CH, Landstrom M, Moustakas A (2009) Mechanism of TGF-beta signaling to growth arrest, apoptosis, and epithelial-mesenchymal transition. Curr Opin Cell Biol 21(2):166–176
23. Massague J (2008) TGFbeta in cancer. Cell 134(2):215–230
24. Akhurst RJ (2004) TGF beta signaling in health and disease. Nat Genet 36(8):790–792
25. Feng XH, Derynck R (2005) Specificity and versatility in tgf-beta signaling through Smads. Annu Rev Cell Dev Biol 21:659–693
26. Park CC, Henshall-Powell RL, Erickson AC, Talhouk R, Parvin B, Bissell MJ et al (2003) Ionizing radiation induces heritable disruption of epithelial cell interactions. Proc Natl Acad Sci U S A 100(19):10728–10733
27. Sankar S, Mahooti-Brooks N, Bensen L, McCarthy TL, Centrella M, Madri JA (1996) Modulation of transforming growth factor beta receptor levels on microvascular endothelial cells during in vitro angiogenesis. J Clin Invest 97(6):1436–1446
28. Henry LA, Johnson DA, Sarrio D, Lee S, Quinlan PR, Crook T et al (2010) Endoglin expression in breast tumor cells suppresses invasion and metastasis and correlates with improved clinical outcome. Oncogene 30(9):1046–1058
29. Wendt MK, Taylor MA, Schiemann BJ, Schiemann WP (2011) Down-regulation of epithelial cadherin is required to initiate metastatic outgrowth of breast cancer. Mol Biol Cell 22(14):2423–2435
30. Ganapathy V, Ge R, Grazioli A, Xie W, Banach-Petrosky W, Kang Y et al (2010) Targeting the Transforming growth factor-beta pathway inhibits human basal-like breast cancer metastasis. Mol Cancer 9:122
31. Sempere LF, Gunn JR, Korc M (2011) A novel 3-dimensional culture system uncovers growth stimulatory actions by TGFbeta in pancreatic cancer cells. Cancer Biol Ther 12(3):198–207
32. Jahchan NS, Wang D, Bissell MJ, Luo K (2012) SnoN regulates mammary gland alveologenesis and onset of lactation by promoting prolactin/Stat5 signaling. Development 139(17):3147–3156
33. Krakowski AR, Laboureau J, Mauviel A, Bissell MJ, Luo K (2005) Cytoplasmic SnoN in normal tissues and nonmalignant cells antagonizes TGF-beta signaling by sequestration of the Smad proteins. Proc Natl Acad Sci U S A 102(35):12437–12442
34. Merlo GR, Basolo F, Fiore L, Duboc L, Hynes NE (1995) p53-dependent and p53-independent activation of apoptosis in mammary epithelial cells reveals a survival function of EGF and insulin. J Cell Biol 128(6):1185–1196
35. Yaswen P, Stampfer MR (2002) Molecular changes accompanying senescence and immortalization of cultured human mammary epithelial cells. Int J Biochem Cell Biol 34(11):1382–1394
36. Debnath J, Muthuswamy SK, Brugge JS (2003) Morphogenesis and oncogenesis of MCF-10A mammary epithelial acini grown in three-dimensional basement membrane cultures. Methods 30(3):256–268
37. Muthuswamy SK, Li D, Lelievre S, Bissell MJ, Brugge JS (2001) ErbB2, but not ErbB1, reinitiates proliferation and induces luminal repopulation in epithelial acini. Nat Cell Biol 3(9):785–792
38. Debnath J, Mills KR, Collins NL, Reginato MJ, Muthuswamy SK, Brugge JS (2002) The role of apoptosis in creating and maintaining luminal space within normal and oncogene-expressing mammary acini. Cell 111(1):29–40
39. Debnath J, Brugge JS (2005) Modelling glandular epithelial cancers in three-dimensional cultures. Nat Rev 5(9):675–688
40. Godde NJ, Galea RC, Elsum IA, Humbert PO (2010) Cell polarity in motion: redefining mammary tissue organization through EMT and cell polarity transitions. J Mammary Gland Biol Neoplasia 15(2):149–168
41. Santner SJ, Dawson PJ, Tait L, Soule HD, Eliason J, Mohamed AN et al (2001) Malignant MCF10CA1 cell lines derived from premalignant human breast epithelial MCF10AT cells. Breast Cancer Res Treat 65(2):101–110
42. Kleinman HK, McGarvey ML, Hassell JR, Star VL, Cannon FB, Laurie GW et al (1986) Basement membrane complexes with biological activity. Biochemistry 25(2):312–318

43. Han J, Chang H, Giricz O, Lee GY, Baehner FL, Gray JW et al (2010) Molecular predictors of 3D morphogenesis by breast cancer cell lines in 3D culture. PLoS Comput Biol 6(2): e1000684
44. Wang F, Hansen RK, Radisky D, Yoneda T, Barcellos-Hoff MH, Petersen OW et al (2002) Phenotypic reversion or death of cancer cells by altering signaling pathways in three-dimensional contexts. J Natl Cancer Inst 94(19):1494–1503
45. Blatchford DR, Quarrie LH, Tonner E, McCarthy C, Flint DJ, Wilde CJ (1999) Influence of microenvironment on mammary epithelial cell survival in primary culture. J Cell Physiol 181(2):304–311
46. Coucouvanis EC, Martin GR, Nadeau JH (1995) Genetic approaches for studying programmed cell death during development of the laboratory mouse. Methods Cell Biol 46: 387–440
47. Boudreau N, Sympson CJ, Werb Z, Bissell MJ (1995) Suppression of ICE and apoptosis in mammary epithelial cells by extracellular matrix. Science 267(5199):891–893
48. Frisch SM, Francis H (1994) Disruption of epithelial cell-matrix interactions induces apoptosis. J Cell Biol 124(4):619–626

Chapter 8

TGF-β Signaling in Stem Cell Regulation

Wenlin Li, Wanguo Wei, and Sheng Ding

Abstract

The transforming growth factor-β (TGF-β) family of cytokines, including TGF-β, bone morphogenic proteins (BMPs), and activin/nodal, is a group of crucial morphogens in embryonic development, and plays key roles in modulating stem/progenitor cell fate. TGF-β signaling is essential in maintaining the pluripotency of human pluripotent stem cells (hPSCs), including both human embryonic stem cells (hESCs) and human induced pluripotent stem cells (hiPSCs), and its modulation can direct lineage-specific differentiation. Recent studies also demonstrate TGF-β signaling negatively regulates reprogramming and inhibition of TGF-β signaling can enhance reprogramming through facilitating mesenchymal-to-epithelial transition (MET). This chapter introduces methods of modulating somatic cell reprogramming to iPSCs and neural induction from hPSCs through modulating TGF-β signaling by chemical approaches.

Key words Pluripotent stem cells, Reprogramming, Neural induction

1 Introduction

The transforming growth factor-β (TGF-β) family contains more than 30 structurally related multifunctional cytokines, including TGF-βs, bone morphogenetic proteins (BMPs), activins, inhibins, nodal, and growth differentiation factors (GDFs) [1]. TGF-β family members are expressed in highly temporal and tissue-specific manner and play pivotal roles in embryonic development, tissue homeostasis, carcinogenesis, and tumor metastasis by regulating diverse cellular functions, such as cell fate specification, cell growth, transformation, or death. Activation of TGF-β signaling is essential for the self-renewal of human pluripotent stem cells (hPSCs). Inhibition of TGF-β signaling, for example by SB431542 (a small molecule inhibitor of TGF-β receptors, including activin like kinase 4, 5, and 7), causes rapid loss of pluripotency and

Xin-Hua Feng et al. (eds.), *TGF-β Signaling: Methods and Protocols*, Methods in Molecular Biology, vol. 1344,
DOI 10.1007/978-1-4939-2966-5_8, © Springer Science+Business Media New York 2016

induces neural induction in part through destabilizing the pluripotency and blocking the mesoendoderm differentiation from hPSCs [2]. We previously demonstrated that combining a GSK3 inhibitor (i.e., CHIR99021) with TGF-β and Notch signaling pathway inhibitors induced a highly efficient conversion of monolayer-cultured hESCs into a homogenous population of primitive neural stem cells (pNSCs) within 1 week. Remarkably, the combination of leukemia inhibitory factor (LIF), CHIR99021 and SB431542 effectively captured and maintained these pNSCs to long-term self-renew in vitro [3]. The underlying mechanisms of this strategy mainly involve the synergy of (1) TGF-β/Activin pathway inhibition that induces neutralization of hESCs by destabilizing pluripotency and blocking mesoendoderm differentiation, and (2) GSK3 inhibition and LIF that promote pNSC self-renewal [4, 5]. Recent studies also demonstrated that inhibition of TGF-β signaling enhances cell reprogramming [6–8], a process converting differentiated somatic cells (e.g., fibroblasts) into the pluripotent state by the ectopic expression of defined transcription factors [9]. Inhibition of TGF-β signaling may enhance reprogramming by facilitating mesenchymal-to-epithelial transition/MET, a process that is entailed during reprogramming fibroblasts into iPSCs. Previous studies also demonstrated that the MAPK and Rho/ROCK signaling can function downstream of TGF-β signaling in epithelial-to-mesenchymal transition/EMT [10, 11]. We had shown that small molecules that target these known MET mechanisms, including inhibition of TGF-β receptors (by SB431542), MEK (by PD0325901), or ROCK (by thiazovivin), significantly enhanced reprogramming efficiency and accelerated reprogramming speed when added individually or in combination [8]. In this chapter, we will introduce methods of modulating somatic cell reprogramming to iPSCs and neural induction from hPSCs through modulating TGF-β signaling and other related signaling pathways by chemical approaches.

2 Materials

2.1 Cells and Cell Culture Reagents

1. Human fibroblasts CRL2097 and BJ (ATCC No. CRL-2097 and CRL-2522).
2. hESCs (HUES9 and HUES1) (Harvard HUES Cell Facility).
3. Viral package cell lines: 293T and PlatE.

All cell culture products were purchased from Invitrogen except where mentioned.

4. Human fibroblast medium: Dulbecco's Modified Eagle's Medium/F12(DMEM/F12) with 10 % fetal bovine serum

(FBS), 1 % penicillin/streptomycin, 1× GlutaMAX, and 0.11 mM 2-mercaptoethanol.

5. hiPSC growth medium: DMEM/F12 with 10 mM Hepes, 20 % Knockout serum replacement, 1 % GlutaMAX, 1 % nonessential amino acids, 1 % penicillin/streptomycin, 0.11 mM β-mercaptoethanol, and 10 ng/mL bFGF.
6. Viral package cell medium for 293T and Plat E: DMEM with 10 % FBS, 0.11 mM 2-mercaptoethanol, and 1 % penicillin/streptomycin.
7. Neural induction medium: Advanced DMEM/F12–Neurobasal (1:1), 1×N2, 1× B27 without vitamin A, 1 % GlutaMAX, 5 μg/mL BSA, and 10 ng/mL hLIF (Millipore).
8. Cell dissociation reagents: 0.05 % trypsin–EDTA and accutase (Millipore).
9. SHH (C24II), FGF8b, BDNF, GDNF, IGF1, and TGF-β3 (R&D Systems).
10. CHIR99021, PD0325901, SB431542, and Thiazovivin (Stemgent). Compound E (γ-Secretase Inhibitor XXI) (EMD Chemicals Inc). Poly-lysine, polybrene, cAMP, retinoic acid (RA), and vitamin C (Sigma-Aldrich). Matrigel (BD Biosciences).

2.2 Plasmids and Cell Transfection

1. pMXs-based retroviral vectors for human Oct4, Klf4, Sox2, and c-Myc and lentiviral package plasmids pMD2.G and psPAX2 (Addgene).
2. pWPXLD-Slc7a1 is generated from pWPXLD (Addgene) by replacing GFP with mouse Slc7a1 ORF (murine ecotropic retroviral receptor).
3. FugeneHD transfection reagent (Roche).

2.3 Cytochemistry and Immunofluorescence Assay

1. Alkaline phosphatase kit (Sigma-Aldrich).
2. Fixation buffer for immunofluorescence: 4 % paraformaldehyde (Sigma-Aldrich) in PBS (Invitrogen).
3. Washing buffer: PBS containing 0.1 % Triton X-100 (Sigma-Aldrich).
4. Blocking buffer: 0.1 % Triton X-100 and 10 % normal donkey serum (Jackson ImmunoResearch Laboratories Inc) in PBS.
5. The primary antibodies: Mouse anti-Oct4 (1:250) and rabbit anti-Nurr1 (1:250) antibodies (Santa Cruz Biotechnology); rabbit anti-Sox2 (1:2000), mouse anti-TRA-1-81 (1:1000), rabbit anti-SSEA4 (1:1000), rabbit anti-tyrosine hydroxylase (1:1000), rabbit anti-FoxA2 (1:1000), goat anti-ChAT (1:500), rabbit anti-NeuN, and rabbit anti-Lmx1b (1:1000)

antibodies (Millipore); goat anti-Brachyury (1:500) antibody (R&D); goat anti-Albumin (1:500), chicken anti-MAP2 (1:50000), and rabbit anti-Nanog (1:500) antibodies (Abcam); rabbit anti-βIII-Tubulin (1:1000), rabbit anti-nestin (1:1500), and rabbit anti-PAX6 (1:1000) antibodies (Covance Research Products); mouse anti-PLZF (1:500) antibody (Calbiochem); mouse anti-N-Cadherin (1:200), mouse anti-Ki-67 (1:200), and Alexa Fluor 555 conjugated mouse anti-TRA-1-81 antibodies (BD Bioscience); rabbit anti-ZO-1 (1:200) antibody (Invitrogen).

6. Secondary antibodies: Alexa Fluor 486/555 donkey anti-mouse, anti-goat or anti-rabbit IgG (1:1000) (Invitrogen).
7. DAPI (Sigma-Aldrich) is dissolved in tissue-culture water at 1 mg/mL, stored in aliquots at −20 °C, and used at 10 μg/mL.
8. Nikon Eclipse TE2000-U microscope.

3 Methods

3.1 Generation of Chemical-Facilitated hiPSCs

1. Thaw a vial of frozen 293T cells from the liquid nitrogen, transfer the cell suspension to a 15-mL tube with 10 mL of viral package cell medium and centrifuge the cells for 5 min at 150 × *g*.
2. Resuspend the cell pellet in 10 mL of viral package cell medium, and seed the cells in poly-lysine coated 100-mm dish (1 million cells per dish).
3. The 293T cells are cultured in a 37 °C, 5 % CO_2 incubator and passaged when approaching 70–80 % confluence with trypsin–EDTA at 1:6 dilution.
4. Before transfection, dissociate the 293T cells by trypsin–EDTA, suspend the cells in appropriate amount of viral package cell medium by gently pipetting. Count the cell number and adjust the concentration to 7×10^5 cells per ml with fresh medium.
5. Seed cells at 100-mm poly-lysine coated culture dishes (10 mL per dish), and incubate overnight at 37 °C, 5 % CO_2.
6. Replace the medium with antibiotics free viral package cell medium (7 mL/dish). The 293T cells are ready for transfection.
7. To make transfection mixture, transfer 0.45 mL of OPTI-MEM into a 1.5-mL tube. Add 4 μg of pWPXLD-Slc7a1 lentiviral plasmid DNA, 3.5 μg psPAX2, and 0.5 μg pMD2.G, and mix gently by finger tapping. Add 30 μL of Fugene HD transfection reagent into the mixture, mix gently by finger tapping, and then incubate for 15 min at room temperature.

8. Add the DNA/Fugene HD complex dropwise into the 293T dishes, and incubate overnight at 37 °C, 5 % CO_2.
9. Replace the transfection reagent-containing medium with 7 mL of human fibroblast medium, and collected the medium at 48 and 72 h post-transfection and filter it through a 0.45 μm pore size cellulose acetate filter.
10. After adding polybrene into the filtrated virus-containing medium at 4 μg/mL final concentration, the medium containing Slc7a1 lentiviruses was used to transduce 20–30 % confluence human fibroblasts CRL2097 or BJ (at passage 2) by two times (5 h each time) by a 24 h interval. The transduced cells were named CRL2097-Slc7a1 and BJ-Slc7a1.
11. Thaw a vial of frozen PlatE cells from the liquid nitrogen, culture and split the cells exactly following the protocol of culturing 293T cells.
12. Transfer 0.45 mL of OPTI-MEM into a 1.5-mL tube. Add 9 μg of pMXs plasmid DNA (encoding human Oct4, Sox2, Klf4, and c-Myc), and mix gently by finger tapping. Add 30 μL of Fugene HD transfection reagent into each of the above tubes, mix gently by finger tapping, and then incubate for 15 min at room temperature.
13. Add the DNA/Fugene HD complex dropwise into the Plat-E dish, and incubate overnight at 37 °C, 5 % CO_2.
14. Replace the transfection reagent-containing medium with 7 mL of human fibroblast medium, and collected the medium at 48 and 72 h post-transfection and filter it through a 0.45 μm pore size cellulose acetate filter.
15. Seed CRL2097-Slc7a1 and BJ-Slc7a1 at 1×10^5 cells per well of a 6-well plate.
16. The next day, add 0.25 mL of each retroviral supernatant containing human Oct4, Sox2, Klf4, and c-Myc to the cells in the presence of 8 μg/mL polybrene overnight for two times by a 24 h interval.
17. Estimate the infection efficiency by fluorescence microscopy on cells transduced in parallel with GFP gene-carrying retroviruses.
18. Seven days after initial transduction, harvest fibroblasts by trypsinization and seed the cells at 1×10^4 cells per well of 6-well plate coated with Matrigel (1:50 dilution).
19. Culture the cells in hiPSC growth medium in the presence of 2 μM SB431542, 0.5 μM PD0325901, and 0.5 μM thiazovivin. Change the media every 2–3 days depending on the cell density.

20. Seven days after compound treatment, about 20 Nanog positive colonies can be generated from 1×10^4 transduced input cells and no Nanog positive colonies should be observed in the culture without compound treatment at this time point.
21. Maintain the cells in the same medium and compound cocktail described above except for the concentrations of PD0325901 (1 μM from day 21 onward) and SB431542 (0.5–1 μM after day 14).
22. On day 30, about 200 Nanog positive colonies (in contrast to ~1 colony in the parallel condition in the absence of small molecules), which are also positive for TRA-1-81 and SSEA4, can be observed from 1×10^4 transduced cells without splitting the cell culture.
23. These colonies can be revealed and picked up after staining the live cells by Alexa Fluor 555 conjugated mouse anti-TRA-1-81 (1:10) under a fluorescence microscope (*see* **Note 1**). These colony-derived hiPSCs can then be long-term maintained in hPSC growth medium on X-ray inactivated mouse embryonic fibroblasts.

3.2 The Pluripotency of hiPSCs

1. To analyze the expression of pluripotent markers by hiPSCs, fix the cells by 4 % paraformaldehyde in PBS.
2. Wash the cells three times with PBS containing 0.1 % Triton X-100.
3. Block the cells with PBS containing 0.1 % Triton X-100 and 10 % normal donkey serum.
4. Incubate the cells with primary antibodies by indicated dilution overnight at 4 °C.
5. Wash the cells with PBS containing 0.1 % Triton X-100 for three times (five minutes each time).
6. Reveal the antigens by appropriate fluorescence conjugated secondary antibodies.
7. Wash the cells with PBS containing 0.1 % Triton X-100 for three times (five minutes each time)
8. These hiPSCs should stably and homogenously express typical pluripotency markers, such as Oct4, Sox2, Nanog, TRA-1-81, and SSEA-4. hiPSCs should also express alkaline phosphatase by cytochemistry assay.
9. For in vitro differentiation of hiPSCs, the cells are dissociated by trypsin and cultured in ultra-low attachment 6-well plate in DMEM medium supplemented with 10 % FBS (with the presence of 0.5 μM thiazovivin for first 12 h) to form embryoid body (EBs). The medium is routinely changed every another day.

10. One week later, the EBs are harvested and transferred into Matrigel-coated 6-well plate in DMEM medium with 10 % FBS.

11. Three to seven days later, the cells are fixed for immunocytochemistry analysis as described. Mesoderm marker Brachyury should be detected at 3 days after transferring EBs on Matrigel-coated plate. Endoderm and ectoderm markers, such as Albumin and βIII-Tubulin, should be detected 7 days later.

3.3 Differentiate hPSCs into Expandable Primitive Neural Stem Cells (pNSCs)

1. hESCs, HUES9 (passages 17–30) and HUES1 (passages 20-30), are cultured in hPSC growth medium on X-ray inactivated mouse embryonic fibroblasts.

2. HUES9 and HUSE1 cells are regularly passaged using Accutase at a dilution of 1:10. The cells at about 20 % confluence are treated with 3 μM CHIR99021, 2 μM SB431542, and 0.1 μM Compound E in neural induction media (Advanced DMEM/F12–Neurobasal (1:1), 1× N2, 1× B27 without vitamin A, 1 % GlutaMAX, 5 μg/mL BSA, and 10 ng/mL hLIF) for 7 days (*see* **Note 2**).

3. After 7 days treatment, the cells should lose expression of pluripotency genes, such as Oct4 and Nanog, but maintain the homogenous expression of Sox2. PAX6 is induced from day 5 and the majority of cells are becoming positive to PAX6 at day 7.

4. The culture is then split 1:3 for the next six passages using Accutase and cultured in neural induction media supplemented with 3 μM CHIR99021 and 2 μM SB431542 on X-ray inactivated MEF feeders or Matrigel-coated surface (*see* **Note 3**).

5. After six passages, the cells are split 1:10 regularly and the cells are able to long-term maintained in neural induction media supplemented with 3 μM CHIR99021 and 2 μM SB431542 (*see* **Note 4**).

3.4 Characterize the Expandable Primitive Neural Stem Cells (pNSCs)

1. By immunocytochemistry analysis as described, pNSCs should uniformly express PAX6, Nestin, PLZF, N-Cadherin, ZO-1, Sox2, and Ki-67, but negative to Oct4 and Nanog. pNSCs are non-polarized neuroepithelia showing even distribution of N-Cadherin and ZO-1.

2. After incubated with fluorescence conjugated primary antibodies on ice for 30 min and washed three times by PBS containing 0.1 % Triton X-100, the cells are analyzed by flow cytometry. More than 95 % pNSCs should be positive for cell proliferation marker Ki-67 and neural stem cell markers PAX6 and Sox2, even after long-term passaging (<30 passages).

3. To induce spontaneous differentiation, pNSCs are plated at 1×10^4 cells per well of Matrigel-coated 6-well plate and culture in DMEM/F12, 1× N2, 1× B27, 300 ng/mL cAMP, and 0.2 mM vitamin C (therefor named differentiation media).
4. The medium is changed every 2–3 days depending on cell density. After 3–4 weeks, more than 70 % cells should be positive to neuronal genes MAP2 or NeuN even from the late passage cells (>30 passages).
5. For dopaminergic neuron differentiation, cells are first treated with 100 ng/mL SHH (C24II) and 100 ng/mL FGF8b in differentiation media for 10 days, and then with 10 ng/mL BDNF, 10 ng/mL GDNF, 10 ng/mL IGF1, 1 ng/mL TGF-β3, and 0.5 mM db-cAMP for another 14–21 days in differentiation media. The differentiated cells are positive to tyrosine hydroxylase and midbrain genes, such as Nurr1, Lmx1b, and FoxA2.
6. For induction of motor neurons, cells are sequentially treated with 1 μM RA in differentiation media for 7 days, then with 100 ng/mL SHH (C24II) and 0.1 μM RA for additional 7 days, and finally with 50 ng/mL SHH (C24II) and 0.1 μM RA for another 7 days. The cells are terminally differentiated in the presence of 10 ng/mL BDNF and 10 ng/mL GDNF in the differentiation media for about 7 days. The differentiated cells are positive to motor neuron markers ChAT and HB9.

4 Notes

1. After live cells staining by Alexa Fluor 555 conjugated mouse anti-TRA-1-81 antibody, the cells should be thoroughly washed by PBS by at least 3 times to remove sodium azide used as preservative in antibody.
2. During the first 7 days of neural induction, if the cells got too confluent and had to be passaged, then passage cells by 1:1 or 1:2. Change medium every day if necessary. High cell density favors neural induction for unknown reason.
3. During the initial a few passages after neural induction, passage at higher cell density (e.g., 1:3) was still required.
4. This protocol to induce expandable pNSCs is not working well with retrovirus/lentivirus generated hiPSCs, probably due to incomplete silence of exogenous pluripotent genes in these hiPSCs.

References

1. Attisano L, Wrana JL (2002) Signal transduction by the TGF-β superfamily. Science 296(5573):1646–1647
2. Smith JR, Vallier L, Lupo G, Alexander M, Harris WA, Pedersen RA (2008) Inhibition of Activin/Nodal signaling promotes specification of human embryonic stem cells into neuroectoderm. Dev Biol 313(1):107–117
3. Li W, Sun W, Zhang Y, Wei W, Ambasudhan R, Xia P et al (2011) Rapid induction and long-term self-renewal of primitive neural precursors from human embryonic stem cells by small molecule inhibitors. Proc Natl Acad Sci U S A 108(20):8299–8304
4. Mao Y, Ge X, Frank CL, Madison JM, Koehler AN, Doud MK et al (2009) Disrupted in schizophrenia 1 regulates neuronal progenitor proliferation via modulation of GSK3beta/beta-catenin signaling. Cell 136(6):1017–1031
5. Kim W-Y, Wang X, Wu Y, Doble BW, Patel S, Woodgett JR et al (2009) GSK-3 is a master regulator of neural progenitor homeostasis. Nat Neurosci 12(11):1390–1397
6. Ichida JK, Blanchard J, Lam K, Son EY, Chung JE, Egli D et al (2009) A small-molecule inhibitor of Tgf-Beta signaling replaces Sox2 in reprogramming by inducing Nanog. Cell Stem Cell 5(5):491–503
7. Maherali N, Hochedlinger K (2009) Tgfbeta signal inhibition cooperates in the induction of iPSCs and replaces Sox2 and cMyc. Curr Biol 19(20):1718–1723
8. Lin T, Ambasudhan R, Yuan X, Li W, Hilcove S, Abujarour R et al (2009) A chemical platform for improved induction of human iPSCs. Nat Methods 6(11):805–808
9. Takahashi K, Yamanaka S (2006) Induction of pluripotent stem cells from mouse embryonic and adult fibroblast cultures by defined factors. Cell 126(4):663–676
10. Bhowmick NA, Ghiassi M, Bakin A, Aakre M, Lundquist CA, Engel ME et al (2001) Transforming growth factor-β1 mediates epithelial to mesenchymal transdifferentiation through a RhoA-dependent mechanism. Mol Biol Cell 12(1):27–36
11. Das S, Becker B, Hoffmann FM, Mertz J (2009) Complete reversal of epithelial to mesenchymal transition requires inhibition of both ZEB expression and the Rho pathway. BMC Cell Biol 10(1):94

Chapter 9

Analysis of Epithelial–Mesenchymal Transition Induced by Transforming Growth Factor β

Ulrich Valcourt, Jonathon Carthy, Yukari Okita, Lindsay Alcaraz, Mitsuyasu Kato, Sylvie Thuault, Laurent Bartholin, and Aristidis Moustakas

Abstract

In recent years, the importance of the cell biological process of epithelial–mesenchymal transition (EMT) has been established via an exponentially growing number of reports. EMT has been documented during embryonic development, tissue fibrosis, and cancer progression in vitro, in animal models in vivo and in human specimens. EMT relates to many molecular and cellular alterations that occur when epithelial cells undergo a switch in differentiation that generates mesenchymal-like cells with newly acquired migratory and invasive properties. In addition, EMT relates to a nuclear reprogramming similar to the one occurring in the generation of induced pluripotent stem cells. Via such a process, EMT is gradually established to promote the generation and maintenance of adult tissue stem cells which under disease states such as cancer, are known as cancer stem cells. EMT is induced by developmental growth factors, oncogenes, radiation, and hypoxia. A prominent growth factor that causes EMT is transforming growth factor β (TGF-β).

A series of molecular and cellular techniques can be applied to define and characterize the state of EMT in diverse biological samples. These methods range from DNA and RNA-based techniques that measure the expression of key EMT regulators and markers of epithelial or mesenchymal differentiation to functional assays of cell mobility, invasiveness and in vitro stemness. This chapter focuses on EMT induced by TGF-β and provides authoritative protocols and relevant reagents and citations of key publications aiming at assisting newcomers that enter this prolific area of biomedical sciences, and offering a useful reference tool to pioneers and aficionados of the field.

Key words Cellular plasticity, Epithelial–mesenchymal transition, Invasion, Motility, Signal transduction, Transcription factor, Transforming growth factor β

1 Introduction

Epithelial-to-mesenchymal transition (EMT) is a developmental cellular process by which an epithelial cell loses cell–cell interactions and apico-basal polarity, changes cell–matrix interactions, meanwhile acquiring mesenchymal features and migratory proper-

Xin-Hua Feng et al. (eds.), *TGF-β Signaling: Methods and Protocols*, Methods in Molecular Biology, vol. 1344, DOI 10.1007/978-1-4939-2966-5_9, © Springer Science+Business Media New York 2016

ties. The important changes in cell phenotype induced by EMT allow the escape of epithelial cells from the well-organized tissue sheet architecture and their subsequent invasion and migration through the extracellular matrix (ECM). EMT was first shown to exert a fundamental role in embryonic morphogenesis [1] occurring in specific developmental stages, such as (1) in gastrulation or neural crest delamination (classified as primary EMTs), (2) in morphogenetic processes contributing to organogenesis (secondary EMTs), and (3) in the development of specialized tissue parts such as the heart valves (tertiary EMTs) [2]. Although originally termed epithelial-to-mesenchymal transformation by Elizabeth Hay [1], the term "transformation" has been subsequently changed to "transition" in accordance with the extraordinary cell plasticity associated with phenotypic reversal of EMT, called the mesenchymal-to-epithelial transition, which occurs naturally during normal development to achieve tissue and organ constructions. EMT also plays important roles in adult skin repair, favoring the re-epithelialization of the wound by keratinocytes. In this context, EMT is partial, as migrating keratinocytes retain some intercellular junctions and migrate as a cohesive cell sheet [3]. Conversely to its "physiological" roles, EMT is also involved in pathological contexts such as chronic inflammation and tissue fibrosis, by stimulating matrix synthesis and deposition [4]. EMT is also implicated in tumor progression, where it participates in carcinoma cell dissemination by favoring tumor cell motility and invasion [2]. Recently, EMT has been shown to impact early stages of tumorigenesis by generating cells with progenitor-like features [5, 6] and by assisting tumor cells to bypass the oncogene-induced senescence [7].

The EMT process is characterized by an extensive nuclear reprogramming of the epithelial cell that supports the aforementioned switch towards a progenitor cell state and the elaborate reorganization of cell–cell junctions, so that the tight associations between well-organized epithelial cells become more flexible or disappear, a necessary event for detachment from the organized epithelium and cell migration [2]. Inside the cell, the reorganization of membrane-based cell–cell or cell–matrix junctions is supported by changes in the dynamics and/or composition of microfilaments and intermediate filaments. More specifically, cortically located actin microfilaments rearrange to generate either motility-associated structures (filopodia or lamellipodia) or stress fibers, as revealed by in vitro studies of cell models. Cytokeratin filaments, initially associated with adherens junctions and hemidesmosomes, will be replaced by the mesenchymal-specific filament vimentin, concomitantly with the dissolution of both junctional systems. Outside the cell, the ECM is also reorganized and becomes richer in a number of protein components and secreted growth factors that can signal back to the cell undergoing EMT or towards neighboring cells. In cancer, at the edge of tumor growth, all these changes prepare the surrounding microenvironment to facilitate local invasiveness, penetration by nearby blood and lymphatic vessels and the intravasation of tumor cells into these vessels [8].

EMT may be induced by environmental conditions, such as ultraviolet radiation, smoke, and hypoxia, and by several growth factors, including transforming growth factor β (TGF-β) [2]. TGF-β has been shown to induce EMT using complementary signaling cues. Indeed, in addition to its common type I receptor, TGF-β receptor I (TβRI) [9] and the canonical Smad pathway [10, 11], TGF-β employs alternative pathways to trigger EMT in epithelial cells. These intracellular cues involve the cleavage and release of the cytoplasmic kinase domain of TβRI, which is capable of translocating to the nucleus [12], the phosphatidylinositol 3-OH kinase (PI3K)/AKT pathways [13], the Rho A small GTPase and the downstream p160 Rho-associated coiled-coil-containing protein kinase ($p160^{ROCK}$) [14], β1-subunit integrin/focal adhesion kinase (FAK) signaling routes [15, 16], the Src tyrosine kinase and p38 MAPK pathways [17, 18], and more [19]. The TGF-β receptor complex also uses the polarity protein Par6 to recruit the E3-ubiquitin ligase Smurf1 at the apical junction complexes to induce the subsequent degradation of RhoA. This event leads to local actin disorganization, hence inducing tight and adherens junction dissolution [20].

During development, EMT is regulated according to the strict spatiotemporal expression of a small group of transcription factors which are re-expressed under pathological conditions and/or upon TGF-β stimulation. These factors include: the zinc finger transcriptional repressors Snail1 (SNAI1) [21, 22], Slug (SNAI2) [23, 24], δEF1/ZEB1 [25], and SIP1/ZEB2 [26], as well as the basic helix-loop-helix (bHLH) E47 [27] and Twist1 [28] transcription factors. Moreover, TGF-β downregulates the inhibitors of DNA binding, ID-1, ID-2, and ID-3 to elicit EMT [29, 30]. Recently, the DNA architectural protein high-mobility group A2 (HMGA2) has been proposed as a major integrator of TGF-β-mediated EMT in human cell lines, coordinating the induction of SNAI1, SNAI2, TWIST and the repression of *ID2* [31]. More precisely, HMGA2 interacts physically with Smad proteins to regulate SNAI1 [32] and TWIST1 [33] gene expression. For a more complete coverage of TGF-β-mediated mechanisms of EMT, the reader is referred to recent authoritative reviews [19, 34–36].

In this chapter, we describe the commonly used experimental approaches to characterize the phenotype of cells undergoing EMT, acquiring motility properties and/or stemness features upon in vitro TGF-β stimulation, as well as strategies to identify and characterize new markers or regulatory proteins involved in TGF-β-mediated EMT.

2 Materials

2.1 Cell Lines

Normal murine mammary gland epithelial cells (NMuMG), immortalized human mammary epithelial cells (MCF-10A), human breast cancer MDA-MB-231 cells and human lung adenocarcinoma A549 cells are obtained from the American Type Culture Collection (ATCC). Human primary mammary epithelial cells (HMEC) and epidermal keratinocytes (HNEK) are purchased from Lonza. Human immortalized mammary epithelial cells (HMLE) are obtained from R. A. Weinberg (Cambridge, MA, USA). Human immortalized lung epithelial cells (HPL1), human immortalized keratinocytes (HaCaT), and α-TN4 murine lens epithelial cells are obtained from T. Takahashi (Nagoya, Japan), N. Fusenig (Heidelberg, Germany) and J. Piatigorsky (Bethesda, MD, USA), respectively. NMe and NMuMG clone 18 cells, clonal epithelial derivatives of NMuMG were developed by K. Vershueren (Leuven, Belgium) and P. ten Dijke (Leiden, The Netherlands), respectively. Mouse EpH4 and Ras-transformed EpRas mammary epithelial cells are obtained from H. Beug (Vienna, Austria). Oncogenically transformed derivatives of MCF-10A cells, Ras-transformed MCF-10AneoT (MIIclone) and a metastatic variant (MCF-10CA1a.cl1, MIV clone) were derived from D. Miller (Seattle, USA) and were recently described [37].

2.2 Cell Culture Reagents

Minimum essential medium (MEM), Dulbecco's modified Eagle's medium (DMEM), Ham's F12 medium, DMEM:F12 medium, calcium- and magnesium-free phosphate buffered saline (PBS), fetal bovine serum (FBS), penicillin/streptomycin sulfate, nonessential amino acids (NEAA), and 0.25 % trypsin/EDTA are from Gibco/Invitrogen.

2.3 Chemicals

Human recombinant TGF-β1 and Epidermal Growth Factor (EGF) are purchased from R&D systems, TβRI inhibitor SB431542 from Calbiochem, protease inhibitor cocktail from Roche, hydrocortisone-HCl, human transferrin, tri-iodothyronine, cholera toxin, EGF, and insulin from Sigma-Aldrich. Recombinant growth factors are dissolved in a 0.1 % BSA-4 mM HCl solution (vehicle).

2.4 Antibodies

Primary antibodies used for Western-blotting analysis and indirect immunofluorescence (IF) staining using fixed cells are listed in Table 1, whereas those used for immunohistochemistry (IHC) and immunohistofluorescence (IHF) staining using fixed tissues are described in Table 2. Table 3 presents secondary antibodies used for the different immunological procedures.

2.5 PCR Primers

Oligonucleotide primers used for real-time reverse-transcription quantitative polymerase chain reaction (RT-qPCR) analyses in human and mouse cells are respectively listed in Tables 4 and 5.

Table 1
Primary antibodies used for Western-blotting (WB) analysis and indirect immunofluorescence (IF) staining of fixed cells

Marker	Antibody clone/ reference	Supplier	Dilution	Antibody species	Expected results	
					Epithelial cells	Mesenchymal cells
E-cadherin	Clone 36 610181	BD Transduction Laboratories	IF; 1:100 WB; 1:500	Mouse	IF; cortical WB; +++	IF; internalization or absence WB; +/– or –
β-catenin	Clone 14 610154	BD Transduction Laboratories	IF; 1:200	Mouse	IF; cortical	IF; internalization
N-cadherin	Clone 32 610921	BD Transduction Laboratories	WB; 1:500	Mouse	WB; +/–	WB; +++
CAR	HPA003342	Atlas Antibodies, Sigma-Aldrich	IF; 1:50 WB; 1:300	Rabbit	IF; cortical	IF; internalization or absence
ZO-1	Clone R40.76 MABT11	Chemicon Intern./ Millipore	IF; 1:100	Rat	IF; cortical	IF; internalization or absence
Claudin-3	341700	Zymed/Invitrogen	WB; 1:500	Rabbit	WB; +++	WB; +/– or –
α-SMA	Clone 1A4 A2547	Sigma-Aldrich	IF; 1:100 WB; 1:1000	Mouse	IF; +/– WB; +/–	IF; +++ WB; +++
Vimentin	Clone V9 V6630	Sigma-Aldrich	IF; 1:200 WB; 1:2000	Mouse	IF; +/– WB; +/–	IF; +++ WB; +++
Fibronectin	F3648	Sigma-Aldrich	IF; 1:500 WB; $1:3\times10^4$	Rabbit	IF; weak ER/ Golgi+ WB; +/–	IF; +++ WB; +++
Twist1	Twist2C1a Sc-81417	Santa Cruz Biotechnology	WB; 1:200	Mouse	WB; –	WB; ++
ZEB1 (δEF1)	H-102 Sc-25308	Santa Cruz Biotechnology	WB; 1:5000	Rabbit	WB; –	WB; ++
ZEB2 (SIP1)	H-260 Sc-48789	Santa Cruz Biotechnology	WB; 1:250	Rabbit	WB; –	WB; ++
Snail1	Ab82846	Abcam	WB; 1:1000	Rabbit	WB; –	WB; ++
β-tubulin	Clone JDR.3B8 T8535	Sigma-Aldrich	WB; 1:1000	Mouse	Housekeeping proteins	
GAPDH	AM43000	Ambion	WB; $1:5\times10^6$	Mouse		

No signal (–), low (+), moderate (++), or high (+++) signal

Table 2
Primary antibodies used for immunohistochemistry (IHC) and immunohistofluorescence (IHF) staining of fixed tissues

Marker	Antibody clone/ reference	Supplier	Dilution	Antibody species	Expected results	
					Epithelial cells	Mesenchymal cells
E-cadherin	Clone NCH-38	Dako	IHF; 1:50	Mouse	Membrane (+++)	+/– or –
	Clone 36 610181	BD Transduction Laboratories	IHC; 1:700	Mouse		
CAR	HPA003342	Atlas Antibodies, Sigma-Aldrich	IHF; 1:50	Rabbit	Membrane (+++)	+/– or –
SMAD4	H-552 Sc-7154	Santa Cruz Biotechnology	IHF; 1:125 IHC; 1:100	Rabbit	Mostly nuclear and cytoplasmic (tissue-dependent)	
SMAD3	ab28379	Abcam	IHF; 1:100	Rabbit	Mostly nuclear and cytoplasmic (tissue-dependent)	
TIF1γ	16G9 MA1-801	Euromedex	IHC; 1:500	Mouse	Nuclear	
SNAIL1	Ab82846	Abcam	IHF; 1:5	Mouse	–	++
α-SMA	M0851 Clone 1A4	Dako	IHC; 1:100	Mouse	–	+++
Vimentin	C-20 Sc-7557	Santa Cruz Biotechnology	IHC; 1:100	Goat	–	++

No signal (–), low (+), moderate (++), or high (+++) signal

2.6 Indirect Immunofluorescence and Direct Fluorescence Staining

Sterile glass coverslips, phosphate-buffered saline solution containing Mg^{2+} and Ca^{2+} (PBS^{++}; PBS with 0.5 mM $MgCl_2$ and 0.9 mM $CaCl_2$, pH 7.0–7.2), 16 % paraformaldehyde solution (Electron Microscopy Sciences), Triton X-100 (Fluka), FBS (Invitrogen), 4′,6-diamidino-2-phenylindole (DAPI, Sigma-Aldrich), Fluoromount-G mounting medium (Interchim), and tetramethylrhodamine B isothiocyanate (TRITC)-conjugated or fluorescein (FITC)-conjugated isothiocyanate-phalloidin (Sigma-Aldrich).

2.7 Immunohistochemical Staining

Superfrost PLUS microscope slides (Thermo Scientific), EasyDip™ slide staining rack (Electron Microscopy Sciences), PBS, H_2O_2 (Fluka), xylene-substitute (Sigma-Aldrich), ethanol (VWR), target retrieval solution, pH 9.0 (Dako) or Vector® antigen unmasking solution (Vector Laboratories), Dako Pen (Dako), antibody diluent (Dako), universal Dako LSAB® + kit-Peroxidase (Dako), glass cov-

Table 3
Secondary antibodies used for the different immunological procedures

Immunological procedure	Antibody (Ig, immunoglobulin)	Supplier	Dilution	Catalogue nr.
Immunohistofluorescence (IHF) Immunofluorescence (IF)	Anti-mouse IgGAlexa Fluor® 488	Molecular Probes/Life Technologies	IHF; 1:700 IF; 1:1000	A-11001
	Anti-mouse IgGAlexa Fluor® 555			A-21424
	Anti-rabbit IgGAlexa Fluor® 568			A-11011
	Anti-rabbit IgGAlexa Fluor® 488			A-11008
	Anti-rabbit IgGAlexa Fluor® 594			A-11012
	Anti-rat IgGAlexa Fluor® 594			A-21203
Western-blotting	Anti-mouse Ig-HRP conjugated	Dako	1:10,000	P0260
	Anti-rabbit Ig-HRP conjugated			P0448
	Anti-rat Ig-HRP conjugated			P0450
Immunohistochemistry	Anti-rat Ig-biotin conjugated	Dako	1:200	E0468
	Universal Biotinylated Link (i.e., biotinylated anti-rabbit, anti-mouse, and anti-goat Ig)		Ready-made solution	K0679 (or K0690)

erslips, DEPEX mounting medium (VWR), Vectashield Mounting Medium with DAPI (Vector Laboratories).

2.8 Immunoblot Assay

Protease and phosphatase inhibitor tablets (Roche), cell lysis buffer (1 % Triton X-100, 150 mM NaCl, 20 mM Tris–HCl, pH 7.5), cell scrapers (Corning®).

2.9 Genome-Wide Transcriptomic Assay

Oligonucleotide- or cDNA-based "home-made" or commercial glass microarray chips, RNA extraction kit (RNeasy kit, Qiagen), a Dual-color labeling kit [e.g., two-color Quick Amp Labeling kit (Agilent), containing Moloney Murine Leukemia Virus-Reverse Transcriptase, T7 Promoter Primer, RNAse inhibitor (RNaseOUT), 5× First Strand Reaction buffer, 4× Transcription buffer, 10 mM

Table 4
Oligonucleotide primers used for quantitative real-time RT-PCR analyses in human cells

Gene	Primer sequence	Strand	Product size (bp)	Accession no./reference	Description
MUC1	5′-TGCTGGTGCTGGTCTGTGTT-3′ 5′-AGACAGCCAAGGCAATGAGATAG-3′	+ –	63	M34089/[11]	Epithelial marker
KRT19	5′-GGTCATGGCCGAGCAGAA-3′ 5′-CGACCTCCCGGTTCAATTC-3′	+ –	77	NM_002276/[11]	
TIMP3	5′-CTATCGGTATCACCTGGGTTGTAA-3′ 5′-CTATCGGTATCACCTGGGTTGTAA-3′	+ –	80	NM_000362/[11]	Mesenchymal marker
HMGA2	5′-CCCAAAGGCAGCAAAAACAA-3′ 5′-GCCTCTTGGCCGTTTTTCTC-3′	+ –	81	NM_003483/[11]	EMT regulator
ZYX	5′-GGACAAGAACTTCCACATGAAGTG-3′ 5′-GGGAAGCAGCCATTGTCATC-3′	+ –	87	NM_003461/[11]	
MSN	5′-CCCGGACTTCGTCTTCTATGC-3′ 5′-CCATGCACAAGGCCAAGAT-3′	+ –	65	M69066/[11]	
GAPDH	5′-GGAGTCAACGGATTTGGTCGTA-3′ 5′-GGCAACAATATCCACTTTACCAGAGT-3′	+ –	78	BC023632/[11]	Housekeeping gene

Table 5
Oligonucleotide primers used for quantitative real-time RT-PCR analyses in mouse cells

Gene	Primer sequence	Strand	Product size (bp)	Accession no./reference	Description
E-cadherin	5′-CTGCGCTGGATAGTGTGTGT-3′ 5′-TGGCATGCACCTAAGAATCA-3′	+ –	70	NM_009864.2/[33]	Epithelial marker
Pai-1	5′-GGGAAAAGGGGCTGTGTGAC-3′ 5′-GTACACGGTGTGTGGCTGTC-3′	+ –	406	M33960/[31]	Mesenchymal marker
Fibronectin-1	5′-CCCAGACTTATGGTGGCAATTC-3′ 5′-AATTTCCGCCTCGAGTCTGA-3′	+ –	200	NM_010233/[31]	
N-cadherin	5′-GAGAGGCCTATCCATGCTGA-3′ 5′-CGCTACTGGAGGAGTTGAGG-3′	+ –	120	NM_007664.4	
Snail1	5′-CCACTGCAACCGTGCTTTT-3′ 5′-CACATCCGAGTGGGTTTGG-3′	+ –	66	NM_011427/[32]	EMT regulator
Snail2 (*Slug*)	5′-CTCACCTCGGGAGCATACAGC-3′ 5′-TGAAGTGTCAGAGGAAGGCGGG-3′	+ –	146	NM_011415/[31]	
Zeb1	5′-ACAAGACACCGCCGTCATTT-3′ 5′-GCAGGTGAGCAACTGGGAAA-3′	+ –	121	NM_011546/[32]	
Zeb2	5′-CACCCAGCTCGAGAGGCATA-3′ 5′-CACTCCGTGCACTTGAACTTG-3′	+ –	101	NM_015753/[32]	
Twist1	5′-CGGGTCATGGCTAACGTG-3′ 5′-CAGCTTGCCATCTTGGAGTC-3′	+ –	196	NM_011658/[31]	
E47	5′-CACAGAGACCTCCCGACTCCTA-3′ 5′-TGCGATTTCCTCATCCTCTTTC-3′	+ –	101	NM_011548/[32]	
E2-2	5′-CTGCCTTAGGGACGGACAAA-3′ 5′-CGCCAAAGAAGTTGGTCCAT-3′	+ –	101	NM_013685/[32]	
Hmga2	5′-AGCAAAAACAAGAGCCCCTCTA-3′ 5′-ACGACTTGTTGTGGCCATTTC-3′	+ –	100	NM_010441/[31]	
Gapdh	5′-TGTGTCCGTCGTGGATCTGA-3′ 5′-CCTGCTTCACCACCTTCTTGA-3′	+ –	76	NM_001001303/[31]	Housekeeping gene
β2-microglobulin	5′-TTCTGGTGCTTGTCTCACTGA-3′ 5′-CAGTATCTTCGGCTTCCCATTC-3′	+ –	306	NM_009735/[33]	

deoxynucleoside triphosphate (dNTP) mix, NTP mix, 0.1 M DTT, T7 RNA Polymerase, 50 % Polyethylene glycol, Cyanine (Cy) 3-CTP, and Cyanine 5-CTP], Agilent 2100 Bioanalyzer (Agilent Technologies), Slide scanner, Spot analysis, and software for data analysis.

2.10 QuantitativeReal Time Reverse-Transcription-Polymerase Chain Reaction

dNTPs (Invitrogen), SuperScript II reverse-transcriptase (Invitrogen), RNAse inhibitor (RNAseOUT, Invitrogen), SYBR® Green quantitative PCR master mix (Applied Biosystems, containing SYBR® Green I dye, AmpliTaq Gold® DNA polymerase, dNTPs with dUTP, passive reference, and optimized buffer components), real-time PCR instruments.

2.11 Mammosphere (Stem Cell) Culture Assay

Stem cell medium (DMEM/F12, supplemented with B27 (Invitrogen), 25 ng/mL EGF, 25 ng/mL bFGF (Peprotech Inc.), and 4 μg/mL heparin (Sigma-Aldrich)), methylcellulose (Becton Dickinson & Co) added to 1 % in the stem cell medium to prevent cell aggregation, 24-well ultra-low attachment multi-well plates (Corning), 40 μm sieve (Becton Dickinson & Co), cell counter.

2.12 Invasion Assay of Three-Dimensional Cell Culture

Methocel solution (*see* **Note 19** for recipe), 96-well plates (flat and round bottom) (Becton Dickinson Falcon or Corning), atelopeptidic and acid-soluble type I collagen (PureCol®; Inamed, store at 4 °C), pH-indicator strips, 0.1 N HCl, and 0.1 N NaOH (filter-sterilized, store at 4 °C), 10× PBS (sterile, store at 4 °C).

2.13 Invasion Assay in Matrigel

BioCoat Matrigel invasion chambers with culture plates (BD Biosciences; we use 24-well plates in this protocol), culture medium such as DMEM, chemoattractant such as 5 % FBS in the culture medium, bovine serum albumin, trypsin inhibitor of choice, cell counter, cotton swabs, methanol, and Wright-Giemsa solution (Sigma-Aldrich) (or other suitable fixative and staining solution).

2.14 In Vitro Wound-Healing (Scratch) Assay

Culture medium with supplements (serum, antibiotics), PBS, growth factor (TGF-β1, BMP-7), culture dishes (60 mm, 35 mm, or 6-well dishes work well), sharpie marker, p200 or p1000 pipette tips (or other appropriate scratching device), phase-contrast microscope with camera, image analysis software.

3 Methods

In vitro analysis of EMT is done by characterizing specific changes to the epithelial cell phenotype upon TGF-β treatment. More specifically, it principally includes the analysis of loss of epithelial features and acquisition of mesenchymal, fibroblastic-like traits. This includes (1) the dissolution of the cell–cell junctions and the subsequent loss and/or delocalization of molecular components of cell junctions (E-cadherin, zonula occludens 1 (ZO-1), coxsackie and adenovirus receptor (CAR), β-catenin), which can be monitored by immunodetection followed by chemical or fluorescent revelation performed on fixed cells and tissue sections, (2) the loss of cell polarity and the associated drop in the barrier function of tight junctions, which can be analyzed using impedance assays, (3) the reorganization of actin microfilaments, usually assessed by direct staining using fluorescently labeled phalloidin as well as the appearance of the alpha-smooth muscle actin by IF, (4) the decrease in total expression levels of specific epithelial proteins (E-cadherin, claudins, occludins) and the increase in expression levels of mesenchymal proteins (N-cadherin, vimentin), which can be analyzed at the mRNA (RT-qPCR) and protein (Western-blotting) levels, (5) the gain of cell motility, which can be characterized by cell migration assay in 2D culture system such as wound healing experiments, or by invasion assay (Boyden chamber or transwell Matrigel invasion assays) and cell behavior in 3D collagen gels, and (6) the acquisition of stem-like features associated with EMT as assessed using spheroid culture assays in defined stem cell media.

This chapter assumes the readers are familiar with several common laboratory techniques including cell culture, SDS-PAGE and Western blot analysis, RNA isolation and RT-qPCR experiments.

3.1 Cell Types That Undergo EMT In Vitro

As can be seen from a deep curation of the literature, virtually all epithelial cell types may undergo a full or partial EMT upon TGF-β stimulation, on the condition that cells (1) exhibit a strong or moderate basal epithelial phenotype and (2) have a functional TGF-β signaling pathway. The full EMT process is achieved between 2 and 7 days, depending on the cell-type and the dose of active recombinant TGF-β used. For this reason, early studies in human epithelial cell lines cultured in vitro concluded that EMT rarely occurred in response to TGF-β, when examined during a 36 h time period [38]. We have shown that NMuMG and their epithelial clonal derivative (NMe cells, [26]), normal human primary (HMEC) and immortalized (MCF-10A) mammary epithelial cells, and immortalized human lung epithelial cells (HPL1, [39]) all undergo a full EMT at 36–48 h following stimulation by any of the TGF-β isoforms (TGF-β1, -β2, or -β3) at a dose range of 0.1–5 ng/mL [11]. α-TN4 murine lens epithelial cells are also able to undergo EMT in the presence of TGF-β1 [29]. TGF-β has also been shown to potently induce EMT in epithelial keratinocytes [40, 41], including normal human primary epidermal (HNEK) and immortalized (HaCaT) keratinocytes [11]. In other laboratories, Madin-Darby canine kidney epithelial (MDCK) cells and mouse mammary epithelial cells EpH4 and their Ras-transformed derivatives, EpRas, are extensively used as a model for studying epithelial cell architecture and polarity, as well as cell plasticity during EMT upon exposure to TGF-β [1,

42]. It is worth noting though that the MDCK response to TGF-β towards EMT is clone-dependent (MDCK type II cells show better EMT response [43]) and the EpRas response is dependent on proper 3D culture conditions [44]. Immortalized HMECs (HMLE) cells also undergo TGF-β-mediated EMT, albeit after longer stimulation (7–15 days) [45, 46].

3.2 Indirect Immunofluorescence and Direct Fluorescence Staining

As illustrated in Figs. 1 and 2a, indirect IF is a suitable technique to detect the loss and/or delocalization of molecular components of cell junctions (E-cadherin, ZO-1, CAR, β-catenin) as well as the induction of certain molecular markers of the mesenchymal cell phenotype (fibronectin, α-smooth muscle actin, vimentin) upon TGF-β stimulation in cell culture. In contrast, actin microfilaments are usually detected by direct fluorescent staining thanks to the high-affinity binding of the phalloidin toxin to filamentous actin (F-actin).

Experimental procedures:

For indirect immunofluorescence staining:

1. Place sterile 22 mm-diameter round or 22 mm-side square glass coverslips in 35 mm-diameter or 6-well culture dishes (one coverslip per dish or well). Alternatively, cells can be seeded on 2-, 4-, or 8-well multi-well chambers on glass slides (Becton Dickinson).
2. Seed the appropriate amount of cells (e.g., 2×10^5 NMuMG cells per 6-well) with 2 mL of culture medium in dishes/wells and culture cells up to 70–80 % confluency in a humidified incubator at 37 °C with 5 % CO_2.
3. Replace the conditioned medium with 1 mL of fresh medium containing 3 % FBS. Treat cells with human recombinant TGF-β1 (0.1–5 ng/mL) or the corresponding volume of vehicle for 36–48 h, or until signs of morphological changes are observed by phase-contrast microscopy.
4. Remove culture medium and carefully rinse cells twice with 2 mL of PBS^{++}.
5. Fix cells by adding 1 mL of 4 %(vol/vol) paraformaldehyde-PBS^{++} in the culture plate and incubate for 30 min at room temperature (*see* **Note 1**). Rinse the cells thrice with 2 mL of PBS^{++}.
6. Permeabilize fixed cells with 1 mL of 0.5 % (vol/vol) Triton X-100-PBS^{++} solution for 10 min at room temperature. Rinse the cells thrice with 2 mL of PBS^{++}.

Fig. 1 (continued) fluo rescence using FITC-conjugated phalloidin (*green*). NMuMG cells display characteristic features of EMT in response to TGF-β treatment, including a more elongated morphology and a neat cell scattering, with increased fibronectin expression, a complete dissolution of tight junctions (represented by CAR staining) and a drastic reorganization of the actin cytoskeleton. A bar indicates 10 μm

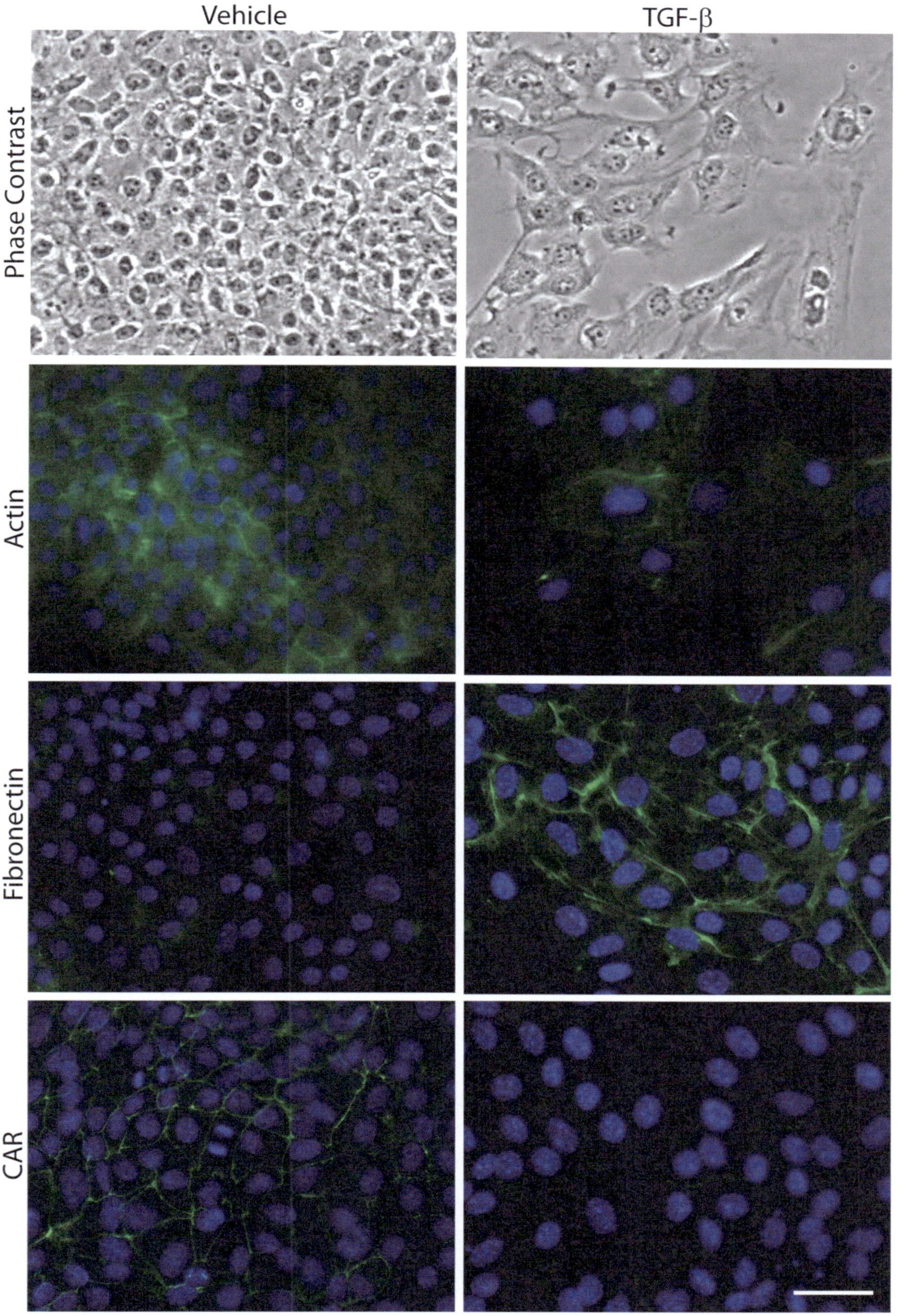

Fig. 1 Representative phase-contrast and fluorescent images of mouse mammary epithelial cells undergoing EMT. NMuMG cells were treated with vehicle or 5 ng/mL TGF-β1 for 48 h prior to fixation and staining. Cells were stained for nuclei (*blue*) and the indicated antibodies (*green*). Actin cytoskeleton is stained by direct

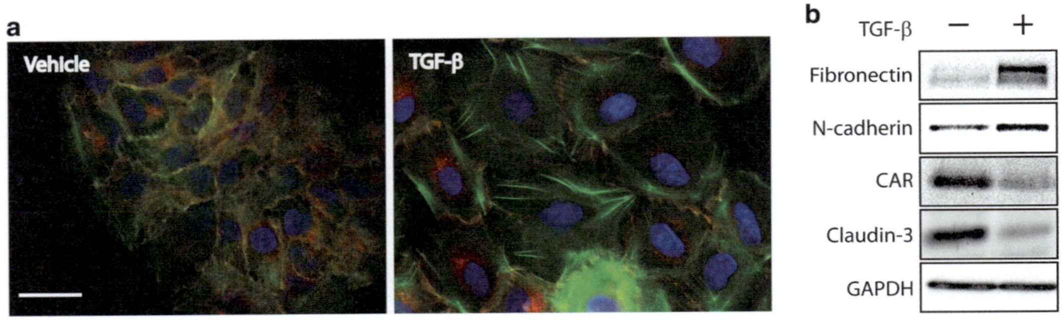

Fig. 2 Representative fluorescent images and western blot analysis in human keratinocytes undergoing EMT. (**a**) E-cadherin (*red*) IF and actin (*green*) direct fluorescence staining of HaCaT cells treated with 5 ng/mL TGF-β1 for 96 h. Nuclei are counterstained using DAPI (*blue*). The TGF-β-treated cells appear larger, have more prominent actin stress fibers, and exhibit partial delocalization and loss of E-cadherin, all hallmarks of the EMT process. A bar indicates 10 μm. (**b**) Western blot analysis of epithelial (CAR and claudin-3) and mesenchymal (fibronectin and N-cadherin) markers in HaCaT cell treated under the same conditions as in (**a**). GAPDH is shown as a loading control. The TGF-β-treated cells demonstrate increased expression of mesenchymal markers and a marked decrease of the levels of epithelial markers

7. Block nonspecific sites with 1 mL of 5 % (vol/vol) FBS-PBS^{++} solution for 1 h at room temperature, or overnight at 4 °C.
8. Remove as carefully as possible any liquid from the coverslips and around the perimeter and place coverslips (cells facing upward) in a humidified dark plastic box containing wet paper sheets.
9. Add 75–100 μL of primary antibody solution on top of each coverslip. Each antibody is diluted in 5 % FBS-containing PBS solution as listed in Table 1. Incubate for 1 h at room temperature (*see* **Note 2**).
10. Remove carefully the antibody solution and place back the coverslips in their original culture support and wash three times with 2 mL of PBS^{++} (5 min each), ideally using an orbital shaker.
11. Repeat **step 8**.
12. Add 75–100 μL of the corresponding fluorescent-labeled secondary antibody solution on top of each coverslip (*see* **Notes 3** and **4**). Antibodies are diluted in 5 % FBS-PBS^{++} solution as described in Table 3. DAPI (0.5–1 μg/mL) is added to the antibody dilution solution to ensure counterstaining of the nuclei (*see* **Note 5**). Incubate for 1 h at room temperature while protecting from light.
13. Remove the antibody solution and wash fixed cells three times with PBS, as in **step 10**.
14. Remove excess liquid from the coverslips and mount the coverslips (cells facing downward) onto glass slides using 5–10 μL

of mounting medium (Fluoromount-G or equivalent). Gently press down on coverslips in order to remove excess mounting medium and air bubbles. For prolonged storage (several months), coverslips can be permanently sealed around the perimeter with plastic sealant or nail polish. Mounted slides are then stored at 4 °C protected from light.

15. Observe staining with either a confocal microscope or an epi-fluorescent microscope equipped with the adapted excitation and emission filters and a digital camera.

For direct staining of F-actin microfilaments with TRITC/FITC-conjugated phalloidin, proceed with sample preparations up to **step 8**. Add 75–100 μL of 0.25 μM TRITC/FITC-labeled phalloidin, diluted in 5 % FBS-PBS^{++} solution containing 0.5–1 μg/mL of DAPI, on top of each coverslip. Incubate for 1 h at room temperature. Then proceed from **step 13** to the end.

3.3 Immunohistochemical Staining

The detection of EMT markers or regulators in tissues, such as tumor samples, and in 3D epithelial cultures, may be performed using IHC or IHF methods. Whereas immunohistochemical methods determine the precise localization of the target antigen in a tumor sample, fluorescence-based immunohistological techniques have the main advantage to allow the simultaneous detection of several antigens and to determine their co-localization, on the condition that primary antibodies are produced in different species [47]. The IHC protocol we describe here uses the Dako LSAB® or LSAB® + HRP-Systems, but other commercial systems may be used. The LSAB® + HRP-system is based on the use of a "universal" horseradish peroxidase (HRP)-conjugated secondary antibody, 3,3′-diaminobenzidine (DAB) substrate, and LSAB enhancer system.

Experimental procedures:

For chromogen-based immunohistochemical staining:

1. Obtain paraffin-embedded sections (4- to 5-μm-thick) using a microtome and place sections on Superfrost PLUS microscope slides.
2. Dissolve paraffin and rehydrate sections by incubating them at 55 °C for a total time of 30 min as follows: twice in xylene-substitute (5 min each), 5 min in ethanol 100 %, 5 min in ethanol 95 % (vol/vol), 5 min in ethanol 70 % (vol/vol), and 5 min in deionized water.
3. Inactivate the endogenous peroxidase activity by incubating the slides with 3 % (vol/vol) H_2O_2 in PBS in a slide staining rack for 30 min in the dark. Rinse in deionized water for 5 min.
4. Perform antigen retrieval using either citrate-based or high pH-based unmasking solutions, depending on the specific

marker and antibody, coupled to high temperature unmasking techniques. As antigen retrieval is an empirical issue for each antigen, the readers should try variations of the indicated methods when using markers of their choice and specific tissues. We successfully detected E-cadherin, cytokeratin-19, α-smooth muscle actin (SMA), vimentin, Smad4, and TIF1γ proteins on mouse pancreatic tumor sections [47, 48] using a citrate-based antigen unmaking solution (Vector Laboratories). Boil slides (in staining rack) in the unmasking solution using a microwave at 1000 W for 5 min then for 10 min at 450 W. Let slides cool down to room temperature for 45 min in the unmasking solution. Rinse three times with PBS (5 min each).

5. Place the slides in a plastic box and mark the area around the sample with a hydrophobic barrier (Dako Pen) to avoid diffusion of the antibody out of the section.
6. Block nonspecific sites with 100 μL of IHC blocking solution or Antibody Diluent solution (Dako) for 1 h at room temperature in a humid atmosphere.
7. Incubate with 100 μL of the selected primary antibody, diluted (Table 2) in IHC blocking solution or Antibody Diluent solution (Dako), overnight at 4 °C in a humid atmosphere.
8. Wash three times with PBS for 5 min each at room temperature, with slow agitation in a slide-staining dish. Place the slides in a dark plastic box.
9. Perform amplification using the biotin-coupled universal secondary antibody as recommended by the manufacturer. Note that the LSAB®+kit (Dako) is only used when primary antibodies are from mouse, rabbit and goat species, otherwise any other biotinylated-secondary antibodies are used with the LSAB® kit (Dako) (Table 3). Add 100 μL of the universal antibody per section and incubate for 30–60 min at room temperature in a humid atmosphere.
10. Wash three times with PBS at room temperature for 10 min each in a staining rack. Place the slides in a dark plastic box.
11. Incubate with streptavidin-HRP for 1 h at room temperature in a humid atmosphere.
12. Prepare fresh DAB-chromogen solution as recommended by the manufacturer (*see* **Note 6**).
13. Add 100 μL of DAB-chromogen solution to each section and incubate for 1–5 min at room temperature until the desired level of staining is achieved (careful checking during revelation).
14. Inactivate the DAB-chromogen solution by immersing the slide in deionized water in a slide-staining rack.
15. Counterstain with hematoxylin as desired. Rinse in running tap water for 2–5 min.

16. Dehydrate the sections through immersion for 5 min each in 70 %, 95 % and twice in 100 % ethanol (all vol/vol solutions) and then incubate in a xylene-substitute for 5 min.
17. Mount the slides with a coverslip and the DEPEX mounting medium. Remove excess mounting medium with a piece of filter paper, and avoid formation of bubbles while mounting.
18. Observe the slides under a light microscope equipped with a digital camera.

For immunohistofluorescence staining:

Double or triple detection of several EMT regulators (Snail1, Smad3 and/or Smad4) or epithelial markers (E-cadherin, CAR) were successfully performed on human breast carcinoma sections [8].

1. Proceed up to **step 8**, except that the retrieval of the antigen is performed by incubating slides in a high-pH antigen unmasking solution (Target Retrieval Solution, pH 9.0; Dako) in a heated (97 °C) water bath for 40 min.
2. Place the slides in a dark plastic box and incubate with 100 μL of the fluorophore-conjugated secondary antibody for 30–60 min in a humid atmosphere (Table 3).
3. Wash three times with 100 mL of PBS at room temperature for 10 min each in a staining rack. Place the slides in a dark plastic box.
4. Mount the slides with a coverslip using Vectashield Mounting Medium with DAPI (Vector Laboratories). Remove excess mounting medium with a piece of filter paper, and avoid bubbles formation while mounting. For prolonged storage, coverslips can be permanently sealed around the perimeter with nail polish. Mounted slides should be stored at 4 °C under light protection.
5. Observe the slides under a confocal or an epifluorescent microscope equipped with the adapted excitation and emission filters and a digital camera.

3.4 Immunoblot Assay

As illustrated in Fig. 2b, Western-blotting analysis offers the main advantage to bring more quantitative results to estimate the level of numerous epithelial and mesenchymal cell protein markers using a single cell lysate.

Experimental procedures:

Cells are cultured either in 35 mm-diameter or 6-well culture dishes and treated (or not, control condition) with TGF-β1 as described in the cell culture method described above.

1. Remove culture medium and carefully rinse the cell monolayer twice with 2 mL of PBS^{++}.

2. Place the culture dish on ice and add 150 μL of ice-cold lysis buffer (containing freshly added protease and phosphatase inhibitors).
3. Homogenize lysis buffer on cell monolayer using a sterile cell scraper and collect cell lysate in a 1.5 mL microfuge tube.
4. Incubate cell lysate for 10 min on ice.
5. Centrifuge cell lysate for 10 min at 16,000 × *g*, 4 °C, and transfer the clarified supernatant containing the soluble cell extract into a new 1.5 mL microfuge tube.
6. Determine protein concentration using a method of your choice.
7. Run 30–50 μg of protein through 8–12 % (wt/vol) SDS-PAGE gels for each sample.
8. Perform Western-blotting using the specific primary and the corresponding HRP-conjugated secondary antibodies (listed Tables 1 and 3 respectively). α-/β-tubulin, glyceraldehyde-3-phosphate dehydrogenase (GAPDH) or any other "house-keeping" protein can be used as loading control (also depending on the cell type) (*see* **Note 7**).

3.5 Genome-Wide Transcriptomic Assay

High-throughput expression analysis based on microarrays can be applied for selection of genes that are related to EMT in response to TGF-β stimulation [49, 50] or downstream signaling or effector molecules, such as Smad4 [29, 51], TIF1γ [45] or JunB [52]. The main goal is the identification of genes and/or pathways whose expression correlates with, and/or regulates the EMT process or the cell scattering in response to TGF-β. The transcriptomic approach has been successful in our hands when comparing the gene expression profiles of NMuMG cells treated, or not (control condition), with recombinant TGF-β1 for various times. Indeed, by focusing on early (2 h), intermediate (8 h) and late (36 h) transcriptional events induced by TGF-β1, we were able to define a list of a limited number of "EMT-related genes" that are shared by epithelial cells of different origins and that have been shown to be either regulators (early and intermediate-responsive genes) or markers (late-responsive genes) of the EMT process [11].

Most of the early high-throughput gene expression analyses were performed using "home-made" or "consortium-made" glass oligonucleotide- or complementary DNA (cDNA)-based chips, but commercial microarray platforms have since emerged offering better reproducibility for hybridization techniques and subsequent analyses and hence bringing more consistent data. Although initially restricted to human and mouse genomes, commercial microarrays have now been extended to other species, such as dog, allowing for the analysis of gene expression in the canine MDCK cell line, a classical model of EMT [53]. Available commercial chip platforms based on either oligonucleotides (Affymetrix, Agilent) or cDNA (CodeLink) often require specific hybridization and detection methods as well as statistical analysis of the data, and the reader has to refer to the recommendations of the microarray's manufacturer. Here we describe a general method to study gene expression using a two-color microarray approach as well as the overall strategy we used to identify EMT-regulatory genes induced by TGF-β1 stimulation.

Experimental procedures:

1. Seed the appropriate amount of cells (e.g., 4×10^6 NMuMG cells) in 150 mm-diameter culture dishes with 20 mL of culture medium and culture cells up to 70–80 % confluency in a humidified incubator at 37 °C with 5 % CO_2.
2. Treat cells with recombinant TGF-β1 (0.1–5 ng/mL) or vehicle (control condition), as described in the cell culture method described above, for the desired period of time.
3. Remove culture medium and carefully rinse the cell monolayer twice with 10 mL of PBS.
4. Isolate total RNA from cell monolayers using the RNeasy kit (Qiagen) according to the manufacturer's protocol. Determine the integrity of the input RNA by running a sample aliquot in a denaturing agarose gel stained with a fluorescent intercalating agent, such as ethidium bromide or SYBR® derivatives. Notably, two clear bands, corresponding to the 18S and 28S ribosomal subunits, should be seen, with no obvious degradation product at the bottom of the gel. Alternatively, use microfluidic instruments such as Agilent 2100 Bioanalyzer (Agilent Technologies) in order to provide detailed information about the condition of RNA samples (such as the ratio between 28S and 18S subunits and the algorithmic "RNA integrity number" parameter) (*see* **Note 8**).
5. Label RNA sample pairs from mock-treated and TGF-β1-treated cells by reverse transcriptase-based methods. The amount of total RNA required (from 500 ng to 25 μg) and the labeling method depend on the microarray platform used; the reader has to refer to the recommended protocol of the chip's manufacturer. The labeling classically involves incorporation of both dCTP-Cy3 and dCTP-Cy5 during the synthesis of cDNA or incorporation of CTP-Cy3 and CTP-Cy5 to create antisense complementary RNA (cRNA) following unlabeled reverse-transcription. The latter method allows the amplification of cRNA samples. For instance, RNA molecules from mock-treated cells are labeled with dCTP-Cy3 (or CTP-Cy3) and those from TGF-β1-treated cells with dCTP-Cy5 (or CTP-Cy5) (*see* **Note 9**).

6. Hybridize equal amounts of labeled cDNA or cRNA probes per pair in at least triplicates (including the dye swap experiment) to the microarray chips of interest. Perform microarray hybridization and washing according to the manufacturer's protocol (e.g., "Two-Color Microarray-Based Gene Expression Analysis" protocol from Agilent Technologies, available on http://www.chem.agilent.com/en-US/Search/Library/_layouts/Agilent/PublicationSummary.aspx?whid=48835).
7. Scan the microarray slides with a 5–10 μm of resolution using a GSI LumonicsScanArray 4000 scanner or the scanner of your choice.
8. Analyze spot intensity by QuantArray (GSI Lumonics) or Feature extraction (Agilent Technologies) software. This step allows the conversion of spot hybridization intensity (Cy3 vs. Cy5) in expression ratio, after performing spot analysis (spot finding and signal optimization), signal to noise subtraction, as well as dye color and scanning variation corrections. In our experiments, raw data were normalized by nonlinear Lowess normalization.
9. Perform statistical analysis of the (replicate/triplicate) data sets using specific data mining software, such as GeneSpring GX (Silicon Genetics/Agilent Technologies). In the different gene profiling analyses that we performed, we selected differentially regulated genes based on the following criteria:
   (a) Average expression ratio values of at least ≥ 1.7 (upregulation) and ≤ 0.57 (downregulation) (*see* **Note 10**). The threshold setting is arbitrary and relies primarily on the quality of the microarray chip, but also of labeling and hybridization efficiency.
   (b) The regulated genes had to be expressed in all three replicates (intensity of the gene spot exceeded the background intensity plus 3 standard deviations based on the intensity of all spots).
   (c) Regulated genes had to score significantly in a *t*-test (P for ratios within triplicates < 0.05).
10. Validate candidate gene expression profiles using independent mock- or TGF-β1-treated cell culture using conventional or quantitative real-time RT-qPCR (Cf. next method on RT-PCR).

Functional classification of highly regulated genes was performed manually based on exhaustive curation of the literature in PubMed (http://www.ncbi.nlm.nih.gov/PubMed/). Candidate genes either stimulated or repressed by TGF-β belong to different functional clusters of genes, such as those involved (1) in the cytostatic response, (2) in the apoptotic program, (3) in the EMT process per se, but also in (4) the feedback negative regulation of the signaling pathway in response to TGF-β. In order to decipher new genes that are specifically involved in the regulation of EMT process, but not necessarily involved in apoptotic or cytostatic responses to TGF-β, we applied additional selection criteria:

1. The regulated genes were clustered according to their time-dependent expression profile (*K*-means method) using GeneSpring GX. Inside each cluster, the genes were classified according to their functional annotation. We postulated that groups of genes regulated at early (2 h) or intermediate (8 h) time points upon TGF-β1 stimulation were potentially new regulators of EMT rather than marker genes of EMT, whose expression were expected to be regulated at the late time point (36 h) by TGF-β1. This clustering analysis also discriminated potentially direct TGF-β-target genes (early time point) to indirect target genes (intermediate and late time-points).
2. As we demonstrated that BMP-7 or BMP type I receptors fail to induce EMT in murine mammary epithelial cells [11], we hypothesized that genes co-regulated by both BMP-7 and TGF-β1 might not be involved in the EMT process. We therefore compared the expression of selected candidate genes in the presence of BMP-7 and excluded genes regulated by this cytokine.
3. We also postulated that target genes involved in EMT should be evolutionarily conserved at least between mouse and human mammary epithelial cells. We therefore validated the expression of selected candidate genes in normal HMEC, as these cells exhibit an EMT response to TGF-β1 similar to the response of murine NMuMG cells.
4. As the EMT process is shared by epithelial cells in different tissues, we compared the kinetic expression pattern of candidate genes between human mammary HMEC and human epidermal HaCaT keratinocytes. EMT related genes were suspected to be expressed in a delayed manner in keratinocytes compared to mammary epithelial cells.

Using this strategy, we selected a small group of genes whose function has later been shown to be involved in the EMT process by our laboratory or others. The selected gene list included, for instance, the cytoskeletal regulatory proteins Zyxin [54] and Moesin [55], and chromatin architectural regulators such as the HMGA2 protein, which has been then shown to be a central regulator of the TGF-β/Smad-induced EMT [31].

Complementary to gene expression profiles obtained by microarray analysis, additional high-throughput methods can be used to study the EMT process triggered by TGF-β signaling. Chromatin immunoprecipitation techniques coupled to massive sequencing (ChIP-seq), which allows the identification of DNA-binding sites of specific transcription factors, was applied to determine the regulatory sequences occupied by nuclear TGF-β-signaling factors, such as Smad4 [56]. Alternatively, the whole transcriptome shotgun sequencing (RNA-Seq) approach can be applied to determine, for instance, alternative splicing events occurring during EMT [57], or non-protein coding RNAs. Transcriptomic analysis of microRNA molecules (small noncoding RNA molecules regulating the activity of specific mRNA targets) has brought a major breakthrough in the understanding of the molecular events controlling epithelial cell plasticity [58].

3.6 Quantitative Real-Time Reverse-Transcription Polymerase Chain Reaction

Following reverse transcription of mRNA into cDNA, real-time quantitative PCR combines PCR amplification and detection into a single step. This method is based on the use of fluorescent dyes that label DNA amplicons during thermal cycling. Real-time PCR instruments measure the accumulation of fluorescent signal during the exponential phase of the reaction in order to get precise quantification of PCR products and objective data analysis. Quantitative PCR reaction products can be fluorescently labeled using the SYBR®Green dye, a molecule emitting fluorescence when specifically bound to double-stranded DNA. As in the PCR reaction, the target-template is the limiting factor, compared to reagents (dNTPs, primers, and DNA polymerase) that are in large molar excess, the fluorescent signal is directly proportional to the amount of target in the input sample.

Alternatively, quantitative PCR reaction products can be labeled using a fluorescent probe-based technology, the TaqMan® technique, using the 5′ nuclease activity of Taq DNA polymerase. The Taqman® approach will not be described here. However, interesting information regarding advantages and drawbacks of these two chemistries may be found at http://www.appliedbiosystems.com/absite/us/en/home/applications-technologies/real-time-pcr/taqman-and-sybr-green-chemistries.html.

Experimental procedures:

1. Cells are cultured in 35 mm-diameter or 6-well culture dishes and treated with TGF-β1 or vehicle (control condition) as described in the cell culture method described above.
2. Remove culture medium and carefully rinse the cell monolayer twice with 5 mL of PBS.
3. Isolate the total RNA from cell monolayers using the RNeasy® Plus kit (Qiagen), or the method of your choice, according to the manufacturer's protocol (*see* **Note 11**).
4. Incubate total RNA (0.5–1 μg) at 65 °C for 5 min in order to denaturate RNA-secondary structures and to ensure optimal

reverse transcription. Maintain RNA samples on ice for 1 min, then pulse and incubate on ice while adding the reagents.

5. Perform reverse-transcription in a 40 μL reaction mixture containing 500 ng of anchored oligo(dT_{17}) primers (5′-AGCT17-3′) or 600 ng of random primers (Invitrogen), at 500 μM concentration of each dNTP, 10 mM dithiothreitol, 40 U of RNaseOUT (Invitrogen), and 200 U of SuperScript II reverse transcriptase (Invitrogen) in the buffer supplied by the manufacturer (First-Strand buffer, Invitrogen). The RT reaction is performed in a thermocycler at 42 °C for 1 h, followed by an enzyme-denaturating step at 70 °C for 15 min (*see* **Note 12**).
6. Perform quantitative real-time PCR from tenfold diluted cDNA in a 40 μL reaction mixture including specific reverse and forward primers (300 nM each, Tables 4 and 5 for human and mouse species, respectively) and reagents supplied in the "Master mix" of your choice.
7. Run the amplification reaction in triplicate in real-time PCR instruments with thermal cycling and fluorescence detection capabilities, using the following conditions: an initial denaturation step consisting of 2 min at 50 °C and 10 min at 95 °C, followed by 40 cycles of 15 s at 95 °C and an annealing/elongation step of 1 min at 60 °C. For each sample, include controls where reverse-transcriptase is omitted during the first-strand cDNA synthesis reaction, as an indicator of genomic DNA contamination. Additionally, for each set of primers, include controls where cDNA is replaced by deionized water in order to exclude possible DNA contamination coming from PCR reagents.
8. Select the threshold line (i.e., level of fluorescence detection above the background signal and thus located in the logarithmic phase of amplification) and determine the corresponding cycle threshold (Ct value) for each sample. Ct value is the PCR cycle at which the sample reaches the level of fluorescence selected by the threshold line (*see* **Note 13**).
9. Levels of target gene expression in each sample can be determined with the comparative Ct method (*see* **Note 14**), using the *GAPDH*, *HPRT* (hypoxanthine phosphoribosyltransferase) or *β2-microglobulin* genes as endogenous controls and relative to a calibrator. For each condition, the ground measurement (for instance without TGF-β1) is set to 1 (Fig. 3). The amount of target is then given by the following arithmetic formula: $2^{-\Delta\Delta Ct}$, where ΔCt represents the difference in threshold cycles for target and reference gene (i.e., *GAPDH*) and ΔΔCt is the difference in threshold cycles for ΔCt (i.e., target normalized to reference) and calibrator (i.e., ground condition).

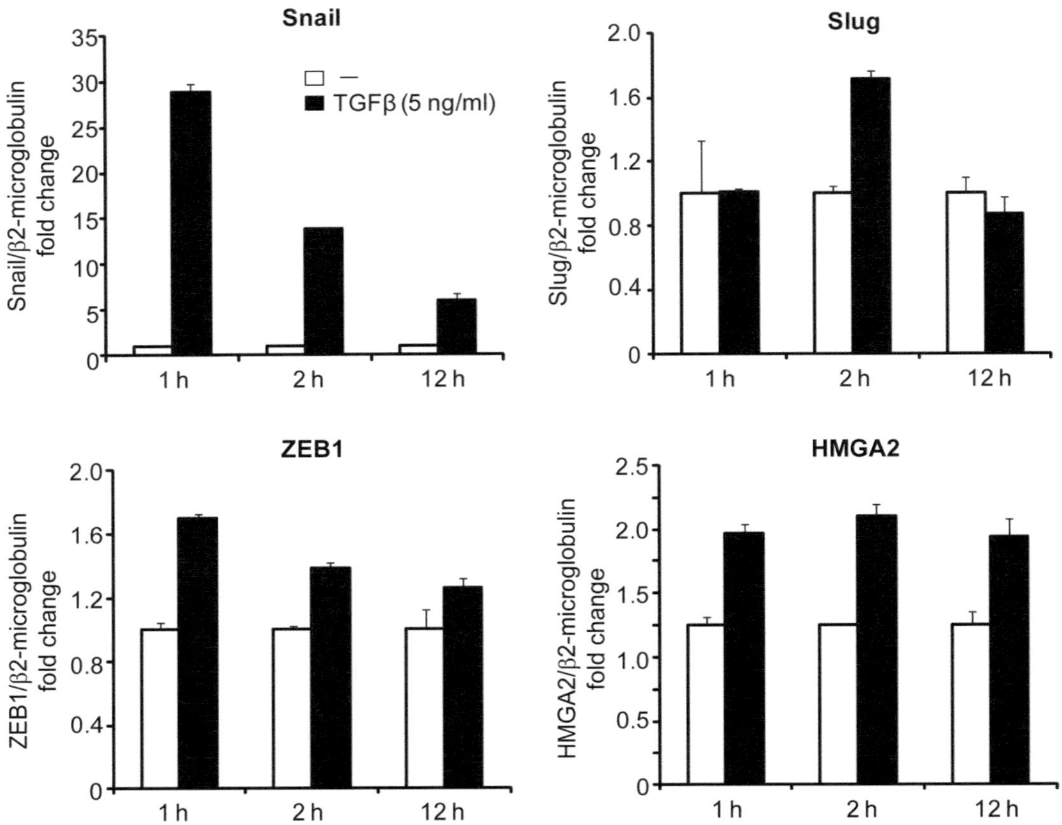

Fig. 3 Representative quantitative real-time RT-qPCR assay in mouse mammary epithelial cells undergoing EMT. NMuMG cells were treated with vehicle or 5 ng/mL TGF-β1 for the indicated time points prior to isolation of total RNA. RT-qPCR analysis was performed for *Snail*, *Slug*, *ZEB1* and *Hmga2* mRNAs using cDNA-specific primers and expression of each gene was normalized to the expression of the housekeeping *β2-microglobulin* gene. Relative expression of each mRNA in the absence of TGF-β1 stimulation at 1 h is set to 1 and all other samples are expressed relative to this reference point as fold-change. Control, unstimulated samples are graphed in *white* and TGF-β1-stimulated samples are graphed in *black* bars. Bars represent average values from triplicate determination and the corresponding standard error is marked

For more information regarding relative quantitation of gene expression, the investigators are invited to read "*ABI PRISM® 7700 Sequence Detection System User Bulletin #2: Relative Quantitation of Gene Expression*," available on the Applied Biosytems website (http://www.appliedbiosystems.com).

3.7 Mammosphere (Stem Cell) Culture Assay

The mammosphere assay is a common method that is used to quantify stem cell activity and self-renewal. It is based on the property of stem cells to survive and produce spherical colonies when cultured in suspension. The mammosphere assay can be used to study stem cell properties in both normal and cancer cells. Most protocols are based on the method developed by Dontu et al. [59].

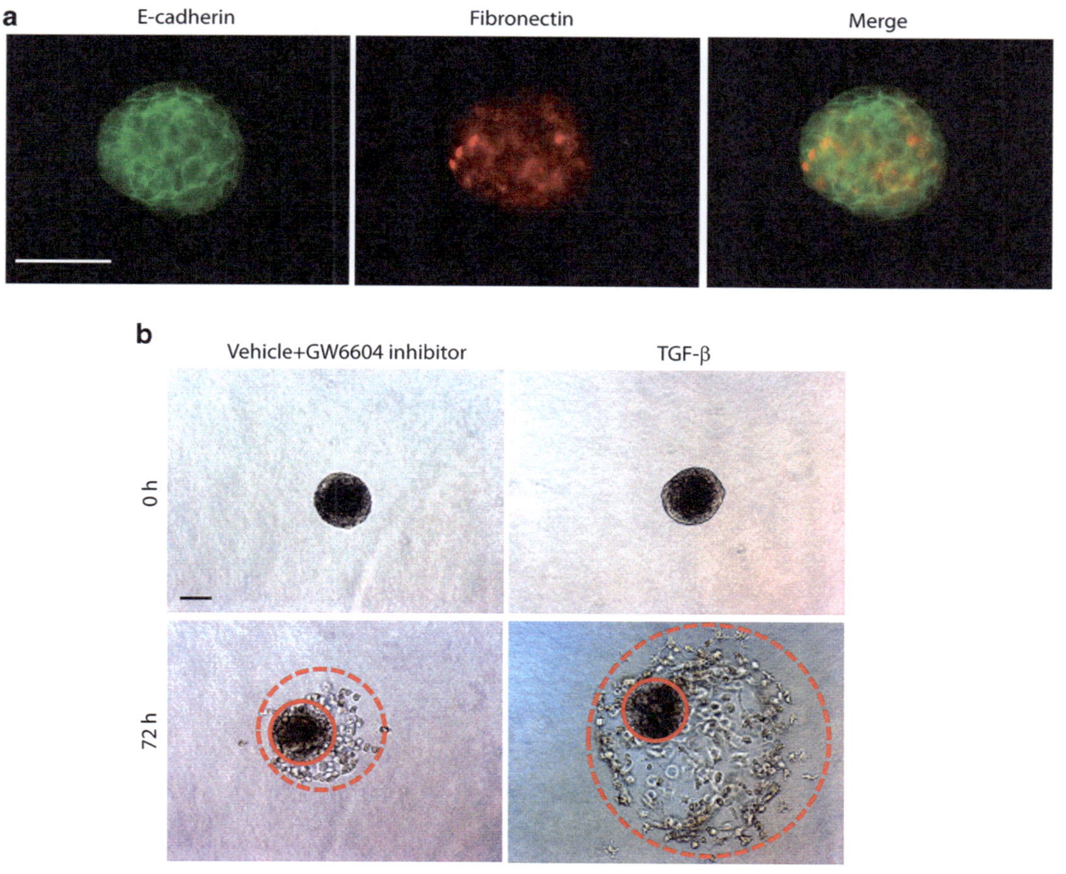

Fig. 4 (a) Representative fluorescent images of human mammary epithelial cells cultured as mammospheres. A single mammosphere is shown after fixation, permeabilization, and staining for E-cadherin (*green*) and fibronectin (*red*). E-cadherin decorates cell–cell junctions and intra-cytoplasmic areas, while fibronectin stains fewer cells, some with strong expression and some with weaker expression. The merged picture shows the overlay of the two protein markers. (b) Representative phase-contrast images of human mammary epithelial cells cultured as organoids embedded into a collagen-based matrix. At time 0 ($t=0$) only tight spherical organoids can be observed. At 72 h ($t=72$ h) the invasion of cells into the matrix is obvious. Cells cultured in the presence of vehicle and a TGF-β type I receptor kinase inhibitor (GW6604, 2 μM) show weak level of invasiveness. Cells cultured in the presence of TGF-β1 show enhanced invasiveness. The degree of invasiveness is calculated by integrating the two-dimensional area of invasion that lies between the two circular lines (contours). Bars indicate 50 μm

A representative figure of a mammosphere generated by Ras-transformed MCF-10AneoT cells (MIIclone) is shown in Fig. 4a.

Experimental procedures:

1. Culture NMuMG, MCF-10A (and oncogenically transformed derivatives) or human breast cancer MDA-MB-231 cells to 70–80 % and detach according to standard protocols.
2. Centrifuge and remove serum-containing medium and trypsin.

3. Add stem cell medium and dissociate cells mechanically by pipetting.
4. Sieve the cell suspension through a 40 μm cell strainer to make a single cell suspension. The single-cellularity of the suspension can be confirmed by looking at the cells under a microscope. If the cells are clumped together, mechanical pipetting and sieving will need to be repeated.
5. Count cells using a cell counter.
6. Seed cells at a concentration of 1–5000 cells/well in the 24-well ultra-low attachment multi-well culture plates (see Notes 15–17)
7. Incubate for 7–14 days in a humidified atmosphere at 37 °C and 5 % CO2.
8. Count the number of spheres larger than 50–100 μm in diameter under the microscope (or by using image analysis software such as ImageJ).
9. Mammospheres are then trypsinized, sieved and single cells are counted and prepared as above to generate secondary mammospheres and successively, tertiary and further passages of mammospheres.

3.8 Invasion Assay of Three-Dimensional Cell Culture

Three-dimensional culture models more accurately reflect in vivo processes than traditional two-dimensional culture systems. The invasion assay of three-dimensional culture model is a method to measure the cell motility and invasive capabilities of cultured multicellular spheroids [37]. In this assay, cells are first plated under non-adherent conditions in round-shaped suspension plates to obtain three-dimensional cell agglomerates. After allowing the cells to aggregate into multicellular organoids, the organoids can be removed and embedded into a collagen matrix in which cells emanating from the organoid are able to invade. A representative figure of MCF-10CA1a.cl1 cells (MIV clone) invading the collagen matrix is shown in Fig. 4b.

Experimental procedures:

1. Prepare the three dimensional culture. Harvest cells from a dish, count and prepare a suspension of 1000 cells per 100 μL (e.g., for one 96-well-plate 120,000 cells in 12 mL of medium with 20 % of Methylcellulose solution). Add 100 μL per well of the suspension to a 96-well plate with round bottom. Check the cells for forming organoids (this takes 1–2 days) and mark the wells containing well-sized organoids.
2. Prepare the neutralized, isotonic collagen solution for coating plates (Day 0). Working on ice, add 1 mL of 10× PBS and 1 mL of the filter-sterilized 0.1 N NaOH to 8 mL of type I collagen (PureCol) to obtain a collagen suspension of 3.3 % (wt/

vol). Use the sterilized 0.1 N HCl to bring the pH of the solution to 7.4. Store the solution on ice prior to using (it will be stable for a few hours).

3. Coat the 96-well-plate with collagen. Add 50 μL of the neutralized collagen solution to each well of a 96-well-plate (flat bottom) and incubate at 37 °C until solidified (30 min) (see Note 18).
4. Prepare the methocel-collagen-ligand/inhibitor solutions to be used for organoid embedding. For each ligand/inhibitor combination to be used, pipette 0.5 mL of collagen solution to a precooled 1.5 mL tube on ice. Add the ligands/inhibitors (keep in mind that the final volume will be 1.0 mL) and mix by vortexing. Add 0.5 mL of methylcellulose and mix by vortexing.
5. Embed the organoids. Take well-sized organoids one-by-one from the 96-well-plate by using a 200 μL pipette (adjusted to 100 μL volume) with a cut Pipetman tip (wide orifice). Pour out the medium to an empty dish (24/48-well-plate) without losing the organoid from the tip. Suck the collagen-methocel-ligand/inhibitor solution with the tip containing the organoid. Carefully pour the solution to a well in a pre-coated 96-well-plate (see Note 19).
6. Place the plates in the incubator and take pictures of the organoids every day (at the lowest magnification) for up to 4 days. Cell invasiveness is measured by calculating the integrated area between the contour that surrounds the original organoid and the contour defined by the most-distant cells away from the center of the organoid (*see* Fig. 4b). Multiple replicates are needed for statistical analysis.

3.9 Invasion Assay in Matrigel

The invasion assay in Matrigel measures the capacity of cells to invade through a barrier of ECM, in vitro [15, 17, 60]. In this experiment, cells are added to the upper chamber of a Matrigel-coated insert and a stimulus is added to the lower chamber. Invasive cells will pass through the Matrigel barrier and the porous membrane to the underside of the insert. The cells that invade through the Matrigel can be quantified by staining and counting the cells under the microscope.

Experimental procedures:

1. Trypsinize the cells and suspend in 5 mL (per 10 cm dish) of medium containing 0.1 % BSA and trypsin inhibitor.
2. Spin down cells and resuspend in 5 mL of medium with 0.1 % BSA.
3. Count the cells.

4. Dilute the cells in 0.1 % BSA to a concentration of 1×10^5 cells/mL (*see* **Note 20**).
5. Add 500 μL per well of chemoattractant medium (such as 5 % FBS) to 24-well culture dishes.
6. Place the BioCoat Matrigel invasion chambers on top of the culture dishes from **step 5**, ensuring no air bubbles are created.
7. Add 100 μL per well of the cell suspension to the top of the Matrigel chambers.
8. Put covers on the plates and incubate at 37 °C for 4–24 h (*see* **Note 20**).
9. Remove the cells from the upper side of the chamber by using a cotton swab and rinse with PBS.
10. Immerse the swabbed inserts into methanol for 10 min to fix the cells that have invaded onto the lower side of the transwell filter.
11. Stain the cells with Giemsa solution for 2–3 min.
12. Wash away excess stain with water and invert and air dry the stained inserts.
13. Use a microscope to visualize (and capture images) and count the cells that have invaded to the bottom side of the insert.

3.10 In Vitro Wound-Healing (Scratch) Assay

The in vitro wound-healing assay is a simple, effective and economical method to measure the migration of cultured cells [13, 14, 18]. The assay involves creating a cell-free area in a confluent monolayer of cells by scraping a line through the cells with a pipette, and then measuring the rate of cell migration into the cell-free area. The protocol below describes a basic approach for measuring cell migration using this technique. In certain experiments, it may be preferred to measure migration following the induction of EMT. In such experiments, cells can be grown to 50–70 % confluence and then be stimulated with TGF-β1 for 24–48 h, prior to performing the scratch assay.

Experimental procedures:

1. Plate cells into culture dishes and allow them to grow until they form a confluent monolayer.
2. Starve cells for 4–6 h (or overnight) in serum-free medium to arrest cell growth.
3. Use a p200 pipette tip to scratch in a straight line through the confluent cell monolayer, making a cell-free area for which the cells will migrate into (*see* **Note 21**).
4. Wash the cells twice in PBS to remove cell debris.

5. Remove PBS and add the appropriate culture medium to be used for the scratch assay (i.e., serum-free medium, or medium that contains a growth factor such as TGF-β1 or BMP-7, *see* **Note 22**).
6. Mark the underside of the culture dish with a sharpie pen so that you can photograph the same area of cells during the migration experiment.
7. Use a phase-contrast microscope with an attached camera to take a photograph of the scratched area.
8. Return cells to the incubator for the appropriate amount of time (12–48 h, *see* **Note 23**). It is best to choose a timeframe that allows the fastest migrating cells to just achieve full closure of the scratch.
9. After incubation for the appropriate amount of time, a second photograph of the same scratch area is taken using the phase-contrast microscope by aligning the reference point created in **step 6**.
10. The rate of migration can be quantified by using image analysis software such as Image Pro-Plus (Media Cybernetics). Simply, the cell-free area created in **step 7** is measured and compared to the cell-free area that remains after the incubation period (**step 9**).

4 Notes

1. For the antibody–antigen reaction of most of the junctional proteins (E-cadherin, ZO-1) and intermediate filaments (vimentin, cytokeratin), an alcoholic precipitation-based fixation is more suitable. Indeed, paraformaldehyde creates intramolecular or intermolecular cross-links that may interfere with the antibody recognition site. Moreover, methanol-induced precipitation induces a partial protein denaturation, thus allowing some cryptic antigens to be unmasked, allowing better reactivity with certain antibodies, whereas it can destroy other more conformational epitopes, leading to inhibition of reactivity with certain antibodies. For methanol fixation, proceed with sample preparations up to **step 4** and fix cells by adding 1 mL of ice-cold methanol and incubate for 5 min at −20 °C. Rinse the cells three times with 2 mL of PBS^{++} and then proceed from **step 7** to the end.
2. Incubation with antibodies should be done in a humid chamber to avoid evaporation.
3. Fluorescent-labeled secondary antibodies and phalloidin are highly light-sensitive. It is strongly recommended to maintain

them protected from light during labeling, washing and storage steps.

4. Specificity controls for the indirect IF should include stain using only the secondary antibody (i.e., in the absence of the primary antibody) to exclude fluorescence due to nonspecific binding of secondary antibodies.
5. Instead of incubation with DAPI, mounting media containing DAPI can be used.
6. DAB chromogen solution is a carcinogenic agent. Use gloves and appropriate protective wear when working. Inactivate the residues with bleach and discard them according to institutional guidelines.
7. Using β-actin protein as a loading control is not a reliable choice as dramatic reorganization of the actin microfilament cytoskeleton occurs during the EMT process and sometimes changes in total β-actin expression have been reported.
8. Polyadenylated tail-containing RNA molecules may be purified to enrich their input in mature messenger RNA.
9. Users may also include a "dye swap" experiment, i.e., an experiment where the sample pair is labeled with the opposite fluorophore (Cy3 versus Cy5) in order to estimate dye bias. Indeed, the users should be aware of the intensity difference between samples labeled with different dyes and whether this is really due to gene expression in the samples but not attributable to the intrinsic fluorescent intensity of the dyes instead.
10. The number of selected genes will be reduced with more stringent conditions (e.g., ≥ 2 and ≤ 0.5).
11. This kit is supplied with "gDNA eliminator" spin columns for effective elimination of genomic DNA. In other cases, DNAse I treatment of RNA samples, followed by heat-inactivation of the enzyme or RNA purification, is highly recommended when PCR reactions are planned, in order to avoid amplification of genomic DNA. The design of primers used in quantitative PCR is a critical parameter for optimal gene expression analysis. Determinant parameters are the primer composition (% of GC content, length, melting temperature), their propensity to form no or very unstable secondary structure, and the size of the amplicon (ideally 50–150 base pairs). To avoid genomic DNA (emanating from RNA samples) amplification during PCR, primers should ideally be designed either to flank a long intron that is not present in the mature mRNA sequence or to span an exon-exon junction. Selected primer pairs should be screened for regions of similarity between biological sequences using a BLAST search (http://blast.ncbi.nlm.nih.

gov/), to verify that the primers do not amplify other sequences than those expected. Finally, the PCR fragment may be sequenced to ensure that the desired cDNA fragment is amplified.

12. A first step of annealing for 10 min at 25 °C is necessary when using random primers.
13. As SYBR® Green is a nonspecific dye that will detect any double-stranded DNA, the users should verify that the PCR produces only the desired amplicon. Perform melt (or dissociation) curve analysis in order to determine the number and approximate size of products. An assay with high specificity should give a single peak at a high temperature (>80 °C) in all reactions and nothing, or very little, should be detected in the no-template controls. Conversely, if several amplicons of different sizes are amplified, their presence is detected by the formation of several peaks at a high temperature (as they should melt at different temperatures).
14. The comparative Ct method relies on the condition that, for each set of primers, every target molecule is copied once at each cycle. This implies that amplification is performed using optimal conditions and highly efficient PCR primers. Consequently, efficiency of amplification should be assessed for each set of primers by realizing reference curves of amplification using serially diluted cDNAs. After having plotted each Ct values versus the log of nucleic acid input (or dilution), the reaction efficiency is calculated from the slope of the line using the following equation: $10^{(-1/\text{slope})} - 1$. If the PCR is 100 % efficient, meaning that the amount of PCR product will double with each cycle, the slope of the standard curve will be −3.33. A slope between −3.9 and −3.0 (i.e., corresponding to 80–110 % efficiency, respectively) is generally acceptable.
15. The number of cells will need to be optimized as it will vary depending on the cell line used.
16. It is advised to avoid using the outside wells of the plate as the medium will evaporate faster. It is also advised to fill the remaining wells with PBS to prevent drying.
17. It is recommended to simultaneously perform a viability assay on day 0 to verify that the initial number of cells used is the same between the different samples. For this purpose MTS (3-(4,5-dimethylthiazol-2-yl)-5-(3-carboxymethoxyphenyl)-2-(4-sulfophenyl)-2H-tetrazolium) assays (CellTiter-96 Aqueous-One-Solution Assay, Promega Inc.) are usually performed as described in the manufacturer's manual.

18. Avoid air bubbles. These can sometimes be removed by using a pipette tip and it is easier to do when the collagen has solidified (30 min after the plate is placed in the incubator).
19. Methocel solution (for 500 mL): autoclave 6 g of methylcellulose (Sigma-Aldrich) in a 500 mL flask containing a magnetic stirrer. Dissolve the methylcellulose in 250 mL of preheated medium w/o additives (60 °C) for 1 h (using a magnetic stirrer). Add 250 mL of medium (room temperature) containing 2× additives and mix overnight (4 °C). Aliquot into 50 mL tubes and clear by centrifugation (5000 × *g*, 2 h, room temperature). Transfer the clear, highly viscous solution to a new tube (about 90–95 % of the stock solution). Store the solution at 4 °C until use.
20. The optimal concentration of cells as well as length of time for migration will vary depending on the cell type and experimental conditions used. This will need to be optimized.
21. It is important to create scratches of equal width. Up to three different scratches can be easily performed in the same well of a six-well culture dish. This is helpful for statistical comparisons when the migration rates are quantified.
22. It is generally advised to use a lower concentration of serum than full growth medium (or use serum-free medium) to minimize the effect caused by cell proliferation.
23. The timeframe for migration will need to be optimized depending on the cell type and serum concentrations. This can be monitored by checking the cells periodically under a microscope.

Acknowledgements

We acknowledge funding by the Ludwig Institute for Cancer Research, the Swedish Cancer Society and the Swedish Research Council (A.M.). We also acknowledge funding by the "Ligue Nationale Contre le Cancer" (LNCC) (L.A.), the "Comité du Rhône de la Ligue Nationale Contre le Cancer" (U.V.) and the "Institut National du Cancer" (INCa) (L.B.).

We thank all past and present members of our groups for their contributions to the scientific work emanating from our laboratories. This chapter summarizes work from a selected number of published papers, however, due to space limitations, we have been unable to include all relevant publications in our discussion. We apologize to those authors whose relevant work has not been included in this review chapter.

We declare no conflict of interest.

References

1. Hay ED (1995) An overview of epithelio-mesenchymal transformation. Acta Anat 154(1):8–20
2. Thiery JP, Acloque H, Huang RY, Nieto MA (2009) Epithelial-mesenchymal transitions in development and disease. Cell 139(5):871–890
3. Savagner P, Kusewitt DF, Carver EA, Magnino F, Choi C, Gridley T et al (2005) Developmental transcription factor slug is required for effective re-epithelialization by adult keratinocytes. J Cell Physiol 202(3):858–866
4. Iwano M, Plieth D, Danoff TM, Xue C, Okada H, Neilson EG (2002) Evidence that fibroblasts derive from epithelium during tissue fibrosis. J Clin Invest 110(3):341–350, PMCID: 151091
5. Mani SA, Guo W, Liao MJ, Eaton EN, Ayyanan A, Zhou AY et al (2008) The epithelial-mesenchymal transition generates cells with properties of stem cells. Cell 133(4):704–715, PMCID: 2728032
6. Morel AP, Lievre M, Thomas C, Hinkal G, Ansieau S, Puisieux A (2008) Generation of breast cancer stem cells through epithelial-mesenchymal transition. PLoS One 3(8), e2888, PMCID: 2492808
7. Ansieau S, Bastid J, Doreau A, Morel AP, Bouchet BP, Thomas C et al (2008) Induction of EMT by twist proteins as a collateral effect of tumor-promoting inactivation of premature senescence. Cancer Cell 14(1):79–89
8. Vincent T, Neve EP, Johnson JR, Kukalev A, Rojo F, Albanell J et al (2009) A SNAIL1-SMAD3/4 transcriptional repressor complex promotes TGF-β mediated epithelial-mesenchymal transition. Nat Cell Biol 11(8):943–950
9. Miettinen PJ, Ebner R, Lopez AR, Derynck R (1994) TGF-β induced transdifferentiation of mammary epithelial cells to mesenchymal cells: involvement of type I receptors. J Cell Biol 127(6 Pt 2):2021–2036, PMCID: 2120317
10. Piek E, Moustakas A, Kurisaki A, Heldin C-H, ten Dijke P (1999) TGF-β type I receptor/ALK-5 and Smad proteins mediate epithelial to mesenchymal transdifferentiation in NMuMG breast epithelial cells. J Cell Sci 112(Pt 24):4557–4568
11. Valcourt U, Kowanetz M, Niimi H, Heldin C-H, Moustakas A (2005) TGF-β and the Smad signaling pathway support transcriptomic reprogramming during epithelial-mesenchymal cell transition. Mol Biol Cell 16(4):1987–2002, PMCID: 1073677
12. Mu Y, Sundar R, Thakur N, Ekman M, Gudey SK, Yakymovych M et al (2011) TRAF6 ubiquitinates TGFβ type I receptor to promote its cleavage and nuclear translocation in cancer. Nat Commun 2:330, PMCID: 3113296
13. Bakin AV, Tomlinson AK, Bhowmick NA, Moses HL, Arteaga CL (2000) Phosphatidylinositol 3-kinase function is required for transforming growth factor β-mediated epithelial to mesenchymal transition and cell migration. J Biol Chem 275(47):36803–36810
14. Bhowmick NA, Ghiassi M, Bakin A, Aakre M, Lundquist CA, Engel ME et al (2001) Transforming growth factor-β1 mediates epithelial to mesenchymal transdifferentiation through a RhoA-dependent mechanism. Mol Biol Cell 12(1):27–36, PMCID: 30565
15. Bhowmick NA, Zent R, Ghiassi M, McDonnell M, Moses HL (2001) Integrin β 1 signaling is necessary for transforming growth factor-β activation of p38^{MAPK} and epithelial plasticity. J Biol Chem 276(50):46707–46713
16. Cicchini C, Laudadio I, Citarella F, Corazzari M, Steindler C, Conigliaro A et al (2008) TGFβ-induced EMT requires focal adhesion kinase (FAK) signaling. Exp Cell Res 314(1):143–152
17. Bakin AV, Rinehart C, Tomlinson AK, Arteaga CL (2002) p38 mitogen-activated protein kinase is required for TGFβ-mediated fibroblastic transdifferentiation and cell migration. J Cell Sci 115(Pt 15):3193–3206
18. Galliher AJ, Schiemann WP (2007) Src phosphorylates Tyr284 in TGF-β type II receptor and regulates TGF-β stimulation of p38 MAPK during breast cancer cell proliferation and invasion. Cancer Res 67(8):3752–3758
19. Wendt MK, Tian M, Schiemann WP (2012) Deconstructing the mechanisms and consequences of TGF-β-induced EMT during cancer progression. Cell Tissue Res 347(1):85–101
20. Ozdamar B, Bose R, Barrios-Rodiles M, Wang HR, Zhang Y, Wrana JL (2005) Regulation of the polarity protein Par6 by TGFβ receptors controls epithelial cell plasticity. Science 307(5715):1603–1609
21. Batlle E, Sancho E, Franci C, Dominguez D, Monfar M, Baulida J et al (2000) The transcription factor snail is a repressor of E-cadherin gene expression in epithelial tumour cells. Nat Cell Biol 2(2):84–89
22. Cano A, Perez-Moreno MA, Rodrigo I, Locascio A, Blanco MJ, del Barrio MG et al (2000) The transcription factor snail controls epithelial-mesenchymal transitions by repressing E-cadherin expression. Nat Cell Biol 2(2):76–83

23. Hajra KM, Chen DY, Fearon ER (2002) The SLUG zinc-finger protein represses E-cadherin in breast cancer. Cancer Res 62(6):1613–1618
24. Bolos V, Peinado H, Perez-Moreno MA, Fraga MF, Esteller M, Cano A (2003) The transcription factor Slug represses E-cadherin expression and induces epithelial to mesenchymal transitions: a comparison with Snail and E47 repressors. J Cell Sci 116(Pt 3):499–511
25. Eger A, Aigner K, Sonderegger S, Dampier B, Oehler S, Schreiber M et al (2005) ΔEF1 is a transcriptional repressor of E-cadherin and regulates epithelial plasticity in breast cancer cells. Oncogene 24(14):2375–2385
26. Comijn J, Berx G, Vermassen P, Verschueren K, van Grunsven L, Bruyneel E et al (2001) The two-handed E box binding zinc finger protein SIP1 downregulates E-cadherin and induces invasion. Mol Cell 7(6):1267–1278
27. Perez-Moreno MA, Locascio A, Rodrigo I, Dhondt G, Portillo F, Nieto MA et al (2001) A new role for E12/E47 in the repression of E-cadherin expression and epithelial-mesenchymal transitions. J Biol Chem 276(29):27424–27431
28. Yang J, Mani SA, Donaher JL, Ramaswamy S, Itzykson RA, Come C et al (2004) Twist, a master regulator of morphogenesis, plays an essential role in tumor metastasis. Cell 117(7):927–939
29. Kowanetz M, Valcourt U, Bergstrom R, Heldin CH, Moustakas A (2004) Id2 and Id3 define the potency of cell proliferation and differentiation responses to transforming growth factor β and bone morphogenetic protein. Mol Cell Biol 24(10):4241–4254, PMCID: 400464
30. Kondo M, Cubillo E, Tobiume K, Shirakihara T, Fukuda N, Suzuki H et al (2004) A role for Id in the regulation of TGF-β-induced epithelial-mesenchymal transdifferentiation. Cell Death Differ 11(10):1092–1101
31. Thuault S, Valcourt U, Petersen M, Manfioletti G, Heldin C-H, Moustakas A (2006) Transforming growth factor-β employs HMGA2 to elicit epithelial-mesenchymal transition. J Cell Biol 174(2):175–183, PMCID: 2064178
32. Thuault S, Tan E-J, Peinado H, Cano A, Heldin C-H, Moustakas A (2008) HMGA2 and Smads co-regulate SNAIL1 expression during induction of epithelial-to-mesenchymal transition. J Biol Chem 283(48):33437–33446, PMCID: 2662269
33. Tan E-J, Thuault S, Caja L, Carletti T, Heldin C-H, Moustakas A (2012) Regulation of transcription factor Twist expression by the DNA architectural protein high mobility group A2 during epithelial-to-mesenchymal transition. J Biol Chem 287(10):7134–7145, PMCID: 3293571
34. Moustakas A, Heldin C-H (2012) Induction of epithelial-mesenchymal transition by transforming growth factor β. Semin Cancer Biol 22(5–6):446–454
35. Massagué J (2012) TGFβ signalling in context. Nat Rev Mol Cell Biol 13(10):616–630
36. Nieto MA (2011) The ins and outs of the epithelial to mesenchymal transition in health and disease. Annu Rev Cell Dev Biol 27:347–376
37. Sundqvist A, Zieba A, Vasilaki E, Herrera Hidalgo C, Söderberg O, Koinuma D et al (2013) Specific interactions between Smad proteins and AP-1 components determine TGFβ-induced breast cancer cell invasion. Oncogene 32(31):3606–3615
38. Brown KA, Aakre ME, Gorska AE, Price JO, Eltom SE, Pietenpol JA et al (2004) Induction by transforming growth factor-β1 of epithelial to mesenchymal transition is a rare event in vitro. Breast Cancer Res 6(3):R215–R231, PMCID: 400675
39. Masuda A, Kondo M, Saito T, Yatabe Y, Kobayashi T, Okamoto M et al (1997) Establishment of human peripheral lung epithelial cell lines (HPL1) retaining differentiated characteristics and responsiveness to epidermal growth factor, hepatocyte growth factor, and transforming growth factor β1. Cancer Res 57(21):4898–4904
40. Cui W, Fowlis DJ, Bryson S, Duffie E, Ireland H, Balmain A et al (1996) TGFβ1 inhibits the formation of benign skin tumors, but enhances progression to invasive spindle carcinomas in transgenic mice. Cell 86(4):531–542
41. Portella G, Cumming SA, Liddell J, Cui W, Ireland H, Akhurst RJ et al (1998) Transforming growth factor β is essential for spindle cell conversion of mouse skin carcinoma in vivo: implications for tumor invasion. Cell Growth Differ 9(5):393–404
42. Oft M, Heider KH, Beug H (1998) TGFβ signaling is necessary for carcinoma cell invasiveness and metastasis. Curr Biol 8(23):1243–1252
43. Peinado H, Quintanilla M, Cano A (2003) Transforming growth factor β-1 induces snail transcription factor in epithelial cell lines: mechanisms for epithelial mesenchymal transitions. J Biol Chem 278(23):21113–21123
44. Huber MA, Kraut N, Beug H (2005) Molecular requirements for epithelial-mesenchymal transition during tumor progression. Curr Opin Cell Biol 17(5):548–558
45. Hesling C, Fattet L, Teyre G, Jury D, Gonzalo P, Lopez J et al (2011) Antagonistic regulation of EMT by TIF1γ and Smad4 in mammary epithelial cells. EMBO Rep 12(7):665–672, PMCID: 3128966

46. Scheel C, Eaton EN, Li SH, Chaffer CL, Reinhardt F, Kah KJ et al (2011) Paracrine and autocrine signals induce and maintain mesenchymal and stem cell states in the breast. Cell 145(6):926–940
47. Vincent DF, Yan KP, Treilleux I, Gay F, Arfi V, Kaniewski B et al (2009) Inactivation of TIF1γ cooperates with Kras to induce cystic tumors of the pancreas. PLoS Genet 5(7), e1000575, PMCID: 2706992
48. Vincent DF, Gout J, Chuvin N, Arfi V, Pommier RM, Bertolino P et al (2012) Tif1γ suppresses murine pancreatic tumoral transformation by a Smad4-independent pathway. Am J Pathol 180(6):2214–2221, PMCID: 3378851
49. Xie L, Law BK, Aakre ME, Edgerton M, Shyr Y, Bhowmick NA et al (2003) Transforming growth factor β-regulated gene expression in a mouse mammary gland epithelial cell line. Breast Cancer Res 5(6):R187–R198, PMCID: 314403
50. Jechlinger M, Grunert S, Tamir IH, Janda E, Ludemann S, Waerner T et al (2003) Expression profiling of epithelial plasticity in tumor progression. Oncogene 22(46):7155–7169
51. Levy L, Hill CS (2005) Smad4 dependency defines two classes of transforming growth factor β (TGF-β) target genes and distinguishes TGF-β-induced epithelial-mesenchymal transition from its antiproliferative and migratory responses. Mol Cell Biol 25(18):8108–8125, PMCID: 1234333
52. Gervasi M, Bianchi-Smiraglia A, Cummings M, Zheng Q, Wang D, Liu S et al (2012) JunB contributes to Id2 repression and the epithelial-mesenchymal transition in response to transforming growth factor-β. J Cell Biol 196(5):589–603, PMCID: 3307698
53. Moreno-Bueno G, Cubillo E, Sarrio D, Peinado H, Rodriguez-Pinilla SM, Villa S et al (2006) Genetic profiling of epithelial cells expressing E-cadherin repressors reveals a distinct role for Snail, Slug, and E47 factors in epithelial-mesenchymal transition. Cancer Res 66(19):9543–9556
54. Mori M, Nakagami H, Koibuchi N, Miura K, Takami Y, Koriyama H et al (2009) Zyxin mediates actin fiber reorganization in epithelial-mesenchymal transition and contributes to endocardial morphogenesis. Mol Biol Cell 20(13):3115–3124, PMCID: 2704162
55. Haynes J, Srivastava J, Madson N, Wittmann T, Barber DL (2011) Dynamic actin remodeling during epithelial-mesenchymal transition depends on increased moesin expression. Mol Biol Cell 22(24):4750–4764, PMCID: 3237619
56. Koinuma D, Tsutsumi S, Kamimura N, Imamura T, Aburatani H, Miyazono K (2009) Promoter-wide analysis of Smad4 binding sites in human epithelial cells. Cancer Sci 100(11):2133–2142
57. Shapiro IM, Cheng AW, Flytzanis NC, Balsamo M, Condeelis JS, Oktay MH et al (2011) An EMT-driven alternative splicing program occurs in human breast cancer and modulates cellular phenotype. PLoS Genet 7(8), e1002218, PMCID: 3158048
58. Gregory PA, Bert AG, Paterson EL, Barry SC, Tsykin A, Farshid G et al (2008) The miR-200 family and miR-205 regulate epithelial to mesenchymal transition by targeting ZEB1 and SIP1. Nat Cell Biol 10(5):593–601
59. Dontu G, Abdallah WM, Foley JM, Jackson KW, Clarke MF, Kawamura MJ et al (2003) In vitro propagation and transcriptional profiling of human mammary stem/progenitor cells. Genes Dev 17(10):1253–1270, PMCID: 196056
60. Gal A, Sjöblom T, Fedorova L, Imreh S, Beug H, Moustakas A (2008) Sustained TGF β exposure suppresses Smad and non-Smad signalling in mammary epithelial cells, leading to EMT and inhibition of growth arrest and apoptosis. Oncogene 27(9):1218–1230

Chapter 10

In Vitro Th Differentiation Protocol

Takashi Sekiya and Akihiko Yoshimura

Abstract

$CD4^+$ T cells play central roles in adaptive immunity, driving appropriate immune responses to invading pathogens of diverse types. Four major $CD4^+$ T cell subsets, Th1, Th2, Th17, and Treg cells are differentiated from naïve $CD4^+$ T cells upon ligation of their T cell receptors with antigens, depending on the cytokines they receive. Th1 cells, which are induced by IL-12 and IFN-γ, mediate host defense against intracellular pathogens by exclusively expressing IFN-γ. Th2 cells, which are induced by IL4, secrete IL-4, IL-5, and IL-13, and protect hosts from helminths. IL-6 plus TGF-β induces Th17 cells, another Th subset identified relatively recently, express IL-17 and play important roles in the eradication of extracellular bacteria and fungi. Treg cells, which play central roles in immune suppression, are composed of either thymus-derived Treg cells (tTreg cells), which are directly developed from CD4-single positive (CD4-SP) cells in the thymus, or peripherally derived Treg cells (pTreg cells), which are induced by TGF-β plus IL-2 from naïve $CD4^+$ T cells. Although the regulated induction of Th cells results in proper eradication of pathogens, their excess activation results in various immune-associated diseases. For example, aberrant activation of Th1 and Th17 has been implicated in autoimmune diseases, excess Th2 activity causes atopic diseases, and impaired function of Treg cells due to abrogation of Foxp3 has been shown to cause fatal inflammatory disorders both in human and in mouse. The methods for in vitro differentiation of each Th subset described above are presented here. We hope these methods will facilitate understanding of differentiation and function of $CD4^+$ T cells and pathogenesis of various inflammatory diseases.

Key words Cytokines, Helper T cell subsets, TGF-β, IL-12, IL-23

1 Introduction

Helper T cells have been shown to play a central role in not only acquired immune responses but also in immune regulation. After emigrating from the bone marrow, thymocyte progenitors enter the thymus, and following positive selection, single positive ($CD4^+$ or $CD8^+$) cells migrate to the periphery as mature T cells. Naturally occurring $CD4^+CD25^+$ $Foxp3^+$ regulatory T cells also develop in the thymus from immature CD4-SP T cells (tTreg cells). Although their developmental mechanisms have not been completely clarified, they are thought to develop from progenitor thymocytes that

Xin-Hua Feng et al. (eds.), *TGF-β Signaling: Methods and Protocols*, Methods in Molecular Biology, vol. 1344,
DOI 10.1007/978-1-4939-2966-5_10,

express T cell antigen receptors (TCRs) autoreactive to complex of self-antigen and major histocompatibility complex presented on thymic antigen-presenting cells [1]. Recently, Nr4a family of orphan nuclear receptors have been shown to play a crucial role in tTreg development, possibly by being induced in the autoreactive thymocytes, whereby they transactivate *Foxp3* gene expression [2, 3]. After exiting the thymus, naïve T cells are activated by antigen-presenting cells and differentiate into effector or memory T cells.

Upon antigen stimulation, $CD4^+$ helper T (Th) cells follow distinct developmental pathways, attaining specialized properties and effector functions. Th cells are traditionally thought to differentiate into Th1 and Th2 cell subsets. Cells of the Th1 lineage, which evolved to enhance eradication of intracellular pathogens (e.g., intracellular bacteria, viruses, and some protozoa), are characterized by their production of interferon (IFN)γ, a potent activator of cell-mediated immunity [4, 5]; cells of the Th2 lineage, which evolved to enhance elimination of parasitic infections (e.g., helminths), are characterized by production of interleukin (IL)-4, IL-5, and IL-13, which are potent activators of B-cell immunoglobulin (Ig)E production, eosinophil recruitment, and mucosal expulsion mechanisms (mucous production and hypermotility) [6].

Recently, a novel helper T cell subset has been described that produces IL-17 (Th17) and has been identified as a subset distinct from Th1 or Th2 cells. Th17 cells secrete a distinctive set of immunoregulatory cytokines, including IL-17A, IL-17 F, IL-22, and IL-21. These cytokines collectively play roles in inflammation and autoimmunity and in elimination of extracellular bacterial and fungal pathogens. Murine autoimmune models, such as experimental autoimmune encephalitis (EAE) and collagen-induced arthritis (CIA), have been shown to be dependent on Th17 cells [7–9].

Th1 polarization is primarily driven by IL-12 and IFN-γ, while Th2 polarization is primarily driven by IL-4. These respective cytokines signal via STAT4, STAT1, and STAT6 to directly control the transcription factors T-bet and GATA3, which, in turn, determine Th1 and Th2 differentiation, respectively [10]. Th1 cells produce IFN-γ, which facilitates their differentiation while inhibiting IL-4-mediated Th2 differentiation [11]. Reciprocally, Th2 cells produce IL-4 and IL-10, which strongly inhibits IL-12/IFN-γ-driven Th1 differentiation.

The Th17 differentiation of naïve T cells is initiated by IL-6 and TGF-β [12–15]. In addition, IL-23, as well as IL-21, is thought to be a key cytokine for the maturation and/or maintenance of Th17 cells [16–18]. IL-6, IL-21, and IL-23 all activate STAT3, which is thought to be essential for Th17 differentiation. It has also been reported that STAT3 plays a critical role in the induction of the orphan nuclear receptor RORγt, which directs Th17 cell differentiation by inducing the IL-23 receptor. The critical role of STAT3 in Th17 differentiation was also confirmed in human patients lacking functional STAT3.

TGF-β also induces differentiation of naïve $CD4^+$ T cells into $Foxp3^+$ Tregs in the peripheral immune compartment (pTreg cells), employing Smad transcription factors as downstream effectors [19]. pTreg cells have been shown to be involved in immune homeostasis especially in intestines.

Considering the substantial contributions of Th differentiation to both physiology and pathophysiology as described above, it is important to reveal the molecular mechanisms which regulate Th differentiation from naïve $CD4^+$ T cells. To better understand the molecular mechanism of Th differentiation, it is important to establish in vitro differentiation systems from primary naïve $CD4^+$ T cells.

2 Materials

Cell Culture reagents:

RPMI1640 medium, 55 mM β-mercaptoethanol, and nonessential amino acids (NEAA) are purchased from Invitrogen. Penicillin/streptomycin sulfate is purchased from Nacalai Tesque.

Antibodies:

Anti-CD3ε (clone 2C11) and Anti-CD28 (clone PV1.17) are purchased from eBioscience. These antibodies are commonly used in all Th differentiation experiments.

2.1 Isolation of Naïve CD4+ T Cells

Antibodies

Biotin-conjugated antibodies against CD8α, B220, CD25, CD11b, CD11c, CD49b, and TER-119 are purchased from eBioscience. Streptavidin microbeads and CD62L microbeads are purchased from Miltenyi.

Reagents

Red blood cell lysis buffer: 150 mM NH_4Cl, 10 mM $KHCO_3$, and 500 μM EDTA

MACS buffer: 2 mM EDTA and 0.5 % BSA in 1× phosphate buffered saline (PBS).

MACS columns: Large depletion columns and large separation columns are purchased from Miltenyi.

2.2 Differentiation of Th1 Cells

Antibodies: IL-4 neutralizing antibodies (clone 11B11) are purchased from eBioscience.

Cytokines: IL-2 and IL-12 are purchased from Peprotech.

2.3 Differentiation of Th2 Cells

Antibodies: IFN-γ neutralizing antibodies (clone R4-6A2) are purchased from eBioscience.

Cytokines: IL-2 and IL-4 are purchased from Peprotech.

Th2 differentiating culture medium: (0.5 μg/mL anti-CD28 and 1 μg/mL anti-IFN-γ antibodies, 5 ng/mL IL-2, and 10 ng/mL IL-4 in RPMI1640 supplemented with 10 % FBS, NEAA, antibiotics, and 55 μM β-mercaptoethanol).

2.4 Differentiation of Th17 Cells

Antibodies: IFN-γ neutralizing antibodies (clone R4-6A2) and IL-4 (clone 11B11) neutralizing antibodies, and IL-2 neutralizing antibodies (clone JES6-1A12) are purchased from eBioscience.

Cytokines: Murine TGF-β1 and human IL-6 are purchased from R&D.

Th17 differentiating culture medium: 0.5 μg/mL anti-CD28, 1 μg/mL anti-IFN-γ, 1 μg/mL anti-IL-2, and 1 μg/mL anti-IL-4 antibodies, 20 ng/mL human IL-6, and 1 ng/mL TGF-β1 in RPMI1640 supplemented with 10 % FBS, NEAA, antibiotics, and 55 μM β-mercaptoethanol.

2.5 Differentiation of In Vitro-Induced Treg Cells (iTreg Cells)

Antibodies: IFN-γ neutralizing antibodies (clone R4-6A2) and IL-4 (clone 11B11) neutralizing antibodies are purchased from eBioscience.

Cytokines: Murine TGF-β1 is purchased from R&D.

iTreg differentiating culture medium: (0.5 μg/mL anti-CD28, 1 μg/mL anti-IFN-γ, and 1 μg/mL anti-IL-4 antibodies, and 2 ng/mL TGF-β1 in RPMI1640 supplemented with 10 % FBS, NEAA, antibiotics, and 55 μM β-mercaptoethanol.)

2.6 Analysis of In Vitro Differentiated Th Cells

Chemicals: Phorbol 12-myristate 13-acetate (PMA) and ionomycin are purchased from Sigma. Brefeldin A solution is purchased from eBioscience.

Reagents: Intracellular cytokine staining kit and Foxp3-staining kit are purchased from eBioscience.

Antibodies: Fluorescence-chemicals conjugated anti-IFN-γ (clone KMG1.2), anti-IL-4 (clone 11B11), anti-IL-17 (clone eBio17B7), and anti-Foxp3 (clone FJK-16 s) are purchased from eBioscience.

3 Methods

In vitro Th differentiation of naïve $CD4^+$T cells is accomplished by stimulating their TCR, in the presence of appropriate cytokines. We first describe a method to purify naïve $CD4^+$ T cells from mice. Then we describe methods to differentiate Th1, Th2, Th17, and iTreg individually. At last, we show a typical method to detect a result of in vitro differentiated Th cells.

3.1 Isolation of Naïve $CD4^+$ T Cells

$CD4^+$ T cells in vivo are typically composed of heterogeneous populations, which include naïve T cells, effector T cells, memory T cells, and regulatory T cells. These subsets can be distinguished by their expression pattern of surface markers including CD25, CD44, and CD62L. Naïve T cells are classified to a $CD4^{+}CD25^{-}CD44^{low}$ $CD62L^{high}$ subset. We first describe how to purify naïve T cells, which are utilized for Th differentiation assay described latter.

Experimental procedures:

1. Cells suspension is prepared from spleen and lymph nodes of 6- to 8-week-old mice, in 5 mL of serum-free RPMI1640 medium. One mouse typically yields $\sim 5 \times 10^7$ of naïve $CD4^{+}$T cells.
2. Add 10 mL of red blood cell lysis buffer, invert five times, then centrifuge for 4 min at $800 \times g$ at room temperature.
3. Resuspend cells with 5 mL of MACS buffer, then centrifuge for 4 min at $800 \times g$ at room temperature.
4. Remove supernatant and repeat the **step 3** once more.
5. Remove supernatant and resuspend cells with mixture of biotin-conjugated antibodies against CD8α, B220, CD25, CD11b, CD11c, CD49b, and TER-119, at a concentration of 1 μg/mL per 1×10^7 cells in 100 μL of MACS buffer. Incubate cells for 15 min on ice.
6. Centrifuge for 1 min at $800 \times g$ at room temperature, then remove supernatant.
7. Resuspend cells with 1 mL of MACS buffer, then centrifuge for 1 min at $800 \times g$ at room temperature.
8. Remove supernatant, then resuspend cells with Streptavidin microbeads at a concentration of 10 μL/mL per 1×10^7 cells in 100 μL of MACS buffer. Incubate cells for 15 min on ice.
9. Centrifuge for 1 min at $800 \times g$ at room temperature, then remove supernatant.
10. Resuspend cells with 1 mL of MACS buffer, then centrifuge for 1 min at $800 \times g$ at room temperature.
11. Remove supernatant, then resuspend cells with 1 mL of MACS buffer.
12. Separate antibody-bound and unbound cells using large depletion columns, according to the manufacturer's protocol. The flow-through contains crude naïve $CD4^{+}$ T cells.
13. Centrifuge the flow-through for 4 min at $800 \times g$ at room temperature.
14. Remove supernatant, then resuspend cells with CD62L microbeads at a concentration of 100 μL/mL per 1×10^7 cells in 100 μL of MACS buffer. Incubate cells for 15 min on ice.

15. Centrifuge for 1 min at 800 × *g* at room temperature, then remove supernatant.
16. Resuspend cells with 1 mL of MACS buffer, then centrifuge for 1 min at 800 × *g* at room temperature.
17. Remove supernatant, then resuspend cells with 1 mL of MACS buffer.
18. Separate CD62L microbeads bound and unbound cells using large separation columns, according to the manufacturer's protocol. The retentate contains naïve $CD4^+$ T cells. Flush out the retentate with 5 mL of MACS buffer. The elute is used as naïve $CD4^+$ T cells in the following Th differentiation experiments. The purity of $CD4^+CD25^-CD62L^{hi}$ cells was consistently greater than 95 %.

3.2 Differentiation of Th1 Cells

Experimental procedures:

1. One day before, prepare a culture dish coated with anti-CD3ε antibodies. Coating a culture dish is done by adding 250 μL of anti-CD3ε antibodies (2 μg/mL in PBS) per a well of 24-well culture dish.
2. Dissolve 4×10^5 naïve $CD4^+$ T cells with 500 μL of Th1 differentiating culture medium (0.5 μg/mL anti-CD28 and 1 μg/mL anti-IL-4 antibodies, 5 ng/mL IL-2, and 10 ng/mL IL-12 in RPMI1640 supplemented with 10 % FBS, NEAA, antibiotics, and 55 μM β-mercaptoethanol.)
3. Plate cells in single well of anti-CD3ε coated 24-well culture dish.
4. Incubate cells for 96 h in 5 % CO_2 incubator at 37 °C.

3.3 Differentiation of Th2 Cells

Experimental procedures:

1. One day before, prepare a culture dish coated with anti-CD3ε antibodies. Coating a culture dish is done by adding 250 μL of anti-CD3ε antibodies (2 μg/mL in PBS) per a well of 24-well culture dish.
2. Dissolve 4×10^5 naïve $CD4^+$ T cells with 500 μL of Th2 differentiating culture medium (0.5 μg/mL anti-CD28 and 1 μg/mL anti-IFN-γ antibodies, 5 ng/mL IL-2, and 10 ng/mL IL-4 in RPMI1640 supplemented with 10 % FBS, NEAA, antibiotics, and 55 μM β-mercaptoethanol.)
3. Plate cells in single well of anti-CD3ε coated 24-well culture dish.
4. Incubate cells for 96 h in 5 % CO_2 incubator at 37 °C.

3.4 Differentiation of Th17 Cells (See Note 1)

Experimental procedures:

1. One day before, prepare a culture dish coated with anti-CD3ε antibodies. Coating a culture dish is done by adding 250 μL of anti-CD3ε antibodies (2 μg/mL in PBS) per a well of 24-well culture dish.
2. Dissolve 4×10^5 naïve $CD4^+$ T cells with 500 μL of Th17 differentiating culture medium (0.5 μg/mL anti-CD28, 1 μg/mL anti-IFN-γ, 1 μg/mL anti-IL-2, and 1 μg/mL anti-IL-4 antibodies, 20 ng/mL human IL-6, and 1 ng/mL TGF-β1 in RPMI1640 supplemented with 10 % FBS, NEAA, antibiotics, and 55 μM β-mercaptoethanol.)
3. Plate cells in single well of anti-CD3ε coated 24-well culture dish.
4. Incubate cells for 96 h in 5 % CO_2 incubator at 37 °C.

3.5 Differentiation of iTreg Cells

Experimental procedures:

1. One day before, prepare a culture dish coated with anti-CD3ε antibodies. Coating a culture dish is done by adding 250 μL of anti-CD3ε antibodies (2 μg/mL in PBS) per a well of 24-well culture dish.
2. Dissolve 4×10^5 naïve $CD4^+$ T cells with 500 μL of iTreg differentiating culture medium (0.5 μg/mL anti-CD28, 1 μg/mL anti-IFN-γ, and 1 μg/mL anti-IL-4 antibodies, and 2 ng/mL TGF-β1 in RPMI1640 supplemented with 10 % FBS, NEAA, antibiotics, and 55 μM β-mercaptoethanol.)
3. Plate cells in single well of anti-CD3ε coated 24-well culture dish.
4. Incubate cells for 96 h in 5 % CO_2 incubator at 37 °C.

3.6 Analysis of In Vitro Differentiated Th Cells

Differentiated Th cells are typically analyzed with detection of cytokines and/or lineage specifying transcription factors by intracellular immuno-staining. IFN-γ, IL-4, IL-17, and Foxp3 expression are used as marker proteins of Th1, Th2, Th17, and iTreg differentiations, respectively. For cytokine staining, cells are typically restimulated with PMA/ionomycin, in the presence of Brefeldin A, for the enhancement of detection.

Experimental procedures:

1. After differentiating Th cells, add PMA, ionomycin, and Brefeldin A to final concentrations of 80 nM, 1 μg/mL, and 10 μg/mL, respectively to the culture media.
2. Incubate cells for 5 h in 5 % CO_2 incubator at 37 °C.
3. Perform intracellular immuno-staining using kits shown above, according to the manufacturer's protocol.
4. Detect cells with FACS.

4 Note

1. IL-17 differentiation has not been firmly established. Some protocols differentiate Th17 cells with combination of IL-6, IL-1β, and IL-23, instead of using TGF-β1 [16]. Other protocols utilize TGF-β3 [20]. These Th17 cells which are differentiated without TGF-β1 have been reported to be more inflammatory than Th17 cells differentiated with TGF-β1 described our method.

References

1. Stritesky GL, Jameson SC, Hogquist KA (2012) Selection of self-reactive T cells in the thymus. Annu Rev Immunol 30:95–114, PMCID: 3518413
2. Sekiya T, Kashiwagi I, Inoue N, Morita R, Hori S, Waldmann H et al (2011) The nuclear orphan receptor Nr4a2 induces Foxp3 and regulates differentiation of CD4+ T cells. Nat Commun 2:269, PMCID: 3104557
3. Sekiya T, Kashiwagi I, Yoshida R, Fukaya T, Morita R, Kimura A et al (2013) Nr4a receptors are essential for thymic regulatory T cell development and immune homeostasis. Nat Immunol 14(3):230–237
4. Abbas AK, Murphy KM, Sher A (1996) Functional diversity of helper T lymphocytes. Nature 383(6603):787–793
5. Agnello D, Lankford CS, Bream J, Morinobu A, Gadina M, O'Shea JJ et al (2003) Cytokines and transcription factors that regulate T helper cell differentiation: new players and new insights. J Clin Immunol 23(3):147–161
6. Mowen KA, Glimcher LH (2004) Signaling pathways in Th2 development. Immunol Rev 202:203–222
7. Harrington LE, Hatton RD, Mangan PR, Turner H, Murphy TL, Murphy KM et al (2005) Interleukin 17-producing CD4+ effector T cells develop via a lineage distinct from the T helper type 1 and 2 lineages. Nat Immunol 6(11):1123–1132
8. Park H, Li Z, Yang XO, Chang SH, Nurieva R, Wang YH et al (2005) A distinct lineage of CD4 T cells regulates tissue inflammation by producing interleukin 17. Nat Immunol 6(11):1133–1141, PMCID: 1618871
9. Puel A, Cypowyj S, Bustamante J, Wright JF, Liu L, Lim HK et al (2011) Chronic mucocutaneous candidiasis in humans with inborn errors of interleukin-17 immunity. Science 332(6025):65–68, PMCID: 3070042
10. Zhu J, Yamane H, Paul WE (2010) Differentiation of effector CD4 T cell populations (*). Annu Rev Immunol 28:445–489, PMCID: 3502616
11. Hwang ES, Szabo SJ, Schwartzberg PL, Glimcher LH (2005) T helper cell fate specified by kinase-mediated interaction of T-bet with GATA-3. Science 307(5708):430–433
12. Veldhoen M, Hocking RJ, Atkins CJ, Locksley RM, Stockinger B (2006) TGFbeta in the context of an inflammatory cytokine milieu supports de novo differentiation of IL-17-producing T cells. Immunity 24(2):179–189
13. Mangan PR, Harrington LE, O'Quinn DB, Helms WS, Bullard DC, Elson CO et al (2006) Transforming growth factor-beta induces development of the T(H)17 lineage. Nature 441(7090):231–234
14. Ivanov II, McKenzie BS, Zhou L, Tadokoro CE, Lepelley A, Lafaille JJ et al (2006) The orphan nuclear receptor RORgammat directs the differentiation program of proinflammatory IL-17+ T helper cells. Cell 126(6):1121–1133
15. Bettelli E, Carrier Y, Gao W, Korn T, Strom TB, Oukka M et al (2006) Reciprocal developmental pathways for the generation of pathogenic effector TH17 and regulatory T cells. Nature 441(7090):235–238
16. Chung Y, Chang SH, Martinez GJ, Yang XO, Nurieva R, Kang HS et al (2009) Critical regulation of early Th17 cell differentiation by interleukin-1 signaling. Immunity 30(4):576–587, PMCID: 2705871
17. Cua DJ, Sherlock J, Chen Y, Murphy CA, Joyce B, Seymour B et al (2003) Interleukin-23 rather than interleukin-12 is the critical cytokine for autoimmune inflammation of the brain. Nature 421(6924):744–748
18. McGeachy MJ, Chen Y, Tato CM, Laurence A, Joyce-Shaikh B, Blumenschein WM et al (2009) The interleukin 23 receptor is essential for the terminal differentiation of interleukin 17-producing effector T helper cells in vivo.

Nat Immunol 10(3):314–324, PMCID: 2945605
19. Takimoto T, Wakabayashi Y, Sekiya T, Inoue N, Morita R, Ichiyama K et al (2010) Smad2 and Smad3 are redundantly essential for the TGF-beta-mediated regulation of regulatory T plasticity and Th1 development. J Immunol 185(2):842–855
20. Lee Y, Awasthi A, Yosef N, Quintana FJ, Xiao S, Peters A et al (2012) Induction and molecular signature of pathogenic TH17 cells. Nat Immunol 13(10):991–999, PMCID: 3459594

Chapter 11

Interrogating TGF-β Function and Regulation in Endothelial Cells

J.A. Maring, L.A. van Meeteren, M.J. Goumans, and Peter ten Dijke

Abstract

Transforming growth factor-β (TGF-β) is a multifunctional cytokine with important roles in embryogenesis and maintaining tissue homeostasis during adult life. There are three isoforms of TGF-β, i.e., TGF-β1, -β2, and -β3, which signal by binding to a complex of transmembrane type I and type II serine/threonine kinase receptors and intracellular Smad transcription factors. In most cell types TGF-β signals via TGF-β type II receptor (TβRII) and TβRI, also termed activin receptor-like kinase 5 (ALK5). In endothelial cells, TGF-β signals via ALK5 and ALK1. These two type I receptors mediate opposite cellular response for TGF-β. The co-receptor endoglin, highly expressed on proliferating endothelial cells, facilitates TGF-β/ALK1 and inhibits TGF-β/ALK5 signaling. Knockout of TGF-β receptors in mice all result in embryonic lethality during midgestation from defects in angiogenesis, illustrating the pivotal role of TGF-β in this process. This chapter introduces methods for examining the function and regulation of TGF-β in angiogenesis in in vitro assays using cultured endothelial cells and ex vivo metatarsal explants.

Key words TGF-β, Endothelial cell, Angiogenesis, VEGF

1 Introduction

Angiogenesis, the growth of new blood vessels from existing vessels, is an important process during development but also plays a role in pathological situations such as tumor progression and metastasis. At a certain point during tumor growth, the existing blood vessels do not provide sufficient nutrients and oxygen to the tumor, and tumor cells will start to produce and secrete proangiogenic factors and reduce the expression of anti-angiogenic factors to stimulate the formation of new blood vessels: the "angiogenic switch" [1]. Besides the major factor known to stimulate angiogenesis, vascular endothelial growth factor (VEGF), many other growth factors such as transforming growth factor-β (TGF-β), bone morphogenetic protein-9 (BMP-9), basic fibroblast growth factor (bFGF), and angiopoietin-2 (ANG-2) are also involved in the angiogenic switch [2]. Of these, TGF-β plays

Xin-Hua Feng et al. (eds.), *TGF-β Signaling: Methods and Protocols*, Methods in Molecular Biology, vol. 1344,
DOI 10.1007/978-1-4939-2966-5_11,

an ambiguous role, either stimulating or inhibiting the formation of new blood vessels [3, 4].

In part the mechanism underlying these differential effects is that TGF-β signals in endothelial cells via distinct receptors. In most cell types TGF-β exerts its cellular effects via TGF-β type I receptor (TβRI), also termed activin receptor-like kinase5 (ALK5), and TβRII. TGF-β signaling via the TβRII/ALK5 complex induces the phosphorylation of Smad2 and 3, which form heteromeric complexes with Smad4. The Smad complex translocates into the nucleus and regulates target gene expression [5, 6]. Endothelial cells selectively express ALK1, another TGF-β type I receptor, and the co-receptor endoglin. Endoglin favors TGF-β/ALK1 signaling, which induces the activation of Smad1, 5, and 8. TGF-β regulates the activation state of endothelial cells by differentially signaling via ALK1 and ALK5; in general, ALK5 inhibits endothelial cell proliferation and migration, and ALK1 induces opposite responses [4]. Depending on the culture conditions and cell types involved, the responses downstream of ALK1 may differ.

To study the complex role of TGF-β signaling or other potential regulators during angiogenesis and the effect of for example proangiogenic or antiangiogenic therapies, many different assays are available that focus on the different phases of angiogenesis. We discuss the endothelial cell spheroid sprouting assay, fetal bone metatarsal explant assay, and the angiogenic differentiation of embryoid bodies from embryonic stem cells, which cover different aspects of the angiogenic process. With the spheroid assay, the capacity of endothelial cells to invade, migrate, and form tubular structures can be studied. Endothelial cells are allowed to adhere and form small balls, “spheroids”, that will be embedded in collagen. The embedded collagen spheroids will form sprouts when proper stimuli are given. This encompasses the basic components for angiogenesis and can therefore provide a complete picture. The spheroid assay is very suitable to be used as a screening tool for investigating for example the effect of inhibitors and stimulators of the TGF-β pathway [7, 8]. However, these spheroids only contain endothelial cells. When investigating the involvement of mural cells other assays, like the metatarsal and embryoid body assay, are more suitable. For the metatarsal assay the primary outgrowth of an endothelial plexus from metatarsals is analyzed, and the organization of the endothelial cells and the coverage by pericytes can be investigated. Interestingly, genetic effects, studying metatarsals isolated from a knockout mouse, as well as the effect of various extracellular physiological cues and compounds can be studied [9]. The metatarsal assay mostly focuses on endothelial cell sprout formation, organization, and stabilization. The embryoid body (EB) assay on the other hand encompasses most processes involved in angiogenesis, because embryonic stem cells (with or without genetic alteration) are allowed to differentiate into endothelial cells and supporting mural cells. When EBs embedded in a 3D collagen gel are stimulated with angiogenic factors, the endothelial cells will invade the surrounding matrix and form endothelial sprouts. Besides these endothelial cells, perivascular cells are also present and provide a supportive network to the endothelial cells, like the in vivo situation [8, 10, 11]. Using ES cells for the EB assay allows the analysis and impact of genetic alterations on vessel development and stability.

2 Materials

2.1 Spheroid Assay

2.1.1 Spheroid Formation

1. Endothelial cell culture, e.g., human umbilical vein endothelial cells (HUVECs).
2. Trypsin–ethylenediaminetetraacetic acid (EDTA).
3. 1.2 % methocellulose (Sigma) in M199.
4. Endothelial cell growth media (EGM)2: EBM-2 with supplemented growth factors (Lonza).
5. U-bottom 96-well plates (Greiner Bio-One, Cat. No. 650185).

2.1.2 Collagen Embedding

1. 96-well plate.
2. PureCol (Advanced Biomatrix).
3. NaOH (1 M).
4. EGM-2.
5. Vascular endothelial growth factor (VEGF).

2.2 Metatarsal Assay

1. Metatarsals from mouse embryo day 17 or 18.
2. 0.1 % gelatin in phosphate buffered saline (PBS).
3. Plastic foil.
4. Medium: alpha MEM with 10 % FBS.
5. 24-well tissue culture plate.
6. Zink Macrodex Formalin: 1 mL of formaldehyde solution with 9 mL of Zink Macrodex solution.
7. mQ water.
8. 40 % methanol/1 % H_2O_2 solution.
9. 0.5 % Boerhringer Milk Powder (BMP)/TBS with Tween (TBST).
10. Antibody of choice, e.g., ERMP12 for CD31 staining.
11. 1 % 3-amino-9-ethylcarbazole (AEC) stock solution.
12. 0.1 M acetate-acetic acid, pH 5.0.
13. 30 % H_2O_2.

2.3 Embryoid Body Assay

2.3.1 EB Formation by Hanging Drops

1. ES cell culture (e.g., R1-ES cells).
2. Trypsin–EDTA.
3. 0.1 % gelatin in PBS.
4. ES medium: Dulbecco's Modified Eagle's Medium (DMEM) (high glucose 4.5 g/Land L-glutamine, Gibco Invitrogen) supplemented with 15 % fetal bovine serum, 25 mM HEPES, 1 mM sodium pyruvate, 0.1 mM β-mercaptoethanol, 1× penicillin/streptomycin.
5. Multichannel pipette p200.
6. Bacterial dishes.
7. Sterile mQ water.

2.3.2 2D and 3D Culture of EBs

1. Collagen solution: PureCol with 34.65 % HAM's F12, 6.25 % NaOH (0.1 M), 6.25 % 10× F12, 1.25 % 4-(2-hydroxyethyl)-1-piperazineethanesulfonic acid (HEPES) (1 M), 0.975 % sodium bicarbonate 7.5 % and 0.625 % GlutaMAX.
2. Tissue culture plates: 6, 12, and 24 wells.
3. VEGF, PeproTech 100 ng/μL.

2.3.3 Staining of EBs

1. 4 % paraformaldehyde in PBS.
2. PBS.
3. Blocking buffer: 3 % BSA in TBST.
4. TBST.
5. Antibody of choice.
6. Hoechst solution 1:25,000 in TBST.

2.4 Spheroid Assay

2.4.1 Spheroid Formation

1. Detach HUVECs by trypsinization and count cell number.
2. Dilute cells to a concentration of 7500 cells/mL in EGM-2 with 0.24 % methylcellulose (1:5 solution of stock methylcellulose) (see Note 1).
3. Pipette 100 μL of cell suspension into suspension U-bottom 96-well plates.
4. Culture cells overnight in humidified incubator at 37 °C and 5 % CO_2.

2.4.2 Collagen Embedding

1. Make cold collagen solution containing 2/3 EGM-2 with 0.5 % 1 M NaOH (e.g., 2 mL EGM-2 with 10 μL NaOH) and 1/3 collagen. Final pH should be approximately 7.0.
2. Keep collagen mixture on ice.
3. Pipette 50 μL of collagen mixture in wells of 96-well plate.
4. Incubate for 1 h at 37 °C to solidify collagen.
5. Gently collect spheroids with p1000 multichannel and collect in 9 cm tissue culture dish.

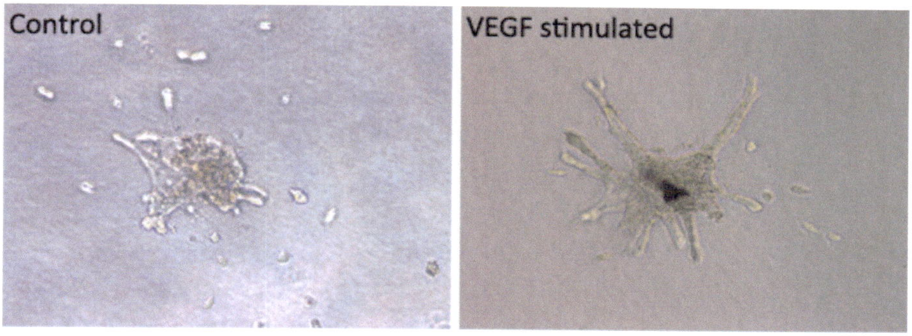

Fig. 1 Sprouting of spheroids after 2 days of culturing in the absence or presence of VEGF

6. Transfer spheroids to 15 mL tube and centrifugate for 5 min at 170 ×*g*.
7. Slowly remove supernatant and gently resuspend spheroids in collagen mixture: 10–12 spheroids per 50 μL.
8. Pipette 50 μL of spheroid/collagen mix on collagen coated wells.
9. Incubate for 1 h at 37 °C to solidify collagen.
10. Add 100 μL of EGM-2 to wells with 50 ng/mL VEGF (final VEGF concentration is 25 ng/mL).
11. Culture for 2 days at 37 °C, 5 % CO_2.
12. Sprouting of HUVECs can be assessed by microscopy (Fig. 1) (*see* **Note 2**).

2.5 Metatarsal Assay

1. Isolate metatarsals from a mouse embryo at day 17 or 18 of gestation (*see* **Note 3**).
2. Coat plates with 0.1 % gelatin for 15 min.
3. Add 180 μL of medium (alpha MEM with 10 % FBS) to each well and place the metatarsals in the middle of the well (one metatarsal per well). Make sure metatarsals do not float by adding a glass coverslip on top of the metatarsal to push the metatarsal down.
4. Wrap the plates with plastic foil and incubate at 37 °C and 5 % CO_2 for 3 days (*see* **Note 4**).
5. On day 3, gently remove the coverslips and old medium and add 250 μL of new medium with desired stimuli.
6. Incubate plates for 7 days.
7. At day 7, remove the medium and gently wash the wells twice with PBS.
8. Fix the bones in Zink Macrodex Formalin for 15 min.
9. Wash twice with mQ water.
10. Preserve the plates with 500 μL of PBS per well for 2 days.

11. Wash wells three times with PBS.
12. Add 500 μL of 40 % methanol/1 % H_2O_2 per well to block endogenous peroxidase activity and incubate at RT for 30 min.
13. Wash twice quickly with PBS and once with TBST for 5 min while gently shaking.
14. Add 245 μL of 0.5 % BMP/TBST and incubate for 1 h at 37 °C.
15. Add first antibody to the well, wrap plates to avoid evaporation of the liquid with plastic foil and incubate overnight at 4 °C (*see* **Note 5**).
16. Wash twice with TBST for 5 min while gently shaking.
17. Add 250 μL of secondary antibody to each well in 0.5 % BMP/TBST and incubate at 37 °C for 45 min.
18. Wash twice with TBST and once with mQ water for 5 min while gently shaking.
19. Add 250 μL of Streptavidin HRP 1:200 in 0.5 % BMP/TBST per well and incubate at 37 °C for 30 min.
20. Repeat **step 18**.
21. Add 500 μL of AEC stock solution + 4500 μL 0.1 M acetate-acetic acid, pH 5.0, mix well.
22. Add AEC solution so metatarsals are immersed.
23. To start staining add 2 μL of H_2O_2 30 % to the AEC solution and incubate for 10 min at room temperature.
24. Stop staining reaction by washing three times with mQ water.
25. Add 500 μL of PBS per well and store at 4 °C for max 14 days (wrap well in plastic foil to avoid evaporation). EC outgrowth can be assessed by microscope analysis (Fig. 2).

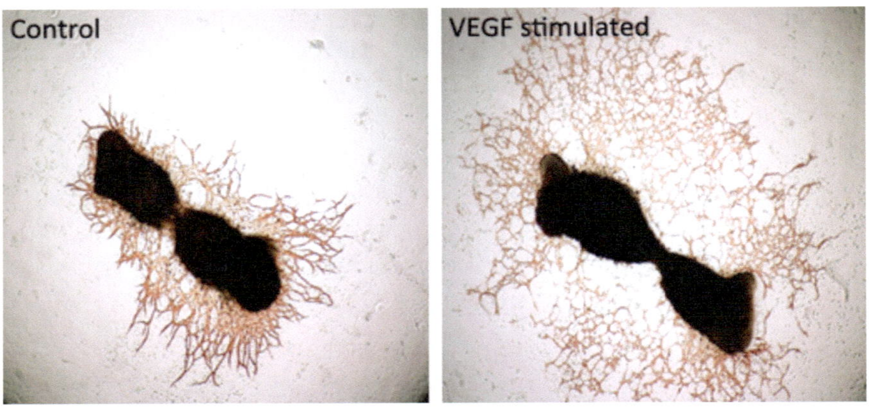

Fig. 2 Outgrowth of endothelial cells from metatarsals cultured with or without VEGF

2.6 EB Assay

2.6.1 EB Formation by Hanging Drops

1. Trypsinize ES cells and wash once with medium. Make sure you have a single cell suspension (*see* **Note 6**).
2. Seed ES cells ±45 min in a gelatin coated 6-well tissue culture plate in ES medium to allow mouse embryonic fibroblasts (MEFs) to adhere to the plate while leaving the ES cells unattached.
3. Collect supernatant containing ES cells and count number of cells.
4. Dilute cell suspension to a concentration of 60 × 103 cells/mL.
5. Use multichannel pipette to pipette drops of 20 μL of the cell suspension onto lid of a 10 cm bacterial dish.
6. Fill bottom dish with sterile water to prevent evaporation.
7. Invert lid and place on top of dish, resulting in hanging drops (Fig. 3a).
8. Incubate for 4 days at 37 °C, 5 % CO_2 (Fig. 3b) (*see* **Note 7**).

2.6.2 3D Culture

EBs can be embedded either as single or multiple EBs per well. Single EBs are usually cultured in 24-well plates while 12 plates are used for culturing multiple EBs.

1. Coat tissue culture plate with collagen: 750 μL for 12-well plate, 400 μL for 24-well plate and allow to polymerize overnight.

Embedding single EB culture

1. Invert lid containing hanging drops carefully.
2. Pick up EB from drop with 200 μL pipette (tip shortened 2–3 mm in order not to damage EB).
3. Place EB in wells of the 24-well plate with a minimum amount of medium.
4. Remove excess medium.
5. Put 350 μL of collagen on top.
6. Allow to solidify for 3 h (see Note 8).
7. Add 750 μL of ES medium containing 60 ng/mL VEGF on top (final concentration 30 ng/mL).

Embedding multiple EB culture

1. Take lid of the dish and keep upside down.
2. Harvest EBs by ticking lid at 45° angle in a clean 9 cm tissue culture dish, thereby collecting EBs and medium in dish.
3. Collect EBs with P1000 (tip shortened 2 mm in order not to damage EBs) and put in 15 mL tube.
4. Allow EBs to settle at the bottom and remove supernatant.

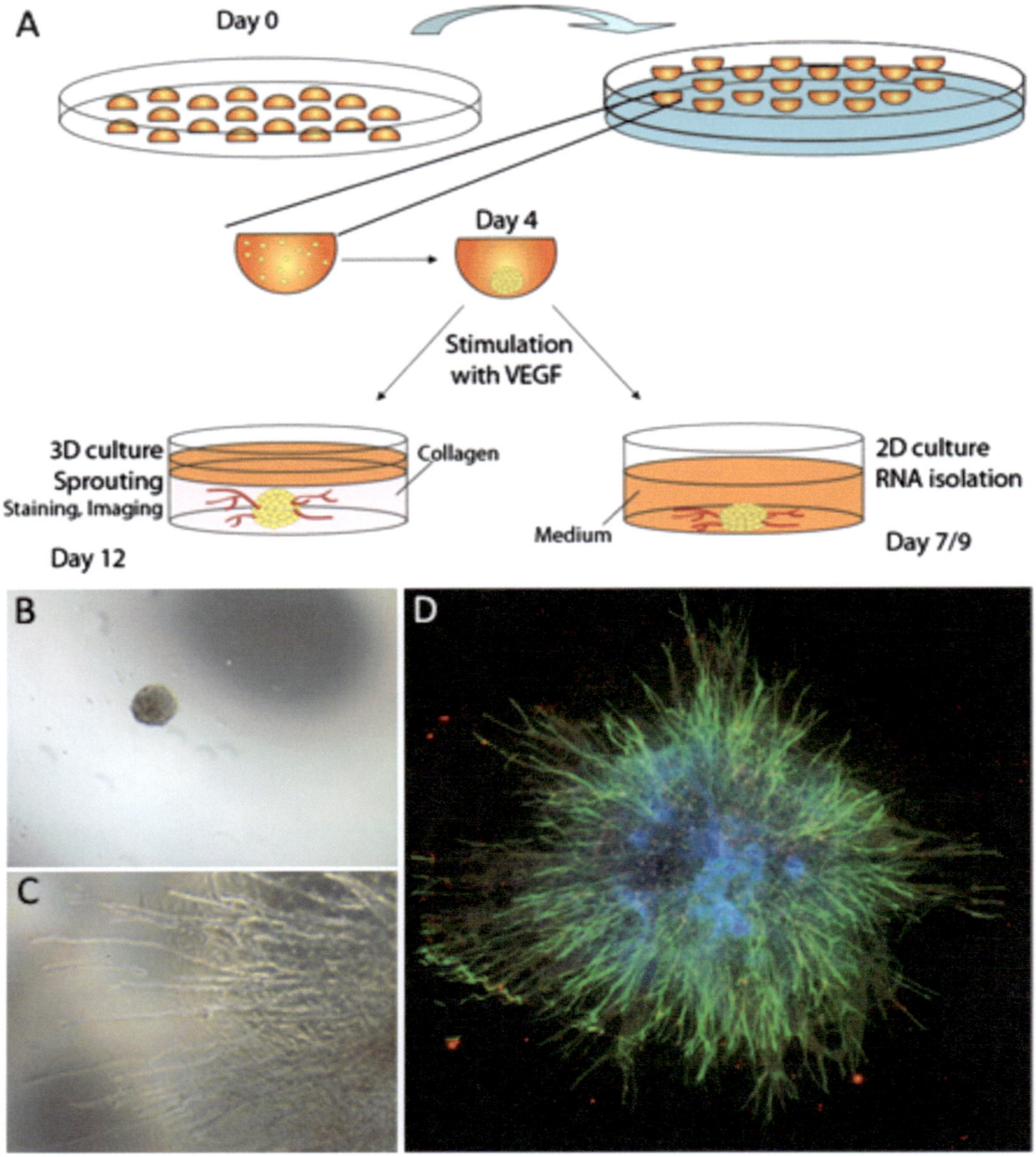

Fig. 3 (**a**) Schematic overview of the EB assay, from hanging drops till 2D and 3D culture. (**b**) EB after 4 days of hanging drop culture. (**c**) Sprouting of an EB from R1-ES cells after 8 days of stimulation with VEGF, bright-field picture. (**d**) EB of R1-ES cells after 8 days of VEGF stimulation, stained for CD31 (*green*), NG2 (*red*), and Hoechst (*blue*)

5. Add collagen to the EBs (10–15 EBs per 750 μL), resuspend carefully and plate EBs in the 12-well plate; 750 μL EB/collagen suspension per well, resulting in 10–15 EBs per well.
6. Allow to solidify for 3 h.

7. Add 1.5 mL of ES-medium containing 60 ng/mL VEGF on top (final concentration 30 ng/mL).

Culture EBs for 8–14 days, depending on sprouting of EBs and desired end-stage. Medium should be changed every 4 days (Fig. 3c).

2.6.3 2D Culture

1. Take lid of the dish and keep upside down.
2. Harvest EBs by ticking lid at 45° angle in a clean 9 cm dish, thereby collecting EBs and medium in the dish.
3. Collect EBs with P1000 (tip shortened 2–3 mm in order not to damage EBs) and put in 15 mL tube.
4. Allow EBs to settle at the bottom and remove supernatant.
5. Resuspend in ES-DMEM containing 50 ng/mLVEGF (10–13 EBs per mL).
6. Add 2 mL of EB suspension to gelatin coated 6-well plate resulting in 20–25 EBs per well.

Culture EBs for up to 21 days, depending on desired end-stage. 2D cultures can be used for RNA, DNA, and protein isolation as well as FACS analyses and stainings.

2.6.4 Staining 3D Culture

1. Remove medium completely.
2. Wash collagen three times with PBS for a total of 20 min at room temperature.
3. Fix with 4 % paraformaldehyde for 30 min at room temperature.
4. Wash with PBS.
5. Cut collagen loose of rim of well.
6. Remove collage carefully from well and place in lid of petri dish.
7. Localize EBs under a microscope and cut away excess collagen without disrupting EBs and sprouts.
8. Put the (clusters of) EBs in a 24-well plate containing PBS.
9. Block with blocking buffer for 2 h or overnight at room temperature.
10. Add primary antibody in blocking solution (concentration depending on antibody) and incubate overnight at 4 °C.
11. Wash four times with TBST.
12. Add secondary antibody in blocking solution and incubate overnight at 4 °C.
13. Wash two times with TBST.
14. Stain with Hoechst for 10 min.
15. Wash two times with TBST.
16. Store at 4 °C with PBS.
17. Sprouting of EBs can be assessed by microscopy (Fig. 3d) (*see* **Notes 9** and **10**).

3 Notes

1. In the spheroid assay, the methocellulose solution should be free of any contaminating particles, which will interfere with the formation of spheroids.
2. The spheroid assay protocol is optimized to ensure best sprouting and is very useful to study inhibition of angiogenesis. When investigating stimulation of angiogenesis, sub-optimal conditions should be used.
3. It is important to isolate the embryos for the metatarsal assay from the uterus and keep them in PBS with calcium and magnesium on ice until isolating the metatarsals. It is better to do one mouse at a time for optimal condition of the bones, max 2.
4. It is best to place the plate with the metatarsals in the back of the incubator and especially the first 2–3 days. Do not slam the door, otherwise the bones will float.
5. Staining for CD31 on the metatarsals should be done using the ERMP12 antibody to get the best results.
6. The ES cells should be in optimal condition before making embryoid bodies, i.e., all colonies should be undifferentiated.
7. Make sure that the hanging drops are not disturbed during the 4 day culture, otherwise the EBs might not form tight clusters. Ideally the EBs should have a compact, round appearance.
8. The second layer of collagen with the EBs should contain no air bubbles and should polymerize at least 3 h, to ensure the layers will not be split when the medium is added to the wells. It will result in reduced sprouting if the layers are not properly attached.
9. The extent to which the EBs sprout varies with the genetic background of the ES cells used. This is important to take into account and test when using this assay. Specifically great care should be taken when knockout ES cells are compared to wild type ES cells. This should always be compared to wild type cells with the same background, preferentially isolated in the same procedure as the mutant ES cells.
10. Besides angiogenesis, the EB assay encompasses the differentiation from ES cell to endothelial cells. Genetic alterations, e.g., knockdown or overexpression, that interfere with the differentiation of ES cells can result in aberrant angiogenesis and confound the results.

Acknowledgements

Our studies on the role of TGF-β receptors in tumor angiogenesis and other diseases are supported by the TAS ZonMW grant, the Dutch Cancer Society, Netherlands Research Council, Le Ducq Foundation, The Netherlands Institute for Regenerative Medicine (NIRM), Cancer Genomics Centre Netherlands, and Centre for Biomedical Genetics.

References

1. Hanahan D, Folkman J (1996) Patterns and emerging mechanisms of the angiogenic switch during tumorigenesis. Cell 86:353–364
2. Carmeliet P, Jain RK (2011) Molecular mechanisms and clinical applications of angiogenesis. Nature 473:298–307
3. Goumans M-J, Valdimarsdottir G, Itoh S et al (2002) Balancing the activation state of the endothelium via two distinct TGF-beta type I receptors. EMBO J 21:1743–1753
4. Goumans M-J, Liu Z, ten Dijke P (2009) TGF-beta signaling in vascular biology and dysfunction. Cell Res 19:116–127
5. Derynck R, Zhang YE (2003) Smad-dependent and Smad-independent pathways in TGF-beta family signalling. Nature 425:577–584
6. ten Dijke P, Hill CS (2004) New insights into TGF-beta-Smad signalling. Trends Biochem Sci 29:265–273
7. Cunha SI, Pardali E, Thorikay M et al (2010) Genetic and pharmacological targeting of activin receptor-like kinase 1 impairs tumor growth and angiogenesis. J Exp Med 207:85–100
8. Jakobsson L, Kreuger J, Claesson-Welsh L (2007) Building blood vessels–stem cell models in vascular biology. J Cell Biol 177: 751–755
9. Liu Z, Kobayashi K, van Dinther M et al (2009) VEGF and inhibitors of TGFbeta type-I receptor kinase synergistically promote blood-vessel formation by inducing alpha5-integrin expression. J Cell Sci 122:3294–3302
10. Zwijsen A, Goumans MJ, Lawson KA et al (1999) Ectopic expression of the transforming growth factor beta type II receptor disrupts mesoderm organisation during mouse gastrulation. Dev Dyn 214:141–151
11. Goumans MJ, Ward-van Oostwaard D, Wianny F et al (1998) Mouse embryonic stem cells with aberrant transforming growth factor beta signalling exhibit impaired differentiation in vitro and in vivo. Differentiation 63:101–113

Chapter 12

Isolation and Manipulation of Adipogenic Cells to Assess TGF-β Superfamily Functions

Maria Namwanje, Juan C. Bournat, and Chester W. Brown

Abstract

A variety of TGF-β superfamily members affect adipocyte differentiation and function with consequential effects on energy metabolism. There has been a growing interest in this area because of the apparent influence of the BMP subgroup on brown adipose characteristics and potential application to the treatment of human obesity. In this chapter we describe methods that are useful in allowing one to assess the roles of specific members of the superfamily on adipocyte differentiation and mature adipocyte function, including the isolation and differentiation of mouse embryo fibroblasts (MEFs) to examine cell autonomous effects and the efficient transfection of two commonly used (but difficult to transfect) adipogenic cell lines, C3H/10T1/2 and 3T3-L1.

Key words Adipogenesis, Neon electroporation, Adipogenic cell lines

1 Introduction

Obesity is a growing epidemic worldwide. In the USA 35 % of adults are classified as obese [1]. Several TGF-β superfamily members have been identified as stimulators or inhibitors of adipogenesis. Bone morphogenetic proteins (BMPs) 2, 4, and 7 stimulate adipogenesis, with BMP4 promoting a white adipocyte lineage while BMP7 drives commitment to a brown adipocyte lineage and contributes to mature cell function [2–5]. TGF-β1 inhibits adipocyte differentiation, while activin A enhances proliferation of human preadipocytes and inhibits their differentiation [6–8]. Thus, members of the superfamily play supportive and inhibitory roles in adipocyte development.

Adipocyte development in vitro is characterized by a multistep process in which multipotent mesenchymal stem cells (MSCs) first become determined to differentiate, then committed to the adipocyte lineage and they ultimately differentiate into mature adipocytes (reviewed in ref. [9]). Undifferentiated MSCs can develop into several mesodermal cell types: osteoblasts, chondrocytes, myoblasts,

Xin-Hua Feng et al. (eds.), *TGF-β Signaling: Methods and Protocols*, Methods in Molecular Biology, vol. 1344,
DOI 10.1007/978-1-4939-2966-5_12, © Springer Science+Business Media New York 2016

and adipocytes [2, 10–12]. Commitment to the preadipocyte stage can be strongly influenced by BMP signaling [2, 3, 13], the hypomethylation of presumably lineage-specific gene(s) [11], and cell shape [14]. The committed preadipocyte population proliferates to confluence, resulting in growth arrest and subsequent mitotic clonal expansion, followed by a second growth arrest stage and terminal adipocyte differentiation. Throughout adipocyte differentiation, temporally precise transcriptional events are required (reviewed in ref [9]). Early differentiation is initiated when hormonal cues stimulate the production of transcription factors that subsequently increase the expression of peroxisome proliferator-activated receptor (PPAR)-gamma (*Pparg*) and C/EBP-alpha (*Cebpa*), ultimately leading to terminal adipocyte differentiation [15–18].

C3H/10T1/2 and 3T3-L1 cell lines are widely used in adipocyte biology. C3H/10T1/2 is a multipotent line that can efficiently differentiate into mature adipocytes when pretreated with BMPs followed by an adipogenic cocktail composed of insulin, dexamethasone, isobutylmethylxanthine (IBMX), +/− a thiazolidinedione (PPAR-gamma agonist, *see* Subheading 3.2, **step 1**). 3T3-L1 is a committed preadipocyte cell line that differentiates efficiently (>90 %) into mature adipocytes when treated with the adipogenic cocktail, without a requirement for BMPs.

Previous work from several laboratories confirms the roles of many BMPs in regulating adipogenesis [19, 20]. BMP4's role in directing commitment to the adipocyte lineage can be achieved by treating C3H/10T1/2 cells exogenously with BMP4 [2]. If treated during the early proliferative stage, a high frequency of MSCs differentiate into adipocytes and subcutaneous injection of these cells into athymic mice results in the development of adipose tissue [2]. During proliferation, C3H/10T1/2 cells normally have very low *Bmp4* mRNA levels, which rapidly increase as the cells approach confluence. These cells differentiate poorly (<10 %) under standard adipogenic conditions. However, if BMP4 is autonomously expressed during early proliferation, up to 90 % differentiation occurs, indicating that the timing of BMP4 expression is important for commitment to the adipocyte lineage [13, 21].

BMP7 initiates adipogenesis in high-density micro-mass cultures of human MSCs [22] and supports brown adipogenesis in brown preadipocytes [3]. In C3H/10T1/2 cells, pretreatment with BMP7 results in brown adipogenesis with lipid accumulation and expression of *Ucp1* [3] and in mature brown adipocytes, BMP7 enhances the uptake, mitochondrial transport and subsequent oxidation of fatty acids leading to increased energy expenditure, indicating dual roles for BMP7 in brown adipocyte differentiation and mature function [5].

Although BMP7 increases phosphorylation of Smads 1/5/8 in 3T3-L1 cells and brown preadipocytes, enhanced phosphorylation of p38 MAPK and ATF2, as well as increased expression of several gene products that participate in mitochondrial biogenesis and function are only observed in cells of brown adipocyte lineage [3, 5]. BMP7 also has a substantial effect on the adipogenic cascade, suppressing the expression of adipogenic inhibitors while upregulating proadipogenic factors including *Prdm16*, an early marker of brown adipogenesis. In summary, BMP7 promotes brown adipogenesis in MSCs as well as committed brown preadipocytes and supports the functions of mature brown adipocytes. This provides a transcriptional environment that augments the expression of genes that promote the phenotypic characteristics and functional activity of mature brown adipocytes.

Growth differentiation factor 3 (GDF3) is expressed in adipose tissue and expression is upregulated under high fat diet (HFD) conditions [23]. *Gdf3* knockout mice are protected against diet-induced obesity compared to wild type mice fed a regular or a HFD [23, 24]. Furthermore, overexpression of GDF3 by adenoviral transduction results in obesity, increased adipocyte size and adipose tissue mass in mice under HFD conditions [25]. The *Gdf3*-null mice are protected from diet-induced obesity due at least in part to increased basal metabolic rate and increased expression of genes involved in mitochondrial function selectively in white adipose tissue [23]. In addition, fatty acid binding protein (Fabp4/aP2) knockout mice under HFD conditions show a 30-fold upregulation of *Gdf3* in adipose tissue relative to wild type suggesting a possible link between nutrient signaling, GDF3 and fatty acid transport [26]. These results also suggest possible roles for GDF3 in energy metabolism by regulating mature adipocyte function.

GDF3 can directly interact with BMP4 and inhibit its signaling during embryogenesis [27]. As such, it may employ a similar mechanism to regulate adipocyte differentiation and/or function. Consistent with this notion, exogenous treatment of mature human adipocytes or fully differentiated 3T3-L1 cells with recombinant GDF3 protein resulted in increased *Pparg* mRNA levels relative to human preadipocytes and undifferentiated 3T3-L1 cells [25].

Our lab studies GDF3 and other TGF-beta superfamily ligands in the context of adipocyte differentiation and mature adipocyte function and is determining the stages during which these effects occur. Cell-based approaches used in combination with genetically engineered mouse models can provide important insights into the mechanisms by which the superfamily members affect differentiation and energy expenditure at the cellular level and on the physiology of an entire organism. In this chapter we describe the isolation of primary mouse embryo fibroblasts (MEFs) and the method used to induce their subsequent differentiation into adipocytes. We also describe the efficient, inexpensive transfection of 3T3-L1 and C3H/10T1/2 cells, adipogenic cell lines that are notoriously difficult to transfect using conventional methods. MEFs from genetically engineered mice are valuable for studies of the cell autonomous effects of specific mutations on adipogenesis and mature adipocyte function. C3H/10T1/2 cells and primary mouse embryo fibroblasts share developmental characteristics with multipotentmesenchymal precursors, whereas 3T3-L1 cells are fully committed to the adipocyte lineage [2, 28]. These properties allow one to also study the stage-selective effects of exogenous TGF-β superfamily ligands.

2 Materials

BMPs positively affect adipocyte differentiation and TGF-β1 is mostly inhibitory, whereas the roles of other members of the TGF-β superfamily are not as well established. In addition to understanding the effects of the specific ligands on adipogenesis, it is also important to determine whether these effects are stage dependent (multipotent, determined, committed, differentiating, mature) and if there are cell autonomous effects during these stages.

C3H/10T1/2 and 3T3-L1 cell lines are inefficiently transfected (<10 %) using standard protocols, such as liposome-mediated transfer, calcium phosphate, DEAE dextran and conventional electroporation [29, 30]. Efficient alternatives include viral transduction with adenoviruses, retroviruses, or lentiviruses, but this can be expensive, labor intensive and require special accommodations in the tissue culture laboratory.

In Subheadings 3.1 and 3.2 we describe the isolation and differentiation of primary mouse embryo fibroblasts. In Subheading 3.3, we describe a highly efficient, reproducible, and simple method to introduce DNA into these cell lines using the small volume (10 μL) Neon Electroporation System [31]. Using this method we have achieved high transfection efficiencies of 60–80 % using plasmid DNA within a highly flexible range of cell numbers ranging from 2×10^4 to 5×10^6.

2.1 Isolation and Differentiation of Mouse Embryonic Fibroblasts

1. Pregnant mice (C57BL/6J) with embryos at E13.5.
2. 0.25 % Trypsin–EDTA (*Invitrogen*: *25200056*).
3. Dulbecco's Modified Eagle's Medium—high glucose (DMEM) (*Invitrogen*: *11965092*).
4. Fetal Bovine Serum—Qualified (*Invitrogen*: *10437028*).
5. 1× Phosphate buffered saline without Calcium and Magnesium (PBS) (*Invitrogen*: *14190044*).
6. 100× Penicillin/streptomycin sulfate (*Invitrogen*: *15140122*).
7. 100× Glutamine (*Invitrogen*: *25030081*).
8. 100 mM Sodium pyruvate (*Invitrogen*: *11360070*).

9. Insulin (*Sigma*: *I5500*).
10. Dexamethasone (*Sigma*: *D4902*).
11. 3-Isobutyl-1-methylxanthine (IBMX) (*Sigma*: *I7018*).
12. Pioglitazone (*Sigma*: *E6910*).
13. Sterile scalpel blades and tissue dissection tools (scissors and forceps).
14. 15- and 50-mL polypropylene tubes (*VWR*: *89039-066 and 89039-658, respectively*).
15. 12-well tissue culture plates (*VWR*: *62406-165*).
16. 37 °C water bath.
17. Complete cell culture medium (10 % FBS in DMEM supplemented with 1 % glutamine, 1 % penicillin/streptomycin, and 1 mM sodium pyruvate).
18. Induction cocktail: 0.5 mM 3-isobutyl-1-methylxanthine (IBMX), 167 nM insulin, and 1 mM dexamethasone in complete cell culture medium.
19. RNA STAT-60 (*Tel-Test, Inc.*, *CS-110*).

2.2 Transfection of C3H/10T1/2 and 3T3-L1 Cells by Small Volume Electroporation

1. Dulbecco's Modified Eagle Medium (DMEM), 4500 mg/L glucose (*Invitrogen*: *11965092*).
2. Opti-MEM (*Invitrogen*: *31985070*).
3. Glutamine (100×) (*Invitrogen*: *25030081*).
4. Penicillin/Streptomycin (100×) (*Invitrogen*: *15140122*).
5. Sodium Pyruvate (100×) (*Invitrogen*: *11360070*).
6. Fetal bovine serum (FBS)-Qualified (*Invitrogen, 10437028*).
7. 0.05 % Trypsin–EDTA (1×) (*Invitrogen*: *25200072*).
8. Plasmid DNA (QIAfilter plasmid Maxi prep kit) (*Qiagen, 12243*).
9. Resuspension buffer R, E electrolytic buffer, electroporation tubes, 10 μL reaction tips [Neon transfection system 10 μL kit] (*Invitrogen*: *MPK1096*).
10. Neon Device, pipette station and pipette (*Invitrogen*: *MPK5000*).
11. 10 cm (surface area 55 cm^2) (*VWR*: *25382-166*) and 12-well (surface area 3.8 cm^2/well) (*VWR*: *62406-165*) tissue culture dishes/plates, respectively.
12. 15-mL polypropylene tubes (*VWR*: *89039-666*).
13. 1.5-mL polypropylene microfuge tubes (*VWR*: *890000-028*).
14. Hemocytometer (*American Optical*).
15. Inverted phase contrast microscope (*Nikon TMS-F*).
16. Inverted phase contrast fluorescence microscope (*Nikon Diaphot*).

17. Spot RT color digital camera (*Diagnostic Instruments, Inc.: 2.2.0*).
18. Tissue culture incubator, 37 °C, 5 % CO_2 (*Forma Scientific: Thermo Forma Series II, 3110*).
19. Cell culture centrifuge (*Eppendorf: 5702 R*).
20. Complete growth medium: DMEM with 10 % FBS, 1× glutamine, 1× Pen/Strep, 1× sodium pyruvate.
21. C3H/10T1/2 cells (*American Type Tissue Collection, CCL-226*).
22. 3T3-L1 cells (*American Type Culture Collection, CL-173*).

3 Methods

3.1 Isolation of Mouse Embryo Fibroblasts

1. Set up timed matings for embryos to be collected at embryonic day 13.5 (E13.5) from C57BL/6J pregnant mice [32]. Euthanize with isofluorane followed by cervical dislocation and dissect out the uterus with sterile forceps and scissors (*see* **Note 1**).
2. In a 10-mm dish containing ice-cold PBS, dissect out the embryos from the uterus and remove the yolk sacs. Transfer the embryos to a fresh 10 mm dish containing ice-cold PBS and remove the head, liver, limbs, and tail. These discarded tissues (including yolk sacs) can be used to genotype the embryos (*see* **Notes 2** and **3**).
3. Transfer the remaining embryonic tissue to a 12-well plate (one embryo/well) with 1 mL of cold PBS to completely wash off remaining blood.
4. Remove the tissues from PBS and place in 15-mL conical tubes (one tube per embryo) so that they adhere to the side of the tube near the opening. To minimize loss, mince the tissue directly in the tube using a sterile, pointed scalpel blade. Wash the minced tissue into the bottom of the tube using 2 mL of 0.25 % trypsin–EDTA per embryo (*see* **Note 4**).
5. Incubate at 37 °C for 30 min with shaking every 10 min to ensure even disruption and complete digestion of the tissue.
6. Add 5 mL of complete cell culture media after incubation to stop trypsinization. Disintegrate the tissue by pipetting up and down several times with a 5 mL pipette to form a uniform cell suspension.
7. Let the cell suspension stand for 30 min so that the remaining pieces of tissue sink to the bottom. Pool and transfer the supernatant from all embryos of the same genotype to a 50 mL polypropylene tube and pellet the cells by centrifugation at 200 × *g* for 5 min at room temperature.

8. Aspirate the supernatant and resuspend the cell pellet in 20 mL of DMEM/FBS complete media. Count the cells and plate 100,000 cells per well (2.5×10^4 cells/cm^2) into a 12-well culture dish in addition to 1.5 mL of culture media per well (*see* **Note 5**).
9. Change media approximately 16 h after initial plating to get rid of the unattached and dead cells.
10. Culture the cells in 10 % FBS/DMEM supplemented with penicillin/streptomycin at 37 °C/5 % CO_2 with media changes every 2 days until they reach confluence in preparation for differentiation (*see* **Note 5**).

3.2 Differentiation of MEFs into Adipocytes

1. Add induction cocktail to the MEFs 2 days post confluency (designated Day 0) to induce differentiation. The induction cocktail is composed of 1.15 g/mL IBMX, 167 nM insulin, and 1 μM dexamethasone in complete growth medium (10 % FBS in DMEM supplemented with 1 % glutamine, 1 % penicillin/streptomycin, and 1 mM sodium pyruvate). To increase differentiation efficiency of the cells, pioglitazone, a PPARγ agonist, can be added to the induction cocktail at 5 μM final conc (*see* **Note 6**).
2. On Day 3, change to complete growth medium containing 167 nM insulin.
3. On day 5, change to complete cell culture medium (without IBMX, dexamethasone, or insulin) and change media every other day until day 8 when cells are fully differentiated.
4. Compare adipocyte differentiation efficiency by Oil Red O staining (Fig. 1) [33].

Additionally, differentiation efficiency can be quantified by measuring gene expression of early and late adipocyte markers such as Pparg, C/ebpα, and Fabp4 by using quantitative real-time PCR. Total RNA is isolated from the cells following the RNA STAT-60 purification protocol.

3.3 Low Volume Electroporation of Plasmid DNA into 3T3-L1 and C3H/10T1/2 Cells

This is a highly efficient, reproducible, simple, and inexpensive method to introduce DNA into these cell lines using the small volume (10 μL) Neon Electroporation System (Fig. 2). In approximately 2 h, one can comfortably process eight 12-well plates (96 wells total) in a single electroporation experiment. For this volume, approximately 1×10^8 cells are required. All experiments should be carried out in cells that have not exceeded ten passages beyond the passage number of the frozen vials from the ATCC.

1. Grow cells in complete growth media (10 % FBS in DMEM supplemented with 1 % glutamine, 1 % penicillin/streptomycin, and 1 mM sodium pyruvate) until 70 % confluency in 10-cm tissue culture dishes.

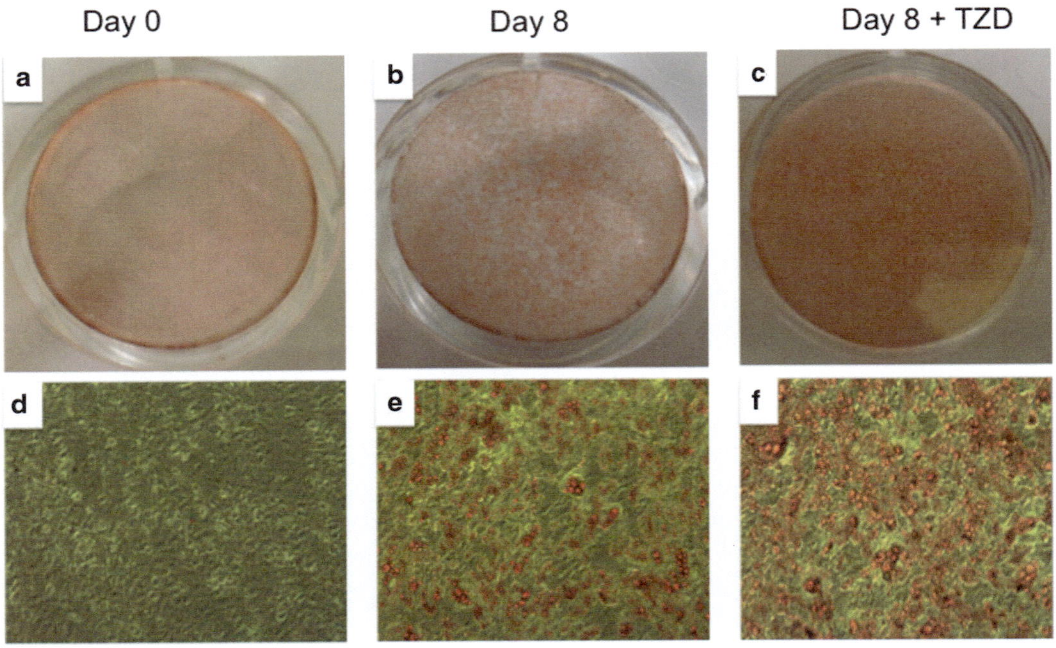

Fig. 1 Oil Red O staining of wild type mouse embryo fibroblasts differentiated into adipocytes. The addition of pioglitazone to the differentiation cocktail improves the differentiation efficiency. (**a** and **d**) Day 0; (**b** and **e**) Day 8 cells differentiated with IDMX, the standard differentiation cocktail mix; (**c** and **f**) Day 8 cells differentiated with IDMX in addition to pioglitazone, a PPARγ agonist

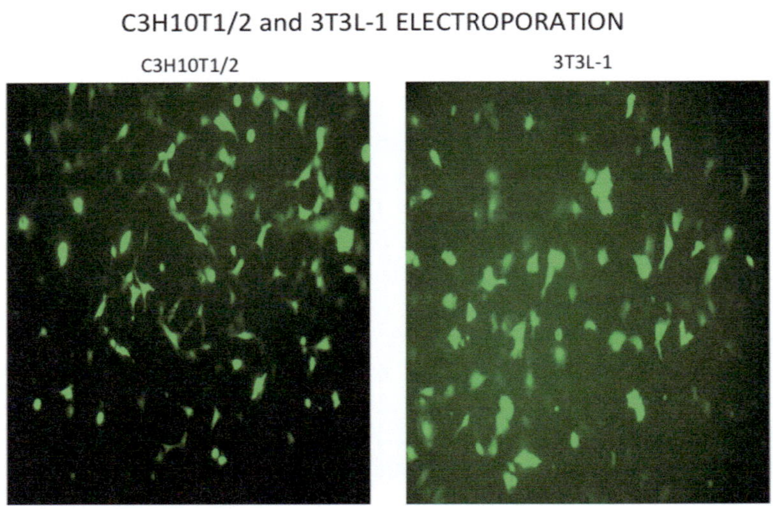

Fig. 2 Electroporation C3H/10T1/2 and 3T3L1 cells. 1×10^6 cells were electroporated using 1 μg pCDNA6.2/emGFP (Invitrogen) and the 10 μL Neon System Kit (Invitrogen). Photomicrographs taken at 40× using a Nikon fluorescence microscope. Several cells with lesser fluorescence intensity than shown in these digital images were evident in the microscopic field, representing 60–80 % of total cells

2. Remove growth media and incubate the cells with 2 mL trypsin–EDTA for 5 min at 37 °C or until cells are fully detached.
3. Add 8 mL growth medium/plate and transfer cells into 15 mL tubes.
4. Centrifuge for 3 min at $1000 \times g$ at room temperature. Discard supernatant.
5. Carefully resuspend cell pellets using sterile 1 mL pipette tips in Opti-MEM and combine cells in a 15-mL tube, 10 mL total volume. Gently mix and count cells with a hemocytometer.
6. Transfer the total desired number of cells ($0.5–1 \times 10^6$ cells/well of a 12-well plate) into a new 15-mL tube. Do not exceed 1×10^6 cells/well for the 10 μL Neon pipet tip. If a greater number of cells is needed, use the Neon System 100 μL kit (up to 5×10^6 cells/electroporation). The same Neon pipette is used for 10 or 100 μL tips.
7. Centrifuge for 3 min at $1000 \times g$ at room temperature. Discard supernatant.
8. Resuspend cells in the appropriate volume of buffer R (10 mL final volume/10^6 cells) and aliquot the suspension into sterile 1.5 mL tubes. Be sure to account for the contributions of both the DNA and cells to the 10 mL final volume. Do not leave the resuspended cells for more than 15–30 min at room temperature, which can reduce cell viability and transfection efficiency
9. Add 1–3 μg DNA to each cell suspension aliquot. Plasmid DNA should be of high quality (260/280 spectrophotometric ratio 1.8 or higher) with a concentration of 1–5 mg/mL. The DNA added to the cells should not exceed 10 % of the total volume. Gently mix the tubes containing cells and DNA by tapping. Carefully avoid introducing bubbles when drawing the cell–DNA mixture into the 10 μL Neon pipet tips (*see* **Note 7**).
10. Fill the Neon tube with 3–4 mL of buffer E. To avoid contamination, use a fresh tube when switching to a different plasmid or cell type. Changing the tube after ten electroporations is recommended, irrespective of plasmid/cell type (*see* **Note 8**).
11. Add 1 mL Opti-MEM to each well of 12-well plates and prewarm in a tissue culture incubator at 37 °C for 10–15 min (*see* **Note 9**).
12. Electroporate cells using 10 μL Neon tips and the following settings: 1400 V/20 ms/2 pulses (Neon System, program 16). Use of the same Neon tip more that two times is not recommended (*see* **Note 10**).
13. Pipette the electroporated cells into the Opti-MEM, prewarmed 12-well plates from **step 11**, and gently rock the plates

two or three times to ensure even distribution of the cells. Place plated cells in incubator.

14. 12–16 h later, change the media to fresh, pre-warmed Opti-MEM. Cells are now ready to be treated with pharmacological agents, recombinant proteins and/or other specific treatments.
15. Incubate cells for an additional 24–72 h in Opti-MEM (serum free) under the desired experimental conditions at 37 °C/5 % CO_2/95 % humidity.

4 Notes

1. Before excising the embryos, generously swab the abdomen with 70 % ethanol as a disinfectant and to prevent cutting through the dry fur, as this creates a bacterial aerosol that could later contaminate the cultured cells. When removing the uterus/embryos, take care not to touch the fur.
2. Prepare two 10-cm dishes containing ice-cold sterile PBS. Wash the embryos once in each dish to ensure complete removal of blood.
3. The extra tissues dissected from the embryo can be used to isolate DNA but not for making MEFs. Procedures after the removal of the extraembryonic tissues are done in the tissue culture hood.
4. The concentration of the trypsin used to dissociate tissue (0.25 %) is higher than that which is routinely used for cell culture (0.05 %). Do not leave the tissue in trypsin longer than 20 min as this results in cell death.
5. Approximately $8–9 \times 10^6$ cells are obtained per embryo. Cells from one embryo can be plated into four dishes (10-cm). Plating the cells directly into the 12-well plates without passaging prior to differentiation minimizes the number of cell divisions needed before the cells are induced. We have observed that MEFs plated directly differentiate into adipocytes with greater efficiency than cells that are passaged three times prior to induction. At the plated cell density, the cells take 3 days to reach confluency.
6. Thiazolidinediones are PPARγ agonists and are used to increase the adipocyte differentiation efficiency of preadipocyte and mesenchymal cell lines. The three commercially available thiazolidinediones (troglitazone, rosiglitazone, and pioglitazone) can all be used to improve adipocyte differentiation efficiency [34, 35]. The pioglitazone is only added in the induction cocktail at day 0 of differentiation. Differentiation efficiency increases to 80–90 %

in the presence of pioglitazone compared to 60–70 % in cells differentiated in absence of pioglitazone (Fig. 1).

7. To ensure plasmid DNA quantity and quality, we recommend large-scale preparation using a Maxiprep kit (Qiagen). We have found that pyrogen-free reagents have no substantial impact on experimental outcome.
8. For convenience in working with a 12-well format, we have found that the chamber buffer/Neon tube can be changed after each 12-well plate has been completed without problems with transfection efficiency or contamination.
9. Opti-MEM is a "reduced serum" medium that supports viability and basal levels of cell activity with much lower concentrations of serum (and absence in some short term applications) than usual. This is advantageous in cases for which assessment of the bioactivity of pharmacological agents or recombinant proteins is desired. C3H/10T1/2 and 3T3-L1 cells can be maintained for up to 72 h in Opti-MEM in absence of serum.
10. We have used the 10 mL Neon tips up to four times with no appreciable change in transfection efficiency.

References

1. Flegal KM, Carroll MD, Kit BK, Ogden CL (2012) Prevalence of obesity and trends in the distribution of body mass index among US adults, 1999–2010. JAMA 307(5):491–497
2. Tang QQ, Otto TC, Lane MD (2004) Commitment of C3H10T1/2 pluripotent stem cells to the adipocyte lineage. Proc Natl Acad Sci 101(26):9607–9611
3. Tseng YH, Kokkotou E, Schulz TJ, Huang TL, Winnay JN, Taniguchi CM, Tran TT, Suzuki R, Espinoza DO, Yamamoto Y, Ahrens MJ, Dudley AT, Norris AW, Kulkarni RN, Kahn CR (2008) New role of bone morphogenetic protein 7 in brown adipogenesis and energy expenditure. Nature 454(7207): 1000–1004
4. Huang H, Song TJ, Li X, Hu L, He Q, Liu M, Lane MD, Tang QQ (2009) BMP signaling pathway is required for commitment of C3H10T1/2 pluripotent stem cells to the adipocyte lineage. Proc Natl Acad Sci 106(31): 12670–12675
5. Townsend KL, An D, Lynes MD, Huang TL, Zhang H, Goodyear LJ, Tseng Y-H (2013) Increased Mitochondrial Activity in BMP7-Treated Brown Adipocytes, Due to Increased CPT1- and CD36-Mediated Fatty Acid Uptake. Antioxidants & Redox signaling 19(3):243–257
6. Choy L, Derynck R (2003) Transforming growth factor-beta inhibits adipocyte differentiation by Smad3 interacting with CCAAT/enhancer-binding protein (C/EBP) and repressing C/EBP transactivation function. J Biol Chem 278(11):9609–9619
7. Tsurutani Y, Fujimoto M, Takemoto M, Irisuna H, Koshizaka M, Onishi S, Ishikawa T, Mezawa M, He P, Honjo S, Maezawa Y, Saito Y, Yokote K (2011) The roles of transforming growth factor-β and Smad3 signaling in adipocyte differentiation and obesity. Biochem Biophys Res Commun. 407(1):68–73
8. Zaragosi LE, Wdziekonski B, Villageois P, Keophiphath M, Maumus M, Tchkonia T, Bourlier V, Mohsen-Kanson T, Ladoux A, Elabd C, Scheideler M, Trajanoski Z, Takashima Y, Amri EZ, Lacasa D, Sengenes C, Ailhaud G, Clement K, Bouloumie A, Kirkland JL, Dani C (2010) Activin A Plays a Critical Role in Proliferation and Differentiation of Human Adipose Progenitors. Diabetes 59(10): 2513–2521
9. Otto TC, Lane MD (2005) Adipose development: from stem cell to adipocyte. Crit Rev Biochem Mol Biol 40(4): 229–242
10. Taylor SM, Jones PA (1979) Multiple new phenotypes induced in 10T1/2 and 3T3 cells treated with 5-azacytidine. Cell 17(4):771–779
11. Konieczny SF, Emerson CP, Jr. (1984) 5-Azacytidine induction of stable mesodermal stem cell lineages from 10T1/2 cells: evidence

for regulatory genes controlling determination. Cell 38(3):791–800
12. Young HE, Mancini ML, Wright RP, Smith JC, Black AC, Jr., Reagan CR, Lucas PA (1995) Mesenchymal stem cells reside within the connective tissues of many organs. Dev Dyn 202(2): 137–144
13. Bowers RR, Kim JW, Otto TC, Lane MD (2006) Stable stem cell commitment to the adipocyte lineage by inhibition of DNA methylation: role of the BMP-4 gene. Proc Natl Acad Sci U S A 103(35):13022–13027
14. McBeath R, Pirone DM, Nelson CM, Bhadriraju K, Chen CS (2004) Cell shape, cytoskeletal tension, and RhoA regulate stem cell lineage commitment. Dev Cell 6(4):483–495
15. Wu Z, Bucher NL, Farmer SR (1996) Induction of peroxisome proliferator-activated receptor gamma during the conversion of 3T3 fibroblasts into adipocytes is mediated by C/EBPbeta, C/EBPdelta, and glucocorticoids. Mol Cell Biol 16(8):4128–4136
16. Rosen ED, Spiegelman BM (2000) Molecular regulation of adipogenesis. Annu Rev Cell Dev Biol 16:145–171
17. Timchenko NA, Wilde M, Nakanishi M, Smith JR, Darlington GJ (1996) CCAAT/enhancer-binding protein alpha (C/EBP alpha) inhibits cell proliferation through the p21 (WAF-1/CIP-1/SDI-1) protein. Genes Dev 10(7): 804–815
18. Morrison RF, Farmer SR (1999) Role of PPARgamma in regulating a cascade expression of cyclin-dependent kinase inhibitors, p18(INK4c) and p21(Wafl/Cip1), during adipogenesis. J Biol Chem 274(24): 17088–17097
19. Ahrens M, Ankenbauer T, Schroder D, Hollnagel A, Mayer H, Gross G (1993) Expression of human bone morphogenetic proteins-2 or -4 in murine mesenchymal progenitor C3H10T1/2 cells induces differentiation into distinct mesenchymal cell lineages. DNA Cell Biol 12(10):871–880
20. Kang Q, Song WX, Luo Q, Tang N, Luo J, Luo X, Chen J, Bi Y, He BC, Park JK, Jiang W, Tang Y, Huang J, Su Y, Zhu GH, He Y, Yin H, Hu Z, Wang Y, Chen L, Zuo GW, Pan X, Shen J, Vokes T, Reid RR, Haydon RC, Luu HH, He TC (2009) A comprehensive analysis of the dual roles of BMPs in regulating adipogenic and osteogenic differentiation of mesenchymal progenitor cells. Stem Cells Dev 18(4): 545–559
21. Bowers RR, Lane MD (2007) A role for bone morphogenetic protein-4 in adipocyte development. Cell Cycle 6(4):385–389
22. Neumann K, Endres M, Ringe J, Flath B, Manz R, Haupl T, Sittinger M, Kaps C (2007) BMP7 promotes adipogenic but not osteo-/chondrogenic differentiation of adult human bone marrow-derived stem cells in high-density micro-mass culture. J Cell Biochem 102(3): 626–637
23. Shen JJ, Huang L, Li L, Jorgez C, Matzuk MM, Brown CW (2009) Deficiency of growth differentiation factor 3 protects against diet-induced obesity by selectively acting on white adipose. Mol Endocrinol 23(1):113–123
24. Andersson O, Korach-Andre M, Reissmann E, Ibanez CF, Bertolino P (2008) Growth/differentiation factor 3 signals through ALK7 and regulates accumulation of adipose tissue and diet-induced obesity. Proc Natl Acad Sci 105(20):7252–7256
25. Wang W, Yang Y, Meng Y, Shi Y (2004) GDF-3 is an adipogenic cytokine under high fat dietary condition. Biochem Biophys Res Commun 321(4):1024–1031
26. Witthuhn BA, Bernlohr DA (2001) Upregulation of bone morphogenetic protein GDF-3/Vgr-2 expression in adipose tissue of FABP4/aP2 null mice. Cytokine 14:129-135
27. Levine AJ, Brivanlou AH (2006) GDF3, a BMP inhibitor, regulates cell fate in stem cells and early embryos. Development 133(2): 209–216
28. Rebbapragada A, Benchabane H, Wrana JL, Celeste AJ, Attisano L (2003) Myostatin signals through a transforming growth factor beta-like signaling pathway to block adipogenesis. Mol Cell Biol 23(20):7230–7242
29. Robinson KA, Ball LE, Buse MG (2007) Reduction of O-GlcNAc protein modification does not prevent insulin resistance in 3T3-L1 adipocytes. American journal of physiology Endocrinology and metabolism 292(3): E884–890
30. Stroh T, Erben U, Kühl AA, Zeitz M, Siegmund B (2010) Combined pulse electroporation—a novel strategy for highly efficient transfection of human and mouse cells. PloS one 5(3):e9488
31. Kim JA, Cho K, Shin MS, Lee WG, Jung N, Chung C, Chang JK (2008) A novel electroporation method using a capillary and wire-type electrode. Biosensors & bioelectronics 23(9):1353–1360
32. Hogan B, Beddington R, Constantini F, Lacy E (1994) Manipulating the Mouse Embryo: A Laboratory Manual, Second edition. Cold Spring Harbor Laboratory Press, Cold Spring Harbor, NY
33. Choy L, Skillington J, Derynck R (2000) Roles of autocrine TGF-beta receptor and

Smad signaling in adipocyte differentiation. J Cell Biol 149(3):667–682

34. Sugii S, Olson P, Sears DD, Saberi M, Atkins AR, Barish GD, Hong S-H, Castro GL, Yin Y-Q, Nelson MC, Hsiao G, Greaves DR, Downes M, Yu RT, Olefsky JM, Evans RM (2009) PPARgamma activation in adipocytes is sufficient for systemic insulin sensitization. Proc Natl Acad Sci 106(52):22504–22509
35. de Souza CJ, Eckhardt M, Gagen K, Dong M, Chen W, Laurent D, Burkey BF (2001) Effects of pioglitazone on adipose tissue remodeling within the setting of obesity and insulin resistance. Diabetes 50(8):1863–1871

Chapter 13

Imaging TGFβ Signaling in Mouse Models of Cancer Metastasis

Yibin Kang

Abstract

Metastatic spread of cancer cells from the primary tumors to distant vital organs, such as lung, liver, brain, and bone, is responsible for the majority of cancer-related deaths. Development of metastatic lesions is critically dependent on the interaction of tumor cells with the stromal microenvironment. As a multifunctional paracrine signaling factor that is abundantly produced by both tumor and stromal cells, TGFβ has been well established as an important mediator of tumor–stromal interaction during cancer metastasis. Imaging the in vivo dynamic of TGFβ signaling activity during cancer metastasis is critical for understanding the pathogenesis of the disease, and for the development of effective anti-metastasis treatments. In this chapter, I describe several xenograft methods to introduce human breast cancer cells into nude mice in order to generate spontaneous and experimental metastases, as well as the luciferase-based bioluminescence imaging method for quantitative imaging analysis of TGFβ signaling in tumor cells during metastasis.

Key words TGFβ, Tumor stroma, Metastasis, Xenograft, Mammary fat pad injection, Intravenous injection, Intracardiac injection, Animal model, In vivo imaging, Bioluminescence, Luciferase

1 Introduction

Metastasis represents the most devastating stage of cancer progression and has attracted intense research efforts in recent years [1–5]. In order to successfully colonize a distant target organ, metastatic carcinoma cells need to overcome several rate-limiting hurdles, including loss of epithelial polarity, invasion and migration through extracellular matrix and basement membrane, intravasation and survival in blood circulation, and finally, extravasation and adaptation to a foreign microenvironment [6]. Throughout the multistep process of cancer metastasis, the important contribution of the stromal microenvironment has been increasingly recognized [7]. In particular, the TGFβ pathway has been shown to convert from a tumor-suppressive pathway in early tumorigenesis to a metastasis-promoting pathway in late stage of cancer progression [8–10]. TGFβ signaling has been shown to promote various

Xin-Hua Feng et al. (eds.), *TGF-β Signaling: Methods and Protocols*, Methods in Molecular Biology, vol. 1344,
DOI 10.1007/978-1-4939-2966-5_13, © Springer Science+Business Media New York 2016

steps of metastasis, including angiogenesis, epithelial–mesenchymal transition, intravasation and extravasation, immunosuppression, and metastatic colonization in bone, lung and various other organs [9, 11–18]. While tumor cells are known to express elevated level of TGFβ, stromal cells, such as carcinoma-associated fibroblasts, platelets, endothelial cells, macrophages, and other immune cells [8, 9, 19–21], as well as other tumor microenvironment components, such as the bone matrix [16, 22, 23], are also rich source of TGFβ. It is therefore crucial to understand the in vivo dynamics of tumor–stroma interactions mediated by TGFβ-signaling during the initiation and progression of metastasis, which is critical for identifying the optimal therapeutic windows for testing TGFβ pathway inhibitors in preclinical and clinical trials.

The development of luciferase-based in vivo bioluminescence imaging (BLI) technology [24] has dramatically improved our ability to prove signaling pathway dynamics in various pathological processes at real time in living animals [25–28]. BLI imaging has been used to analyze 26S proteasome degradation [29], E2F and Cdk2 activities in tumor proliferation [30, 31], NFκB [32], Akt [31], hypoxia [33], and many other signaling pathways. Luciferase reporters under the control of TGFβ responsive elements have been used in stable cell lines and genetically engineered mice for imaging TGFβ activity during cancer, fibrosis, and other physiological and pathological condition [34–37]. We recently coupled the system with a Renilla luciferase internal control reporter for quantitative analysis of TGFβ signaling analysis in vivo during breast cancer metastasis (Fig. 1) [16]. In this system, the status of TGFβ-Smad signaling was monitored by placing firefly luciferase (F_{LUC}) under the control of a TGFβ-responsive promoter containing multiple Smad binding elements (SBEs). Constitutive expression of a *renilla*-luciferase (R_{LUC}) reporter under the CMV promoter was used to quantify tumor burden during metastasis development and to additionally serve to normalize F_{LUC} activity to tumor size in order to obtain a readout of relative strength of TGFβ signaling activity. Combing with mouse xenograft models of bone and lung metastasis, this system allowed us to obtain unprecedented insights into the TGFβ signaling dynamics during metastasis and its response to therapeutic regiments that directly or indirectly disrupt TGFβ signaling in vivo [16].

Among transplantable mouse models of cancer metastasis, injection of tumor cells into anatomically relevant sites (orthotopic injections) has the advantage of generating physiological relevant "primary tumors" that lead to spontaneous metastases to different distant sites. For example, breast cancer cells are often injected into the mammary fat pad to generate primary tumors in the mammary gland. Alternatively tumor cells can be introduced directly into blood circulation, a method that is often termed "experimental metastasis." This method bypasses the early steps of the metastasis

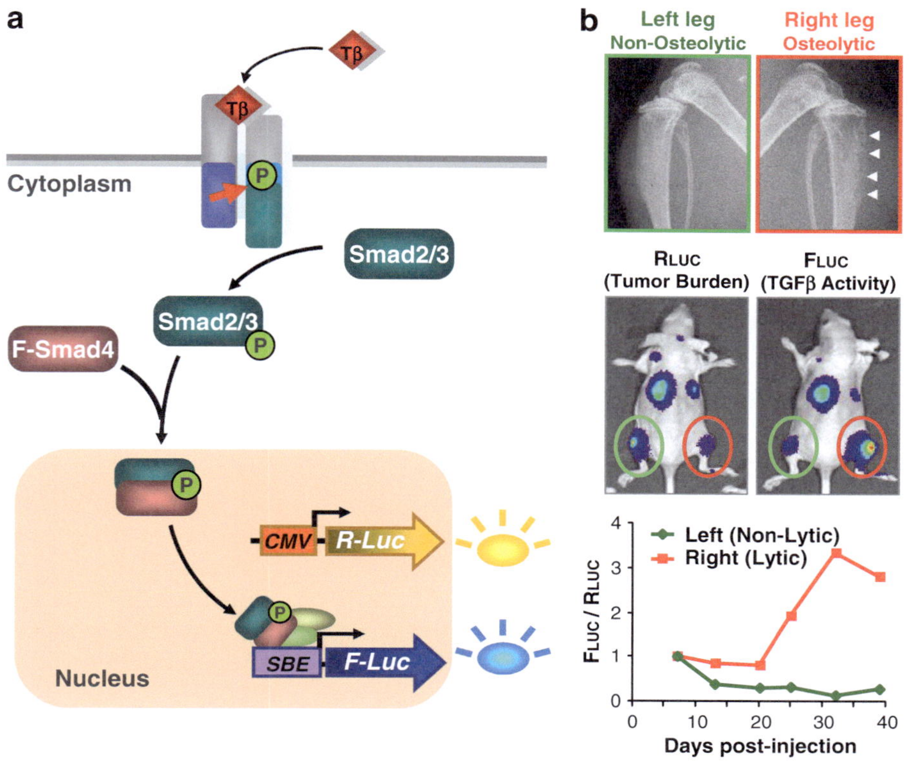

Fig. 1 The dual-luciferase reporter system for quantitative imaging of TGFβ activity in vivo. (**a**) A schematic diagram of the dual-luciferase reporter system. A stably integrated Renilla luciferase gene (R_{LUC}) under the control of the constitutively active CMV promoter allows for the detection of tumor burden in vivo by BLI. A firefly luciferase gene (F_{LUC}) under the control of multiple Smad-binding elements (SBEs) is the used for functional imaging of the TGFβ pathway activity in tumor cells during metastasis. (**b**) Tumor-induced osteolysis enhances signaling activity of the TGFβ pathway in tumor cells in vivo. Osteolytic lesion in the right leg induces significant bone destruction of the tibia and fibula (indicated by white *arrow heads*), resulting in enhanced signaling activity of the TGFβ-Smad pathway as indicated by increasing normalized F_{LUC} activity (*bottom panel, red line*). In contrast, non-osteolytic lesions at comparable sites on opposite legs show little sign of bone destruction and persistent low level of TGFβ signaling activity (*green line*). Figure adapted from [16].

cascade and can generate metastasis much more efficiently to facilitate meaningful statistical analysis. Additionally it has the advantage of the shorter incubation time for metastasis development and the higher possibility of generating metastases in organs that may be difficult or even impossible to produce using orthotopic injections. However, experimental metastasis models do not encompass the initial stages of primary tumor growth, invasion and metastasis from an orthotopic site. The route of inoculation greatly influences the metastasis pattern. For example, lateral tail vein injection results primarily in pulmonary metastases because the lung is the first capillary bed that tumor cells encounter following tail vein injections. Intracardiac injection of cells allows widespread distribution of tumor cells throughout the body of the animals and may produce metastases in multiple sites.

In this chapter, I describe three commonly used xenograft methods for the analysis of breast cancer metastasis, including mammary fat pad injection to generate spontaneous metastasis of breast cancer from the mammary gland, and intravenous as well as intracardiac injections to generate experimental metastases in lung, bone and other organs. I also describe methods to image TGFβ activity using luciferase reporters. Similar experimental metastasis models can also be applied to the analysis of tumor cells from other tumor types.

2 Materials

2.1 Mice

For analysis of breast cancer cells, we use female nude mice of about 5–6 weeks old at the beginning of the experiment. We commonly use NCr-nu/nu strain for intracardiac injections and BALB/c-nu/nu strain for tail vein and mammary fat pad injections. Mice can be ordered from the NCI animal production program, The Jackson Laboratory, the Charles River Laboratories, and other vendors (*see* **Note 1**).

2.2 Tumor Cells

Human tumor cell lines, free of mycoplasma and murine pathogenic viruses. Checking the cell lines for these viruses will reduce the risk of introducing pathogens into the animal facility. In addition, contamination of cell lines will dramatically influence their metastatic behaviors in vivo. For imaging TGFβ activity in vivo using BLI, genetically engineered cell lines with stable expression of dual luciferase reporters as previous reported [16] should be generated using standard cell culture and transfection techniques.

2.3 Media and Solutions

1. Cell culture medium with serum and antibiotics (penicillin, streptomycin, and fungizone).
2. PBS without Ca^{2+} and Mg^{2+}.
3. Trypsin–EDTA: 0.25 % w/v trypsin and 0.02 % w/v EDTA in PBS without Ca^{2+} and Mg^{2+}.
4. Anesthetic solution: make 10 mg/mL ketamine and 1 mg/mL xylazine from stock solutions (Sigma, K-113 or K4138). Store the solution at 4 °C. Use 10 μL of anesthetic solutions per 1 g weight of mice.
5. D-Luciferin substrate solution for firefly luciferase bioluminescence imaging: dissolve D-Luciferin (from Caliper, Promega or Nanolight) in PBS to make 15 mg/mL solutions. Aliquots can be stored with protection from light at −80 °C for several months. Use 100 μL/mouse for bioluminescence imaging.

6. Coelenterazine substrate solution for *Renilla* luciferase bioluminescence imaging: dissolve coelenterazine (from Caliper, Promega, or Nanolight) in (in 1:1 ethanol–1,2-propanediol) to make 5 mg/mL solutions. Aliquots can be stored with protection from light at −80 °C for several months. Right before in vivo imaging use, dilute the stock solution by tenfold in PBS to 0.5 mg/mL, vortex and make sure the diluted substrates is completely clear. Use 100 μL/mouse for bioluminescence imaging. The aqueous solution of the substrates is not stable and freshly diluted substrates for in vivo use need to be prepared each time.

2.4 Surgical Equipment

1. Sterile instruments and surgical supplies for surgery, cloth tape, alcohol wipes, scissors, pins, 12 mm wound clips and wound clip applier (Harvard Apparatus).
2. Mouse restraint device for intravenous injections.
3. Sterile 1 mL tuberculin syringes, insulin needles (for retro-orbital injection of luciferin solution), 26 gauge (G) (for intracardiac injections) and 27G (for intravenous injections) × 1/2″ (13 mm) needles. 10 μL Hamilton syringe with a 26G needle for mammary fat pad injections.
4. Dissecting microscope.
5. Warming lamp and warming pad.
6. Calipers for tumor measurements.

3 Methods

3.1 Preparation of Tumor Cell Suspensions

1. Aspirate culture medium from rapidly growing cancer cell culture that is 75 % ~ 90 % confluent. Wash with 10 mL of PBS per 10 cm dish; add 1–2 mL of the trypsin–EDTA solution. Shake the dish to cover the cells with trypsin and incubate until the cells begin to round up under the microscope. Shake or tap the dish to detach the cells (*see* **Note 2**).
2. Resuspend the cells in 10 mL of culture medium and transfer to a centrifuge tube. Determine total cell number by using a hemacytometer. Centrifuge at 200 × *g* for 5 min at 4 °C and resuspend the cell pellet in 10 mL of PBS.
3. Centrifuge again at 200 × *g* for 5 min at 4 °C and resuspend the cells in an appropriate volume of PBS to reach the desired concentration of cells for each injection method (see below). Make sure to have more than twice the amount of cells that are actually needed for the injection into mice.
4. Place the suspension in ice and proceed immediately to inject the cells.

3.2 Mammary Fat Pad (MFP) Injection for Spontaneous Metastasis Assay

1. Make a 10^7/mL cell suspension for MFP injection as in Subheading 3.1. 10 μL of the cell suspension will be injected into each mouse (*see* **Note 3**).
2. Anesthetize mice by using intraperitoneal injection of ketamine at 100 mg/kg and xylazine 10 mg/kg.
3. Invert tube or vortex the cell suspension gently to mix settled cells. Be certain that the cells are free of aggregates to prevent embolic obstruction (*see* **Note 2**). Gently mix the cells periodically and prior to each inoculation.
4. Use scissors to make an incision along the abdominal midline and laterally between the fourth and fifth nipples midway down to right hind leg (ventral side up). Carefully separate the skin flap from the body wall. The separated skin flap should be pinned to the surgery board thus exposing the #4 mammary fat pad.
5. Locate mammary fat pad under a dissection microscope. Use a 10 μL Hamilton syringe with a 26 g needle to 10 μL of cell suspension into the intact fat pad (below the draining lymph node). Close the wound by using 4–5 surgical staples.
6. Perform the same procedure for the other #4 mammary gland, if necessary.
7. Transplanted mice will then be imaged immediately after surgery using bioluminescence imaging as time 0 point (*see* Subheading 3.5).
8. The primary tumor outgrowth should be monitored weekly by taking measurements of the tumor length (L) and width (W). Tumor volume was calculated as $LW^2/6$. For metastasis assays, tumors should be surgically resected when they reached a volume greater than 300 mm^3. After resection, the mice should continue to be monitored by bioluminescent imaging for the development of metastases.

3.3 Intravenous (Tail Vein) Injection for Experimental Lung Metastasis Assay

1. Make a 2×10^6/mL cell suspension for intravenous (I.V.) injection as in Subheading 3.1. 100 μL of the cell suspension will be injected into each mouse.
2. Use a heating lamp to warm up the mice in order to dilate the tail veins (*see* **Note 4**).
3. Immobilize mice by using a mouse restraint (Fig. 2) and without anesthesia.
4. Invert tube or vortex the cell suspension gently to mix settled cells (*see* **Note 2**). Be certain that the cells are free of aggregates to prevent embolic obstruction. Gently mix the cells periodically and prior to each inoculation.
5. Inject 2×10^5 cells (100 μL) into the lateral tail vein through a 27G needle attached to a 1.0 mL syringe. The needle is inserted

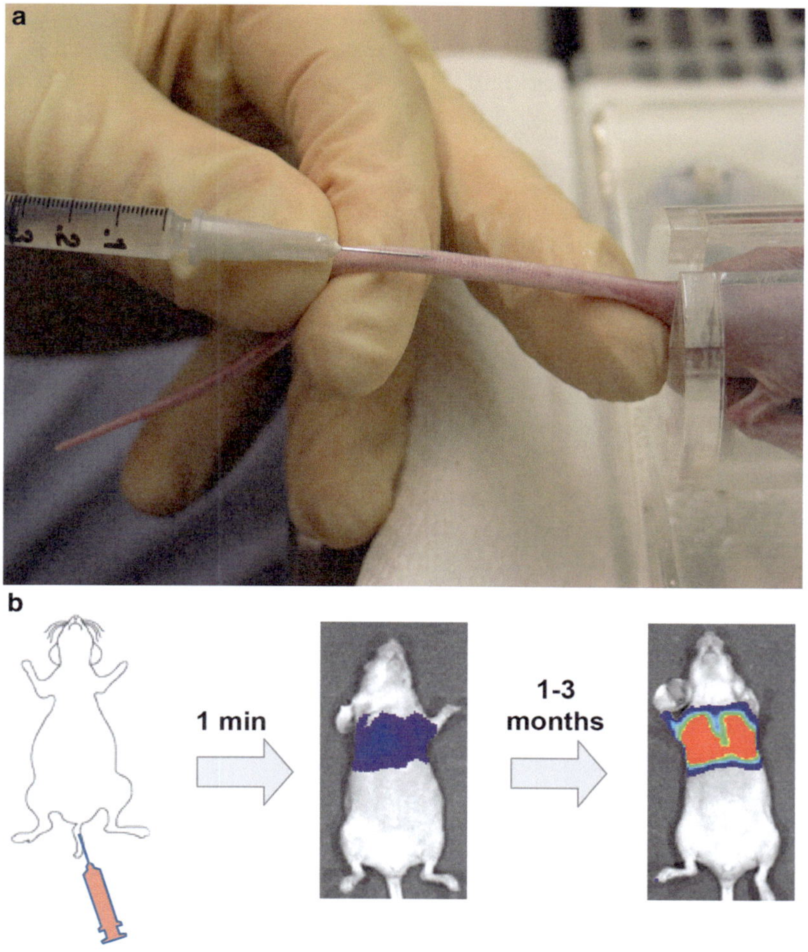

Fig. 2 The intravenous injection method and the development of lung metastasis after injection. (**a**) A mouse was stabilized by a mouse restraint and tumor cells were injected via the lateral tail vein. Note the positioning of needle and the angle of injection. (**b**) Bioluminescence imaging was performed immediately after injection to visualize the localization of luciferase-labeled tumor cells in the lung. The increase of the lung metastasis burden over time can be quantified by bioluminescence imaging

with its bevel pointing downward, at an angle of that is almost parallel to the vein (Fig. 2). Apply a pressure on the needle downward toward the vein while moving the needle straight forward (*see* **Note 5**). Injections should begin close to the tip of the tail in order to preserve injection sites further up the tail if necessary.

6. Following injections, pressure should be applied by gently holding cotton or gauze over the injection site for approximately 30 s to stop bleeding and prevent hematoma formation.

7. Transplanted mice should then be imaged immediately after surgery using bioluminescence imaging as time 0 point (*see* Subheading 3.5). These recipient mice should be imaged weekly to track the development of lung metastasis.

3.4 Intracardiac Injection for Experimental Bone and Brain Metastasis Assay

1. Make a 10^6/mL cell suspension for intracardiac injection as in Subheading 3.1. 100 μL of the cell suspension will be injected into each mouse.
2. Anesthetize mice by using intraperitoneal injection of ketamine at 100 mg/kg and xylazine 10 mg/kg.
3. Invert tube or vortex the cell suspension gently to mix settled cells (*see* **Note 2**). Be certain that the cells are free of aggregates to prevent embolic obstruction. Gently mix the cells periodically and prior to each inoculation.
4. Once the mouse is fully anesthetized, place it on its back with arms and legs extended (Fig. 3).
5. Place a piece of tape (cloth tape is better as it does not irritate the skin of nude mice and is less sticky) across the abdomen to secure the animal firmly on the working surface. Then tape the upper extremities in the extended position (Fig. 3). Do not place tape across the abdomen too tightly or the mouse will not be able to breathe properly. It is imperative that the mouse is symmetrically positioned.
6. When the mouse is firmly secured and not moving, wipe the chest with 70 % ethanol. This will enable you to visualize the landmarks as well as cleaning and sterilizing the inoculation site.
7. Gently mix the cell suspension. Leave an air space or bubble of about 200 μL near to the plunger and draw up the cell suspension into a syringe and attach a 26G needle. Tap the syringe with the needle pointing down to make sure all air bubbles are removed from the hub of the needle while maintaining the air bubble near the plunger. This is an important step as it allows the spontaneous entrance of blood into the hub of the needle when the left ventricle is entered. The arterial blood can not enter the syringe if no air space is left or if the needle is blocked. It is imperative that a new needle is used for each inoculation as blood can coagulate and clog the needle; however, the same syringe can be used.
8. Insert the needle, at an angle that is 45° to the right, 45° to the horizontal plane and pointing toward you. Insert the needle into the second intercostal space 3 mm to the left of the sternum, directing the tip into the center of the chest, to a depth of 6 mm. Pulsatile flow of red blood into the hub of the needle will indicate correct placement of the needle in the left ventricle, and gentle turning of the needle may be needed if red

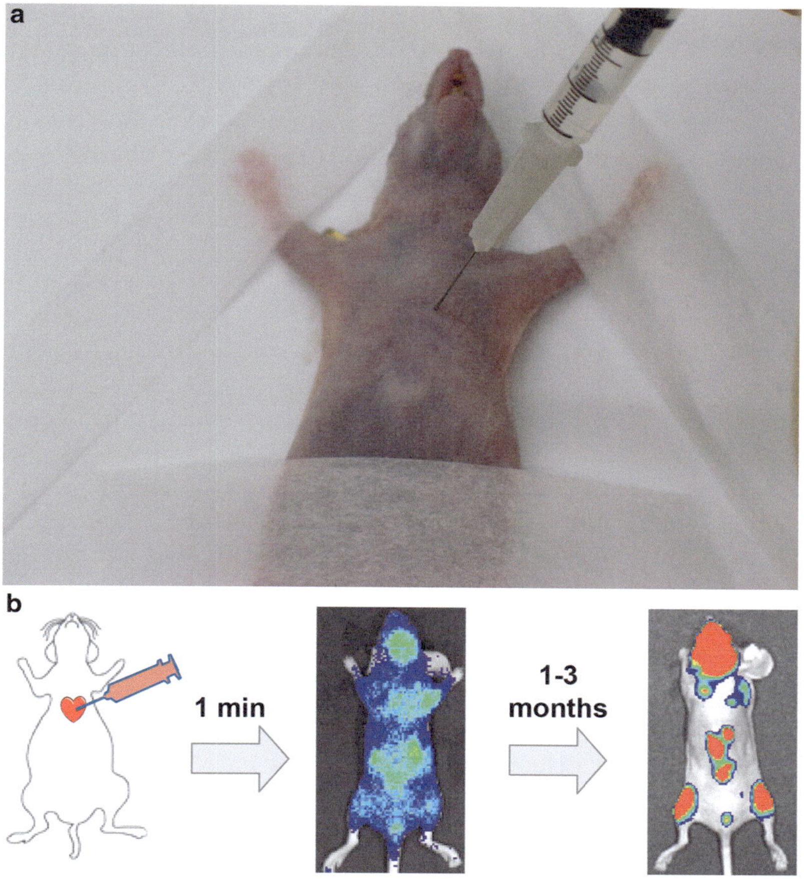

Fig. 3 The intracardiac injection method and the development of bone and brain metastasis after injection. (**a**) A mouse was anesthetized and stabilized with cloth tapes. Tumor cells were injected second intercostal space 3 mm to the left of the sternum, directing the tip into the center of the chest. Note the positioning of needle and the angle of injection. (**b**) Bioluminescence imaging was performed immediately after injection to visualize the diffusive distribution of luciferase-labeled tumor cells throughout the mouse. The development of metastases in the brain, spine, and limbs can be detected and quantified by noninvasive bioluminescence imaging

blood does not appear immediately (*see* **Note 6**). Steady the syringe and slowly inject 0.1 mL of suspension with one hand over 20–30 s. Do not inject the air bubble.

9. Quickly withdraw the needle from the chest to prevent seeding of tumor cells into the heart and lung. Place pressure on the chest with alcohol wipes for about 30 s to stop bleeding. Place the mouse on a heating pad to accelerate recovery and observe for bleeding or other unusual behavior. Return the mouse to a clean cage after it has completely recovered from the anesthetic.

10. Transplanted mice should then be imaged immediately after surgery using bioluminescence imaging as time 0 point (*see*

Subheading 3.5). These recipient mice should be imaged weekly to track the development of bone, brain, and other types of metastases.

11. Monitor the mice 1 h after injection to make sure they fully recover from injection. You can anesthetize one cage (5–6) of mice at a time and perform their injections together. Occasionally, administer warm sterile saline intraperitoneally (about 1–3 mL) to help the recovery if the mouse appears to be hypotensive (cold and pale) after getting anesthesia or an injection.
12. Observe the mice daily for signs of tumor burden, including paralysis, hunched posture, or weight loss. Euthanize when moribund or at predetermined time points, and necropsy and preserve tissue for histology if required. Examination of the skeleton by X-ray radiography can detect skeletal lesions. Soft tissue metastases can be detected and monitored noninvasively by bioluminescence imaging.

3.5 Bioluminescence Imaging of Metastasis with the Caliper (Xenogen) IVIS 200 System

1. Initialize IVIS system and wait for the status light to turn green.
2. Anesthetize mice by using intraperitoneal injection of ketamine at 100 mg/kg and xylazine 10 mg/kg.
3. Inject 100 μL of Luciferin (for firefly luciferase) or Coelenterazine (for Renilla luciferase) solution through the orbital plexus using an insulin needle.
4. Image the mice using the Xenogen IVIS 200 imaging system with appropriate field of view (E for imaging 5 mice at the same time) and binning (small for high resolution). The default imaging time is 1 min. Shorten the exposure time if the image is saturated.
5. To obtain firefly and Renilla luciferase activities from the same mouse, two imaging session using two imaging substrates should be performed on consecutive days.

4 Notes

1. The genetic background and the age of the nude mice may influence the outcome of xenograft metastasis assays. It is important to keep these parameters consistent in different sets of experiments in order to obtain reproducible results. When placing animal orders, specify the desired age (usually 4-weeks old) of mice at the time of delivery. Allow 1 week for the mice to adjust to the environment before the experiment. Place ear tags on mice and weigh them a day before the xenograft experiment. Coordinate the cell culture with the delivery of the mice to achieve the optimal time of the animal experiment. Do

not use mice more than 6 weeks old for xenograft experiments as they are more resistant to the development of metastasis.

2. Using a consistent method to prepare cell suspensions is another important consideration for reducing the variability in metastasis experiments. Use cells that are recently thawed from the frozen stock or freshly obtained from cancer patients. The cells should be in sub-confluent, actively growing cultures. The cells from confluent cultures are more likely to form clumps or aggregates, which can lead to dramatic changes in their metastatic behavior in vivo. In addition, the degree of confluence in vitro has been reported to regulate gene expression, which might also influence the metastasis phenotype. Using the Ca^{2+} and Mg^{2+}-free buffer will reduce the formation of clumps, and gentle vortexing before injection may help to break up aggregates. However, vigorous pipetting or vortexing should not be used in order to avoid the damage to the cells. Avoid over-trypsinizing the cells. Keep the suspension on ice and proceed with inject experiment as soon as possible. Depending on the set-up of animal facility, it may be necessary to arrange the experiment to be performed through cooperation of a team of 2 or 3 researchers.
3. Mammary fat pad injection has also been done through direct injection of tumor cells through nipples. However, such method does not guarantee the delivery of tumor cells into the mammary fat pad and in fact often create subcutaneous tumors. Injection of tumor cell suspension in a volume bigger than 10 μL will likely lead to the leakage of cells out of the mammary fat pad.
4. Dilation of lateral tail vein by heating the mice greatly facilitates intravenous injections. Dilation of the blood vessel allows the insertion of the needle which has a diameter that is often bigger than normal lateral tail vein vessels. Mice can be heated up by a heating lamp. Alternatively, the vein can be heated by immersing the tail in hot water. Such method is less effective and more time consuming than the heating lamp method. When mice are heated by a heating lamp, they need to be monitored closely. Overheating of the animals can lead to heat shock and death.
5. It is very easy to insert the needle "through" the tail vein (instead of "into" the vein) and inject the tumor cells intramuscularly instead of intravenously. To avoid such mistake, make sure that the needle can move up and down the vein without much resistance after the insertion of the needle into the blood vessel. Injection of the tumor cell suspension should meet little resistance if the needle is inserted correctly. If the blood vessel is clogged during the injection, remove the needle immediately and attempt the injection again at a different location up the tail.

6. It is important to inject the tumor cells into the left cardiac ventricle instead of other compartment of the heart to ensure the diffusive distribution of the cancer cells throughout the body of the mice. The appearance of dark color blood is often an indication of misplaced needles. In such circumstance, the needle needs to be retrieved immediately and the injection can be attempted again. However, it is not recommended to attempt intracardiac injection on the same mouse more than three times. Bioluminescence imaging immediately after the injection can be used to ensure the success of each injection.
7. All of the animal procedures need to be approved by the Institutional Animal Care and Use Committee (IACUC). Experimental design should take into account the well being of the mice and use appropriate procedures to reduce pain and suffering. In the context of this chapter this means careful monitoring of mice for development of tumor and metastasis burden, appropriate animal handling and efficient execution of surgical procedures, and the humane and timely use of euthanasia. Usage of bioluminescence imaging helps reducing the number of mice required for each experiment. Using a moribund end-point, rather than a death end point for a study is more practical if the point of the study is to assess the extent of tumor spread. The following end point criteria is routinely used in animal metastasis assays:
8. The animals should be euthanized as soon as possible and not longer than 24 h following detection of any the following symptoms:
   (a) Severe cachexia (weight loss approaching 25 %).
   (b) Inability to obtain food or water. General lack of moving activities.
   (c) Pale appearance with body temperature below 99 °F.
   (d) Breathing problem.
   (e) Pathological bone fracture.
   (f) Bite wound or ulcer.
   (g) Infection.
   (h) Apparent neurological disorder as judged from the pattern of movement.
9. Renilla luciferase signal fades quickly after injection of the substrate. Imaging should be performed within 1 min after substrate injection and the imaging time frame needs to be consistent among all mice in the experimental cohort.
10. The current protocol utilizes tumor cells lines with stable expression of TGFβ-responsive luciferase reporter and an internal CMV-driven Renilla reporter control [16]. To image TGFβ

activity in stromal cells, different lineage of stromal cells derived from TGFβ-responsive luciferase reporter transgenic mice [35] can be used by bone marrow transplantation and other methods in the xenograft models of metastasis.

References

1. Gupta GP, Massague J (2006) Cancer metastasis: building a framework. Cell 127(4):679–695
2. Kang Y (2005) Functional genomic analysis of cancer metastasis: biologic insights and clinical implications. Expert Rev Mol Diagn 5(3): 385–395
3. Steeg PS (2006) Tumor metastasis: mechanistic insights and clinical challenges. Nat Med 12(8):895–904
4. Sethi N, Kang Y (2011) Unravelling the complexity of metastasis—molecular understanding and targeted therapies. Nat Rev Cancer 11(10): 735–748
5. Valastyan S, Weinberg RA (2011) Tumor metastasis: molecular insights and evolving paradigms. Cell 147(2):275–292, PMCID: 3261217
6. Chambers AF, Groom AC, MacDonald IC (2002) Dissemination and growth of cancer cells in metastatic sites. Nat Rev Cancer 2(8): 563–572
7. Hanahan D, Coussens LM (2012) Accessories to the crime: functions of cells recruited to the tumor microenvironment. Cancer Cell 21(3): 309–322
8. Massague J (2012) TGFbeta signalling in context. Nat Rev Mol Cell Biol 13(10):616–630
9. Bierie B, Moses HL (2006) Tumour microenvironment: TGFbeta: the molecular Jekyll and Hyde of cancer. Nat Rev Cancer 6(7): 506–520
10. Kang Y (2006) Pro-metastasis function of TGFbeta mediated by the Smad pathway. J Cell Biochem 98(6):1380–1390
11. Yin JJ (1999) TGF-[beta] signaling blockade inhibits PTHrP secretion by breast cancer cells and bone metastases development. J Clin Invest 103:197–206
12. Padua D, Zhang XH, Wang Q, Nadal C, Gerald WL, Gomis RR et al (2008) TGFbeta primes breast tumors for lung metastasis seeding through angiopoietin-like 4. Cell 133(1):66–77, PMCID: 2390892
13. Kang Y, He W, Tulley S, Gupta GP, Serganova I, Chen CR et al (2005) Breast cancer bone metastasis mediated by the Smad tumor suppressor pathway. Proc Natl Acad Sci U S A 102(39):13909–13914
14. Muraoka-Cook RS, Shin I, Yi JY, Easterly E, Barcellos-Hoff MH, Yingling JM et al (2006) Activated type I TGFbeta receptor kinase enhances the survival of mammary epithelial cells and accelerates tumor progression. Oncogene 25(24):3408–3423
15. Massague J (2008) TGFbeta in Cancer. Cell 134(2):215–230
16. Korpal M, Yan J, Lu X, Xu S, Lerit DA, Kang Y (2009) Imaging transforming growth factor-beta signaling dynamics and therapeutic response in breast cancer bone metastasis. Nat Med 15(8):960–966
17. Sethi N, Dai X, Winter CG, Kang Y (2011) Tumor-derived JAGGED1 promotes osteolytic bone metastasis of breast cancer by engaging notch signaling in bone cells. Cancer Cell 19(2):192–205, PMCID: 3040415
18. Calon A, Espinet E, Palomo-Ponce S, Tauriello DV, Iglesias M, Cespedes MV et al (2012) Dependency of colorectal cancer on a TGF-beta-driven program in stromal cells for metastasis initiation. Cancer Cell 22(5):571–584, PMCID: 3512565
19. Yingling JM, Blanchard KL, Sawyer JS (2004) Development of TGF-beta signalling inhibitors for cancer therapy. Nat Rev Drug Discov 3(12): 1011–1022
20. Dumont N, Arteaga CL (2003) Targeting the TGF beta signaling network in human neoplasia. Cancer Cell 3(6):531–536
21. Labelle M, Begum S, Hynes RO (2011) Direct signaling between platelets and cancer cells induces an epithelial-mesenchymal-like transition and promotes metastasis. Cancer Cell 20(5):576–590, PMCID: 3487108
22. Pfeilschifter J, Mundy GR (1987) Modulation of type beta transforming growth factor activity in bone cultures by osteotropic hormones. Proc Natl Acad Sci U S A 84(7):2024–2028
23. Dallas SL, Rosser JL, Mundy GR, Bonewald LF (2002) Proteolysis of latent transforming growth factor-beta (TGF-beta)-binding protein-1 by osteoclasts. A cellular mechanism for release of TGF-beta from bone matrix. J Biol Chem 277(24):21352–21360
24. Contag PR, Olomu IN, Stevenson DK, Contag CH (1998) Bioluminescent indicators in living mammals. Nat Med 4(2):245–247

25. Gelovani Tjuvajev J, Blasberg RG (2003) In vivo imaging of molecular-genetic targets for cancer therapy. Cancer Cell 3(4):327–332
26. Gross S, Piwnica-Worms D (2005) Spying on cancer: molecular imaging in vivo with genetically encoded reporters. Cancer Cell 7(1):5–15
27. Jenkins DE, Oei Y, Hornig YS, Yu SF, Dusich J, Purchio T et al (2003) Bioluminescent imaging (BLI) to improve and refine traditional murine models of tumor growth and metastasis. Clin Exp Metastasis 20(8):733–744
28. Weissleder R (2002) Scaling down imaging: molecular mapping of cancer in mice. Nat Rev Cancer 2(1):11–18
29. Luker GD, Pica CM, Song J, Luker KE, Piwnica-Worms D (2003) Imaging 26S proteasome activity and inhibition in living mice. Nat Med 9(7):969–973
30. Uhrbom L, Nerio E, Holland EC (2004) Dissecting tumor maintenance requirements using bioluminescence imaging of cell proliferation in a mouse glioma model. Nat Med 10(11):1257–1260
31. Zhang L, Lee KC, Bhojani MS, Khan AP, Shilman A, Holland EC et al (2007) Molecular imaging of Akt kinase activity. Nat Med 13(9):1114–1119
32. Gross S, Piwnica-Worms D (2005) Real-time imaging of ligand-induced IKK activation in intact cells and in living mice. Nat Methods 2(8):607–614
33. Lu X, Yan CH, Yuan M, Wei Y, Hu G, Kang Y (2010) In vivo dynamics and distinct functions of hypoxia in primary tumor growth and organotropic metastasis of breast cancer. Cancer Res 70(10):3905–3914, PMCID: 2872139
34. Luo J, Lin AH, Masliah E, Wyss-Coray T (2006) Bioluminescence imaging of Smad signaling in living mice shows correlation with excitotoxic neurodegeneration. Proc Natl Acad Sci U S A 103(48):18326–18331, PMCID: 1838750
35. Lin AH, Luo J, Mondshein LH, ten Dijke P, Vivien D, Contag CH et al (2005) Global analysis of Smad2/3-dependent TGF-beta signaling in living mice reveals prominent tissue-specific responses to injury. J Immunol 175(1):547–554
36. Zhang J, Lu Y, Dai J, Yao Z, Kitazawa R, Kitazawa S et al (2004) In vivo real-time imaging of TGF-beta-induced transcriptional activation of the RANK ligand gene promoter in intraosseous prostate cancer. Prostate 59(4):360–369
37. Chong AK, Satterwhite T, Pham HM, Costa MA, Luo J, Longaker MT et al (2007) Live imaging of Smad2/3 signaling in mouse skin wound healing. Wound repair and regeneration: official publication of the Wound Healing Society [and] the European Tissue Repair. Society 15(5): 762–766

Chapter 14

Generation and Characterization of Smad7 Conditional Knockout Mice

Yi Pan and Yan Chen

Abstract

Smad7 is an important negative modulator that regulates the duration and strength of TGF-β signaling. Dysregulation of Smad7 is associated with the pathogenesis of many human diseases. Various mouse models are developed to facilitate addressing the physiological functions of Smad7. We have recently demonstrated that loss of Smad7 function by deletion in its MH2 domain leads to multiple cardiac defects and aggravates ethanol-induced liver injury. Here, we describe the procedure to construct and characterize the *Smad7* conditional knockout mice.

Key words Smad7, Mouse model, Gene deletion, Heart development

1 Introduction

Transformation growth factor-β superfamily members (TGF-βs) play important roles in development, immunity, inflammation, and various physiological processes [1]. The transmission of TGF-β signals is initiated by TGF-β family member binding to their serine/threonine kinase receptors. Then the receptor regulated R-Smads (Smad1/5/8 for BMPs and Smad2/3 for TGF-β) are phosphorylated, and form a complex with Co-Smad (Smad4). The complex is translocated into the nucleus and results in the expression of a broad spectrum of target genes [2–4]. On the other hand, I-Smads (Smad6 and Smad7) negatively regulate the TGF-β signaling, in which Smad6 specifically inhibits BMP signaling, while Smad7 antagonizes the signal transduction downstream both BMPs and TGF-β/activins [5, 6].

Smad7 is regarded as a potential effector in auto-regulatory feedback loop in TGF-β signaling and mediates the cross talk between TGF-β pathway and various other signaling pathways [7]. Aberrant regulation of Smad7 is implicated in many human diseases, including malignancy, scleroderma, and chronic inflammatory bowel diseases [8–10]. To illustrate the physiological function of

Xin-Hua Feng et al. (eds.), *TGF-β Signaling: Methods and Protocols*, Methods in Molecular Biology, vol. 1344,
DOI 10.1007/978-1-4939-2966-5_14,

Smad7, multiple *Smad7* mouse models have been developed by either overexpression or deletion of *Smad7* in specific organs or tissues [11].

We have recently constructed a *Smad7* conditional knockout mouse, which can allow deletion of Smad7 in a tissue-specific and temporal-specific manner [12]. The construction procedure of these mice is based on the *E. coli* recombination system reported by Liu [13] with some modifications, which includes *Smad7* genome sequence retrieval from BAC, the first flox targeting to the upstream region of exon 4, and the insertion of the FRT-flox site after the stop codon. With this mouse model, exon 4 that encodes MH2 domain of Smad7 protein can be deleted, leading to complete abolishment of the inhibitory activity of Smad7 on TGF-β signaling. We have obtained *Smad7*$^{\Delta MH2-/-}$-mutant mice by breeding the conditional *Smad7*^{loxp} mice with transgenic mice expressing Cre in germ cells. Additionally, the liver specific Smad7 knockout mice were achieved by crossing the *Smad7*^{loxp} mice with *Albumin-Cre* transgenic mice [14, 15]. With these two mice models, we demonstrate that Smad7 is required for cardiovascular development and that endogenous Smad7 plays an important role in liver function.

2 Materials

Reagent:

1. Ampicillin, chloramphenicol, kanamycin, L (+) arabinose.
2. NotI, SpeI, BamHI, EcoRI, SalI HindIII, XhoI, ScaI, T4 ligase are purchased from NEB.
3. dNTP and high-fidelity Taq polymerase are purchased from TAKARA.
4. Qiagen gel purification kit, QIAquick PCR Purification Kit.
5. LB medium: Dissolve 10 g tryptone, 5 g yeast extract, and 10 g NaCl in 950 mL deionized water, then adjust the volume up to 1 L with ddH_2O. The medium is autoclaved on liquid cycle for 20 min at 15 psi.
6. SOB medium: dissolve 20 g trptone, 5 g yeast extract, 0.5 g NaCl in 900 mL ddH_2O. Add 2.5 mL of 1 M KCl, 10 mL of 1 M $MgCl_2$, and 10 mL of 1 M $MgSO_4$, then adjust the volume up to 1 L with ddH_2O. The medium is autoclaved on liquid cycle for 20 min at 15 psi.

Strains and plasmids: the *E. coli* strain EL350 and EL250 and plasmid PL451, PL452, PL253 are obtained from Frederick National Laboratory. 129S7/AB2 clones are ordered from Sanger. DH10B competent cells.

Primers: the primer sets and amplification targets are listed in the Table 1.

Table 1
Primer sets and amplification targets

Amplification targets	Primer	Primer sequence
5′ Left mini-arm DNA	LR-F	atctgcggccgc ACACATTTGCTGTTCCCACA
	LR-R	aagcactagt GGTATCACCGTGAGCCTGTT
3′ Right mini-arm DNA	RR-F	ctgaactagtAGTTCCCAGGATTTGCAATG
	RR-R	atcggatccGCCAAAAGTGTTGGGTGACT
Mini-arm P1	P1-F	atctgcggccgcGCATCCTCTTTGTGTGTAGCTG
	P1-R	tta ggatccCTCGAGGCCTTTGTTGTTTCTG
Mini-arm P2	P2-F	cag gaattcGACTAGACACATTCCCGGTACA
	P2-R	attc gtcgacTGGAAGTGACACTGGTAAGGTG
Mini-arm P3	P3-F	atct gcggccgcTGGTGTTGTTATTGCCATTCAT
	P3-R	atc ggatccCCAACAAAAATCCGCACCATA
Mini-arm P4	P4-F	Cgt aagcttGTTGTGTGTGTGTGTTTTGTTTTT
	P4-R	gac ctcgagGTGTCCAAAAGGCTAGAGAGACA

3 Methods

3.1 Target Sequence for Smad7 Conditional Knockout Mice

1. Using a mouse genome browser http://www.ensembl.org/Mus_musculus, download the *Smad7* genomic locus, and locate the exon and intron on the genomic sequence, and then design the region to be floxed (*see* **Note 1**).
2. In http://www.ensembl.org/Mus_musculus, BAC clones can be viewed with setting in the *Misc. regions & clones* menu in "configure this page" dialogue at the left of the region in detail or region overview pages under the location tab. Select and order 129S7/AB2 BAC clone from Sanger Institute.
3. Analyze the restriction enzyme sites of the sequence with online tool NEBcutter2 (http://tools.neb.com/NEBcutter2/index.php). And select HindIII as a 3′ end of genomic cutter for *Smad7* knockout genotyping by Southern blotting (*see* **Note 2**).

3.2 Construct of the Retrieval and Targeting Vector

1. Design 5′ left mini-arm and 3′ right mini-arm which are used to retrieve the *Smad7* genome sequence based on the genomic analysis in the above section.
2. The 3′ right mini-arm fragment is amplified in the reaction mixture containing 4 μL of dNTP (2.5 mM each), 0.5 μL of each primers (10 μM of RR-F/RR-R), 1 μL of BAC (10–100 ng), 5 μL 10× PCR buffer, 0.5 μL high-fidelity Taq (5 U/μL), and 38.5 μL of H_2O. The PCR reaction is performed in the

Bio-Rad thermal cycler with the following program: 94 °C for 10 min, 35 cycles of 94 °C for 45 s, 57 °C for 30 s, 72 °C for 90 s, then the additional extension at 72 °C for 10 min, and cool down to 4 °C. 5 μL of PCR product is checked by running in the 1.2 % of gel (*see* **Note 3**).

3. The remaining PCR product is purified with the QIAquick PCR Purification Kit as the follows: the PCR product is mixed with 5 volume of PB and applied to the QIAquick column, and then centrifuged for 1 min. The column is washed twice with 750 μL PE buffer and spins at the maximum speed for 1 min. Discard the flow-through, and then centrifuge for an additional 1 min. The DNA is eluted with 35 μL of ddH_2O and 5 μL of the eluate is loaded on the 1.2 % of gel to be checked.
4. The 30 μL of the purified 3' right mini-arm fragment is mixed with 4 μL of 10× restriction buffer, 0.5 μL of BamHI, 0.5 μL of SpeI, and 5 μL of ddH_2O, and incubated at 37 °C for 1–4 h. And the product is purified with the QIAquick PCR Purification column as described **step 3**.
5. The PL253 is digested in the reaction mixture containing 10 μg of PL253, 5 μL of 10× restriction buffer, 0.5 μL of BamHI, 0.5 μL of SpeI. The reaction mixture is adjusted to 50 μL with ddH_2O, and then incubated at 37 °C for 4 h to overnight.
6. The digested mixture is loaded onto the 0.6 % of gel. After separation, the digested PL253 is subject to purification with the Qiagen gel purification kit as the follows. The gel only with the expected DNA fragment is excised and added with 3 volumes of buffer QC, then incubated at 50 °C until the gel completely dissolved. Add 1 volume of isopropanol to the sample and mix well. Load the mixture onto the column and spin at the maximum speed for 1 min. Then wash the column twice with PE buffer, and discard the flow-through. After 1 min additional centrifugation, the digested PL253 is eluted with 35 μL ddH_2O.
7. 50–100 ng of digested product is mixed with 1 μL of 10× ligation buffer, 1 μL digested PCR fragment from **step 4** (about 1:1–3:1 molar ratio of the insert DNA to the vectors), 1 μL of T4 ligase. Adjust the reaction volume to 10 μL and incubate at 16 °C for overnight.
8. The ligation product is added into 50 μL of DH10B competent cells, mixed well, and kept on ice for 30 min. Then the mixture is heated at 42 °C for 90 s, 1 mL of LB is added and incubated at 37 °C for 1 h, and then spread on the LB plate with 100 μg/mL ampicillin. The positive colony is identified by the restriction digestion and named as PL253-RR.
9. The 5′ left mini-arm DNA fragment is amplified with primers LR-F/LR-R as described in **step 2**, and then purified as **step 3**.

The purified DNA fragment is digested with SpeI/NotI. The reaction is set as the **step 4** and then purified as described as **step 3**.

10. The PL253-RR plasmid is prepared with Qiagen tip-100 midi-prep kit then digested with SpeI/NotI in 50 μL of reaction mixture (*see* **step 5**), followed by gel-purified (*see* **step 6**).
11. The retrieval vector is generated in the 10 μL of ligation mixture containing 50–100 ng PL253-RR cut with SpeI/NotI from **step 10**, 1 μL 5' left mini-arm DNA fragment digested with SpeI/NotI from **step 9** (about 1:1–3:1 molar ratio of the insert DNA to the vectors), 1 μL of 10× ligation buffer, 1 μL of T4 ligase. The ligation mixture is incubated at 16 °C for overnight and then transformed into the DH10B competent cells as **step 8**. The positive retrieval vector is identified by the restriction digestion (*see* **Note 3**).
12. For construction of the first loxp targeting vector, 200–400 bp of min-arm P1 and P2 is amplified separately with primers P1-F/P1-R and P2-F/P2-R by PCR as described in **step 2**, with NotI/BamHI added at 5′/3′ end of P1 and with EcoRI/SalI added at 5′/3′ end of P2., PCR products of P1 and P2 are purified respectively as described in **step 3**.
13. The first ligation reaction is performed with P1 (digested with BamHI/NotI) and PL452 (cut with BamHI/NotI) as described in **step** 7. Then the second ligation is performed between the positive plasmid carrying P1 fragment (EcoRI/SalI digestion) and P2 (EcoRI/SalI digestion). The final positive product is named as P1-loxp-Neo-flox-P2 targeting vector (*see* **Note 3**).
14. For construction of the second loxP targeting vector, 200–500 bp of P3 and P4 are amplified separately with primers P3-F/P3-R and P4-F/P4-R by PCR, with NotI/BamHI added at the 5′/3′ end of the P3 and with HindIII/XhoI added at 5′/3′ end of P4. PCR product is purified as **step 3**.
15. P3, P4 is digested by BamHI/NotI and XhoI/HindIII respectively, and sequentially cloned into PL451 as described in **step 13**. The final positive plasmid is named as P3-flox-FRT-Neo-FRT-P4 targeting vector (*see* **Note 3**).

3.3 Transformation of BAC into EL350

1. The BAC is purified with Qiagen tip-100 midi-prep as following. The DH10B with BAC is inoculated into 100 mL of LB with 20 μg/mL chloramphenicol and grown overnight. Cells are harvested by centrifugation at 5000×*g* for 10 min at 4 °C. The pellets are well resuspended in 4 mL of P1. Then 4 mL of P2 is added, mixed gently and thoroughly by inverting four to six times. Then the mixture is incubated in the room temperature for 5 min. 4 mL of prechilled P3 is added, mixed immediately by inverting four to six times, and incubated on ice for 15 min.

The sample is centrifuged at 20,000 ×*g* for 30 min at 4 °C. The DNA supernatant is transferred promptly to a clean tube and recentrifuged at 20,000 ×*g* for 15 min to make the fluid clear. Equilibrate the QIAGEN-tip 100 with 4 mL of QBT, and allow the buffer drain by gravity flow. Apply the DNA supernatant to the column and let it enter the resin by gravity flow. Wash the QIAGEN-tip 100 with 10 mL of QC twice. Elute the BAC DNA with 5 mL of QF. 0.7 volume of isopropanol is added into the eluate to precipitate the BAC DNA by immediate centrifugation at 15,000 ×*g* for 15 min. Finally, the pellet is washed with 2 mL of 70 % ethanol, air-dried, and then resuspended in 100 μL of H_2O.

2. EL350 cells are grown in 5 mL of SOB at 32 °C overnight (*see* **Note 4**). The next day, the culture is diluted 50-folds into 100 mL of SOB in a 500 mL flask. When OD_{600} of the cells reaches 0.4, the culture is placed on ice for 15 min. The cells are harvested by centrifugation at 4000 ×*g* for 10 min at 4 °C. The pellet is resuspended in 10 mL of ice-cold ddH_2O and collected by centrifugation at 4000 ×*g* for 10 min. This process is repeated once. The supernatant is decanted and the pellet is washed in the prechilled 10 % glycerol and collected by centrifugation twice. Finally, the supernatant is carefully aspirated, and the pellet is resuspended gently in 100 μL of 10 % glycerol by swirling. The cells are aliquoted and stored at −80 °C or can be directly used for immediate electroporation.
3. The electrocompetent EL350 cells frozen at −80 °C are thawed on ice. 1 μL of BAC DNA (0.1–0.5 μg) is added and mixed with the electrocompetent EL350 cells on ice. The mixture is transferred into the 0.1 cm-gap electroporation cuvette. The electroporation is performed with electroporator setting at 1.75 KV, 25 μF and 200 Ω. Then 1 mL of LB is added into the cuvette and transferred into a 1.5 mL tube, which is incubated in the 32 °C for 1 h. Cells are then collected and spread on a plate with 20 μg/mL chloramphenicol.

3.4 Retrieval of Smad7 Sequence from BAC Clone

1. A single colony of the EL350 with BAC is inoculated in the 5 mL of SOB with 20 μg/mL chloramphenicol at 32 °C, and reinoculated into 100 mL of SOB with 20 μg/mL chloramphenicol by 50-fold dilution the next day. When the OD_{600} reaches 0.4, transfer the culture into 42 °C incubator and shake at 250 rpm for 20 min. Then the culture is cooled on ice and prepared for electrocompetent cells (*see* **step 2** in section "Transformation of BAC into EL350").
2. The retrieval vector (1–10 μg) from **step 11** in section "Construct of the Retrieval and Targeting Vector" is digested with 20 unit SpeI (NEB) in 40 μL volume for overnight. 1 μL of linearized product (0.1–0.5 μg) is mixed with EL350/BAC

electrocompetent cells, and electroporated is performed in a 0.1 cm-gap cuvette (1.75 kV, 25 μF and 200 Ω). 1 mL of LB is added into the mixture and incubated at 32 °C for 1 h, and the cells are plated on LB plate with 100 μg/mL ampicillin. The positive colony is diagnosed with the restriction digestion.

3.5 Targeting the Flox Sites into the Subcloned Genome Fragment

1. The EL350 containing the plasmid with the retrieved fragment from **step 2** in section "Retrieval of Smad7 Sequence from BAC Clone" is grown in 100 mL of SOB with 100 μg/mL ampicillin. When the OD_{600} reaches 0.4, the cells are placed in the 42 °C, and kept shaking at 250 rpm for 20 min. Then the cells are put on ice for 15 min to cool down. Then the cells are prepared for the electrocompetent cells.
2. 10 μg of P1-loxp-Neo-flox-P2 vector is digested by SalI/NotI (NEB) to release the P1-flox-Neo-flox-P2 fragment in 40 μL reaction volume. The fragment is recovered by gel purification with Qiagen gel purification kit(*see* **Note 5**).
3. 1 μL of targeting P1-flox-Neo-flox-P2 fragment (0.1–0.5 μg) is added to the electrocompetent EL350 with the retrieved DNA from **step 1**. After electroporation (1.75 kV, 25 μF and 200 Ω), 1 mL of LB is added and the mixture is incubated at 32 °C for 1 h. The cells are collected and spread on the plate with ampicillin/kanamycin (100 μg/mL ampicillin and 50 μg/mL kanamycin). The colony is picked and diagnosed with the restriction digestion (*see* **Note 6**).
4. Before second flox (flox-FRT-Neo-FRT) is targeted into the retrieval vector, the Neo cassette between flox sites in the targeted fragment needs to be removed as **step 5**.
5. A single colony of EL350 is picked and grown in 5 mL of LB at 32 °C overnight, then inoculated into the 100 mL LB at 32 °C. When the OD_{600} reaches 0.4, add l(+) arabinose (Sigma) to the culture to a final concentration of 0.1 % to induce Cre. The culture is grown for another 1 h, and then cooled down on the ice and prepared for the electrocompetent cells. 1 μL of positive plasmid DNA (1–10 ng) with flox-Neo-flox cassette from the **step 3** in this section is mixed with the electrocompetent cells and performed electroporation. The cells are then plated on LB plate with 100 μg/mL ampicillin. The colony with excision of Neo cassette will grow on LB plate with ampicillin but not on LB plate with kanamycin. The colony is picked and diagnosed by the restriction enzyme digestion and confirmed by sequencing. The final product is the plasmid containing the retrieval DNA with the first flox site.
6. The single colony of EL350 containing the plasmid with first flox in retrieval fragment is picked and grown in 5 mL SOB at 32 °C overnight and then transferred to 100 mL of SOB with

100 μg/mL ampicillin by 50-fold dilution. When the OD_{600} reaches 0.4, the cells are placed at 42 °C and shook at 250 rpm for 20 min to induce the recombination proteins. The cells are prepared for the electrocompetent cells.

7. The P3-flox-FRT-Neo-FRT-P4 targeting vector is digested with NotI/XhoI and subjected to the gel purification (*see* **Note 5**). 1 μL of purified P3-flox-FRT-Neo-FRT-P4 fragment (0.1–0.5 μg) is mixed with the electrocompetent cells (from **step 6**) and used for electroporation. The mixture is recovered in LB at 32 °C for 1 h, and then spread on the LB plate with ampicillin/kanamycin (100 μg/mL ampicillin and 50 μg/mL kanamycin).
8. Pick the single colony to perform the restriction digestion (*see* **Note 6**) and sequence confirmation for the locus of the second flox sites (*see* **Note 7**). The final product is named as *Smad7* CKOΔMH2 targeting vector.

3.6 Targeting Smad7 into the ES Cells and Generating of Smad7 Mutant Mice

20 μg of *Smad7*CKOΔMH2 targeting vector is linearized by NotI and electroporated into embryonic stem cells that are subjected to G418 selection (400 μg/mL). The targeted clone is identified with 3′ probes by southern screening (*see* **Note 8**). After the chimeric mice are generated by blastocyst injection, they are intercrossed with C57BL/6 mice to obtain heterozygous Smad7 conditional knockout mice (Fig. 1). By crossing with EIIa-Cre mice, systemic deletion of Smad7 MH2 domain is obtained (*see* **Note 9**).

4 Notes

1. *Smad7* genome sequence on chromosome18: 75362768: 75403325 is used for analysis, and the coding region of exon 4 located in 79393827–79394568 is to be floxed. The first flox site is designed to insert at 1423 bp before the fourth exon, and the second flox site is inserted at 202 bp after the stop codon.
2. The restriction recognition site (HindIII in this protocol) used for the southern blot should be located outside of the retrieved DNA region. Before the genomic region is retrieved, a Southern blotting assay needs to be tested with the wild type genome to make sure the expected wild type bands are obtained.
3. PCR primer is designed by primer 3, http://frodo.wi.mit.edu/primer3/. The BAC DNA is used as the template for the current and following PCR reaction as it can give rise to the high yield of products. We use the two-step ligation to construct for the retrieval vector and targeting vectors. It may take some times, but we can get good yield of product. One-step ligation

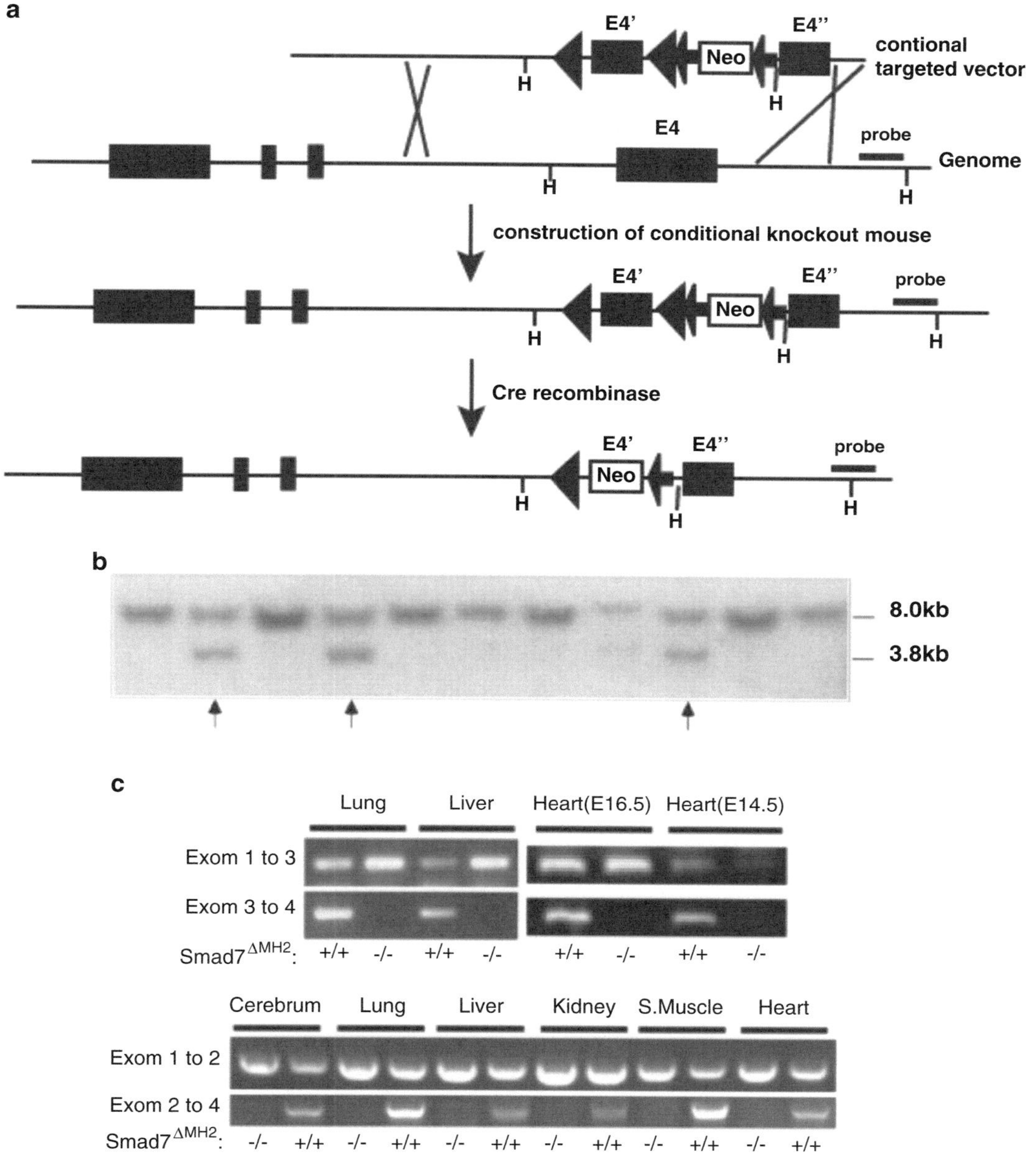

Fig. 1 Generation of *Smad7* conditional knockout mice. The conditional knockout allele is obtained by homologous recombination between the conditional targeted vector and *Smad7* genomic locus. The partial fourth exon of Smad7 is floxed (**a**). A HindIII restriction site is introduced after the right flox site. A 3′ probe is used for screening as indicated in **a**. The correct targeted ES cell will give rise to 3.8 kb HindIII-digested fragment in addition to the 8.0 kb wild type band (**b**, *arrow*). Using the Cre transgenic mice with Cre expressing in the germ cells, systemic Smad7 knockout mice are developed. No MH2 domain transcript is detected in the mutant mice (**c**)

is also used in some protocol [13]. Sequence confirmation is necessary for the construction of the targeting vectors.

4. Always grow EL350 at 32 °C, except for Induction of recombination proteins.
5. Since size of the released fragment is close to the vector backbone, *Sca*I can be used to cut the vector backbone. The P1-flox-Neo-flox-P2 and the P3-flox-FRT-Neo-FRT-P4 fragment can be well separated from the backbone on the gel.
6. When screening the recombination clone by restriction digestion, some strange pattern could be observed in addition to the expected bands. Retransformation of the plasmid into EL350 is needed to obtain purified clones.
7. In order to test the function of FRT sites, EL250 strain can be used to excise the Neo cassette. The excision of the Neo in EL250 is performed the same way as removal of Neo cassette in the first flox sites in EL350 (*see* **step 5** in section "Targeting the Flox Sites into the Subcloned Genome Fragment"). However, the Neo cassette in the flox-FRT-Neo-FRT should be kept in Smad7CKOΔMH2 targeting vector for the ES cells selection for gene targeting.
8. In this procedure, we used the 3′ probe to screening the targeted clone (Fig. 1). However, 5′ probe is recommended because the loss of one loxp site is observed in some targeted ES cells.
9. After crossing with EIIa-Cre mice, we crossed the heterozygous Smad7 mutant mice with wild type mice to exclude the chimera.

Acknowledgements

This work was supported by research grants from National Natural Science Foundation of China (81021002 and 81130077 to Y.C. and 30971660 to Y.P.), and Ministry of Science and Technology of China (2012CB524900 to Y.C. and 2010CB529506 to Y.P. and Z.W.).

References

1. Massague J (1998) TGF-beta signal transduction. Annu Rev Biochem 67:753–791
2. Heldin CH, Miyazono K, ten Dijke P (1997) TGF-beta signalling from cell membrane to nucleus through SMAD proteins. Nature 390(6659):465–471
3. Derynck R, Zhang YE (2003) Smad-dependent and Smad-independent pathways in TGF-beta family signalling. Nature 425(6958):577–584
4. ten Dijke P, Hill CS (2004) New insights into TGF-beta-Smad signalling. Trends Biochem Sci 29(5):265–273
5. Imamura T, Takase M, Nishihara A, Oeda E, Hanai J, Kawabata M et al (1997) Smad6 inhibits signalling by the TGF-beta superfamily. Nature 389(6651):622–626
6. Nakao A, Afrakhte M, Moren A, Nakayama T, Christian JL, Heuchel R et al (1997) Identification of Smad7, a TGFbeta-inducible antagonist of TGF-beta signalling. Nature 389(6651):631–635
7. Nakao A, Okumura K, Ogawa H (2002) Smad7: a new key player in TGF-beta-associated disease. Trends Mol Med 8(8):361–363

8. Kleeff J, Ishiwata T, Maruyama H, Friess H, Truong P, Buchler MW et al (1999) The TGF-beta signaling inhibitor Smad7 enhances tumorigenicity in pancreatic cancer. Oncogene 18(39):5363–5372
9. Monteleone G, Boirivant M, Pallone F, MacDonald TT (2008) TGF-beta1 and Smad7 in the regulation of IBD. Mucosal Immunol 1(Suppl 1):S50–S53
10. Dong C, Zhu S, Wang T, Yoon W, Li Z, Alvarez RJ et al (2002) Deficient Smad7 expression: a putative molecular defect in scleroderma. Proc Natl Acad Sci U S A 99(6):3908–3913
11. Zhu L, Chen S, Chen Y (2011) Unraveling the biological functions of Smad7 with mouse models. Cell Biosci 1:44
12. Kuhn R, Schwenk F (2003) Conditional knockout mice. Methods Mol Biol 209: 159–185
13. Liu P, Jenkins NA, Copeland NG (2003) A highly efficient recombineering-based method for generating conditional knockout mutations. Genome Res 13(3):476–484
14. Chen Q, Chen H, Zheng D, Kuang C, Fang H, Zou B et al (2009) Smad7 is required for the development and function of the heart. J Biol Chem 284(1):292–300
15. Zhu L, Wang L, Wang X, Luo X, Yang L, Zhang R et al (2011) Hepatic deletion of Smad7 in mouse leads to spontaneous liver dysfunction and aggravates alcoholic liver injury. PLoS One 6(2):e17415

Chapter 15

Monitoring Smad Activity In Vivo Using the Xenopus Model System

Marco Montagner, Graziano Martello, and Stefano Piccolo

Abstract

The embryo of the African clawed frog *Xenopus laevis* plays a central role in the field of cell and developmental biology. One of the strengths of Xenopus as model system lies in the high degree of conservation between amphibians and mammals in the molecular mechanisms controlling tissue patterning and differentiation. As such, many signaling cascades were first investigated in frog embryos and then confirmed in mouse and/or human cells. The TGF-β signaling cascade greatly benefited from this model system. Here we review the overall logic and experimental planning for studying Smad activity in vivo in the context of Xenopus embryonic development, and provide a guide for the interpretation of the results.

Key words *Xenopus laevis*, Animal cap explants, TGF-β, BMP, Smad, Mesoderm and neural induction

1 Introduction

1.1 Xenopus as Model System

Xenopus laevis is a pseudotetraploid vertebrate that lives in fresh water and that can be housed and fed easily and inexpensively. Each female can lay eggs up to 3–4 times a year, but, in the laboratory, ovulation is typically induced through a simple injection of human chorionic gonadotropin 15–18 h preceding the experiment. A single female typically lays thousands of eggs per day; embryos are easily maintained in saline medium, without specialized equipment. The dimension of the early frog embryo (approximately 1 mm) makes it easy to microinject and manipulate. Moreover, the rapid developmental rate (8–9 h from fertilization to gastrulation) allows collecting samples from all the relevant embryonic stages.

After fertilization, the zygote undergoes rapid cell divisions without growth (cleavage stages) and generation of a blastula. Up to the 32 cells stage, each blastomere can be individually microinjected with vital dyes or lineage tracers to generate fate maps (for an atlas of fate maps refer to [1]). Moreover, injecting

Xin-Hua Feng et al. (eds.), *TGF-β Signaling: Methods and Protocols*, Methods in Molecular Biology, vol. 1344, DOI 10.1007/978-1-4939-2966-5_15,

dyes or tracers into the zygote at the stage of 1–2 cells, allows labeling of the entire embryo; these marked tissues can then be transplanted into unlabeled embryos to monitor the fate of the injected cells in new environments (see for example [2]).

Within approximately 5 h after the fertilization, the embryos reach the blastula stage and after additional divisions the embryo starts "gastrulation", a complex set of movements and tissue rearrangements that shapes the body plan along the anteroposterior and dorsoventral axes. During the blastula stages, the zygotic genome is largely silent and cellular activities rely on maternal proteins and transcripts, while the genes of the zygote start to be expressed at mid-blastula transition (or MBT, about 4000 cells).

At the blastula stage, the three germ layers of the embryo (ectoderm, mesoderm, and endoderm) are established. Figure 1a shows a schematic view of the three germ layers: mesoderm is a ring of cells localized in the equatorial region; ectoderm and endoderm are localized respectively above and below the mesoderm. In the frog jargon, the "animal pole" is the highly cellularized "cap" of the sphere, while the "vegetal pole" is the bottom of the embryo, where most of the heavy oocyte's yolk resides. The endoderm germ layer is mainly specified by maternal transcription factors, such as Veg-T, with the contribution of zygotically expressed TGF-β ligands emanating from the vegetal hemisphere (Fig. 1a, b) [3]. Historically, the inductive capacity of the vegetal region was discovered by elegant recombination experiments

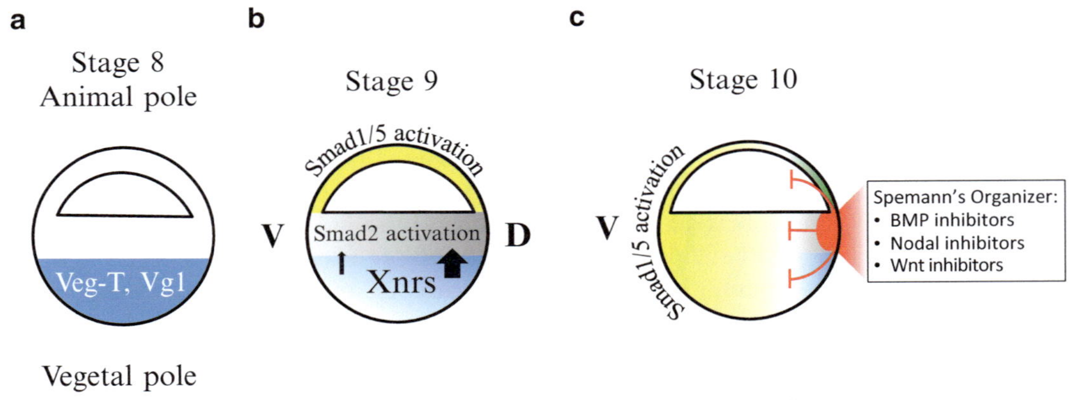

Fig. 1 (**a**) The germ layers of Xenopus embryo at the early blastula stage: endoderm (*blue*) at the vegetal pole, prospective mesoderm in the equatorial region and ectoderm at the animal pole. At this stage, maternal TGF-β activators are present (Vg1 and Veg-T) and will activate transcription of Nodal ligands (Xnr) in endoderm. (**b**) Nodal ligands emanating from the endoderm activate Smad2 (*gray gradient*) on the overlying tissue that will differentiate into mesoderm (*gray*). In addition, β-catenin on the dorsal side contributes to generate a dorsal enrichment of Smad2 activation. At this stage, Smad1/5 is evenly activated in the animal pole (*yellow*). (**c**) At the gastrula stage, endogenous activity of BMP inhibitors (Noggin, Chordin, Cerberus, Follistatin) emanating from the Spemann Organizer cause a shift of Smad1/5 signaling towards the ventral side of the germ layers. Embryos are presented as sections to allow visualization of the blastocoel cavity. V, ventral side; D, dorsal side

placing in contact ectoderm explants (i.e., the animal cap) with the prospective endoderm tissue: under this condition the ectodermal cells differentiated into mesoderm [4]. This experiment highlighted not only the inductive capacity of the endoderm, but also the pluripotency of the animal cap explants. Indeed, during normal development, this tissue is fated to become epidermis, but, on the dorsal side, it is induced to differentiate into neural tissue from signals emanating from the dorsal mesoderm (also called Spemann's Organizer, see below). That said, if removed before this stage, and cultivated under the appropriate stimuli, the cells of the animal cap can be induced to generate a variety of tissues, that, after prolonged culture, self-organize into functioning organs. The differentiation of explanted animal cap cells by exogenous TGF-β ligands has been an invaluable tool to discover the biology and biochemistry of this class of cytokines.

2 TGF-β in Early Embryos

TGF-β is a superfamily of cytokines that can be subdivided into two subfamilies: TGF-β/Activin/Nodal-related ligands, and the BMP family of ligands. The first group is required for endoderm, mesoderm specification and left-right patterning; the BMP family controls neuroectodermal differentiation and the dorsoventral patterning of the germ layers. Both types of ligands signal by causing the heterodimerization of two serine-threonine kinase receptors (named type I and II receptors). This event triggers phosphorylation of type I receptors, that, in turn, phosphorylate the Receptor-Smads (R-Smads), leading to formation of a trimeric complex between two R-Smads and one Smad4 molecule (the Co-Smad). R-Smads thus function as receptor substrate, signal transducer and transcription factors; once formed, the R-Smad/Smad4 complex is mainly nuclear. The affinity of nuclear Smads for cognate sequences is too low to achieve high affinity recognition of their target promoters; in fact, Smads mediate their transcriptional effects by associating to other transcription factors. As such, the output of TGF-β stimulation is extremely dependent on the specific pool of Smad DNA-binding cofactors expressed by a cell in space and time [5].

These transcriptional features explain only in part the pleiotropic effects of TGF-β signaling. Importantly, quantitative differences in TGF-β stimulation can trigger qualitatively distinct transcriptional profiles. This property defines TGF-β ligands as key "morphogens" in early embryos: cells near the source of the ligand (or where the ligand is free from extracellular "traps") will follow a developmental pathway different from the cells experiencing a reduced concentration of the same ligand [6, 7].

This morphogen-type property of TGF-β ligands can be well appreciated in animal cap assays, whereby increasing concentrations of Activin leads to progressive activation of different genes [7].

2.1 Nodal/Activin-Related Ligands

This family of signaling molecules, that includes Xenopus Nodal-related factors (or Xnrs), signals through Smad2 and Smad3 and plays prominent roles in the regulation of endoderm and mesoderm differentiation, gastrulation movement and left-right patterning.

Endoderm specification is initiated by a maternal transcription factor, Veg-T, localized in the vegetal pole of the embryo (Fig. 1a). After MBT, Veg-T activates the transcription of a cascade of Xnrs that act in the overlying marginal zone driving mesoderm differentiation (Fig. 1b) [8, 9]. Nodal activity is not even throughout the marginal zone: a dorsal enrichment of Smad2 activity leads to the formation of the Spemann's Organizer, the primary signaling center of the early embryo (Figs. 1b, c) [10]. The Spemann Organizer is a source of secreted antagonists of Nodal (Cerberus, Lefty), Wnt (Cerberus, DKK), and BMP ligands (Cerberus, Chordin, Noggin, and Follistatin) [10].

2.2 BMP Ligands

BMP ligands signal through Smad1 and Smad5 as R-Smads and regulate two fundamental processes in early embryonic development: (1) the induction of epidermis in the ectodermal region, and (2) ventralization of the mesoderm.

BMP-2, 4 and 7 are uniformly distributed in the mesoderm and ectoderm at the blastula stage (Fig. 1b); during gastrulation their activities and expressions starts being higher in the ventral side compared to the dorsal side (Fig. 1c). Another BMP-related factor, ADMP is expressed dorsally within the Spemann Organizer [11, 12]. Knockdown of all the four members of BMP family leads to the massive transformation of the ectoderm into neural tissue, revealing the central role of these proteins in the inhibition of a default neural fate [12]. In normal embryos, BMP inhibition and neural induction is mediated by the antagonists emanating from the Spemann's Organizer (Chordin, Noggin, Follistatin) generating a ventral-to-dorsal gradient of BMP activity that pattern the germ layers (Fig. 1c) [13].

3 Manipulating and Studying Gene Expression in Embryos of *Xenopus laevis*: The Basics

One of the reasons for the success of the Xenopus embryo as model system is that it is suitable both for gain- and loss-of-function assays. Most studies employ injection of mRNA or oligonucleotides, and are thus temporally transient manipulations; however, long-term overexpression or mis-expression studies with transgenic frogs have been also developed [14].

3.1 Reagents for Gain- and Loss-of-Function Studies

3.1.1 RNA

During the 1970s, Gurdon and colleagues pioneered the possibility to manipulate gene expression manipulation in Xenopus embryos by showing that frog oocytes could translate an injected mRNA [15]. After the development of recombinant DNA techniques, mRNA microinjection opened the possibility to "clone" new genes, not according to their sequence, but to their biological function: this has been carried out by injecting embryos with pools of hundreds (or thousands) of different mRNA species derived from in vitro transcription of cDNA libraries, followed by reiteration of the assay with progressively less and less complex pools, down to the individual, biologically active, cDNA.

There are several important controls to be considered in each mRNA microinjection experiment. First, sibling embryos must be injected with neutral mRNA (such as β-globin or prolactin). Second, a potentially informative control is injection of mRNA encoding for an "inactive" version of the protein of interest, for example a catalytically inactive enzyme or a DNA-binding defective transcription factor. Third, coinjection of the nucleic acids of interest with lacZ or GFP mRNAs as lineage tracers may allow to discriminate cell-autonomous from non-cell autonomous effects (in case of secreted factors), and to exclude potential effects due to toxicity from the injected mRNA [15, 16].

Injected mRNA diffuses rapidly, but is relatively unstable inside frog embryos. The pCS plasmid series [16] allow to surpass this problem as these plasmids contain the SV40 polyadenylation signal, causing polyadenylation of the injected transcripts in vivo, resulting in enhanced stability and translation efficacy. Thus, subcloning into a pCS plasmid is typically very critical and highly recommended step for a successful mRNA overexpression experiment. mRNA can be then obtained by in vitro transcription with commercial kits (i.e., one of the most used one is the Ambion SP6/T7 mMessage mMachine) followed by mRNA purification by standard phenol–chloroform extraction and ethanol precipitation using Sodium Acetate (Note: Ammonium acetate is toxic and teratogenic so this salt should be avoided in the preparation of nucleic acids to be injected in frog embryos. Similarly, DEPC-treated water can be also toxic and commercial non-DEPC RNAse-free water should be used for mRNA resuspension and dilution).

In principle, mRNA is suitable both for gain- and loss-of-function experiments depending on the overexpressed construct. Loss-of-function may be performed with dominant negative proteins. The prerequisite for this approach is that the mutant form should interfere with the wild-type protein (e.g., when this operates as a multimer) or downstream components of the same pathway. An advantage of dominant-negative proteins is that they may interfere with the activity of multiple family members, bypassing the problem of redundancy. Dominant negative proteins can be designed for secreted factors, receptors, intracellular transducers and transcription factors. That said, the specificity of a dominant negative might be questioned and should be controlled, whenever possible, by a rescue experiment, for example by overexpressing a protein acting downstream in the same pathway. Popular dominant negative proteins useful for the study of TGF-β activity are truncated type I BMP2/4 receptor (tBR) [17], and dn-Smad5 [19].

3.1.2 DNA

DNA is suitable both for gain- and loss-of-function experiments. In Xenopus embryos injected DNA is not integrated into the genome and is maintained as an episome. At difference with mRNA injection, expression from injected DNA is very heterogenous, restricted to patches of cells. Other drawbacks are the toxicity of the DNA (the upper limit is 50 pg) and the low amount of protein that could be obtained from DNA compared to RNA. An important advantage of DNA injection is that while injected RNA starts to be translated immediately, the transcription of an injected plasmid can start only after the MBT stage. This offers the chance to test the function of a gene at different developmental stages: a striking example is expression of Wnt ligands at very early stages (by mRNA), that causes the formation of two Siamese embryos, or just prior to gastrulation (by injecting XWnt8 DNA), causing almost the opposite result, that is, loss of head and shortening of the primary axis [20].

3.1.3 Morpholinos

Morpholino oligonucleotides are excellent loss-of-function reagents, allowing the unprecedented opportunity to knockdown multiple genes at once. This has brought back the Xenopus embryo under the spotlight as ideal tool for epistatic and gene-dosage experiments.

Morpholinos (MOs) are synthetic oligonucleotides that sterically block translation initiation complex leading to efficient knockdown of a targeted mRNA. For this reason, morpholinos are usually designed to match and cover the ATG sequence of the messenger RNA (i.e., they are antisense of an mRNA sequence). They are very stable, not toxic (up to 100 ng per embryo is well tolerated) and remain active within the embryo for days (although the concentration of oligos per cell obviously decreases by dilution after the embryo starts growing). Morpholinos can be designed to block splicing and this is exploited to inhibit only certain isoforms of a given target protein. It is important to validate the results with at least two nonoverlapping morpholinos, and to include negative controls, such as injection of the Gene Tools-negative control morpholinos, or the reverse or mutant morpholino sequence.

While protein knockdown by morpholinos is usually very efficient, two main problems need to be discussed. The first is that Xenopus laevis genome is allotetraploid, meaning that this animal might have 2–4 alleles of the same gene. These alleles might differ in their sequences in a way that only one allele might be targeted by a single morpholino. Thus, the best practice is to carefully design the morpholino oligonuclotides (ideally with GeneTool, the main morpholino provider, that offers free design services with a proprietary software), taking into account what are highly conserved regions (by comparing multiple ESTs). That said, sometimes only combinations of morpholinos might be ultimately required for efficient target downregulation (a good example being Chordin or Smad4 knockdown [21]). The second issue is that morpholino-mediated knockdown might be blunted, at very early stages, by the effects of maternal proteins that the embryo inherits from the oocytes. In this case the alternative setups are the use of dominant negative or the inhibition of maternal transcripts in the oocytes [8].

3.2 Xenopus Animal Cap Explants

The use of explants in embryology is crucial for understanding how a cell decides to follow a particular fate or what are the signaling combinations that drive lineage commitment. The explant could be induced to differentiate by (1) the activity of a previously injected mRNA or by (2) treatment of the naive explants with synthetic proteins diluted into the culture medium (Fig. 2).

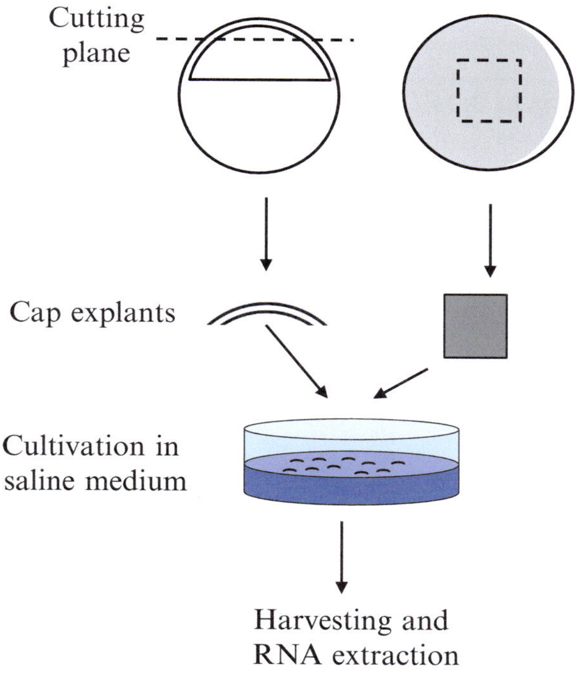

Fig. 2 Animal cap assay. Injected or uninjected embryos are devitellinized and animal poles are removed with the help of an eyebrow knife around stage 9. The explants are then cultivated onto an agarose plate containing saline buffer or on polyHEMA-treated tissue culture plate (the latter in case of added TGF-β ligands or other growth factors) until the siblings reach the desired stage. Note that the embryo on the left (lateral view) is presented as a section to allow visualization of the blastocoel cavity

The results are then assayed using molecular markers that define the identity of different tissues.

As mentioned above, animal caps are pluripotent, and different TGF-β family members could trigger distinct differentiation routes. Thus, the animal cap assay provided an outstanding background for the identification of Smad regulators. Activation of the Activin/Nodal-Smad2/3 leads to differentiation of the ectoderm into dorsal mesoderm or endoderm (depending on the degree of activation), whereas activation of the BMP-Smad1/5 pathway induces ventral mesoderm. Conversely, inhibition of the BMP pathway by treatment (or mRNA injection) with Chordin or Noggin (secreted BMP antagonists) differentiates the caps into neural tissue.

Note

Complete and straightforward protocols and reference pictures documenting the expression of molecular markers can be found in the books "Early development of Xenopus laevis: a laboratory manual" [18] and from several issues of Methods in Molecular Biology series [19–22]. Moreover, www.xenbase.org website offers updated versions of atlases and protocols, as well as explicative videos and links to the most recent publications of Xenopus field.

4 Analysis of the Results

Widely diffused assays to monitor the effects of gain- or loss-of-function of a given gene—or set of genes—are: whole embryo phenotypic analyses, identification of molecular markers by in situ hybridization or qPCR, and reporter assays.

4.1 Phenotype Analysis in Whole Embryos

This approach includes the careful examination of the overall external anatomy of the embryo and histology complemented by in situ hybridization studies to visualize molecularly the differentiation pattern of mRNA-injected or "morphant" embryos. For example, in absence of molecular markers it would be impossible to discriminate if an embryo fails gastrulation because mesoderm has not been induced by TGF-β ligands or if the cell movements of gastrulation are impaired.

4.2 In Situ Hybridization (ISH)

Whole-mount in situ hybridization (ISH) is a very powerful technique that allows the spatial detection of the expression domains of markers specific for a particular lineage, cell-specific differentiation route or histotype. The protocols for ISH are now quite straightforward (for example see http://www.hhmi.ucla.edu/derobertis/index.html); the limits of this technique is that it is not quantitative, requires some optimization to set-up, and that only one or two markers could be studied at once in the same embryo.

4.3 RT-PCR (or Quantitative PCR)

qPCR is a straightforward assay that allows the detection of many targets from the same samples. After RNA isolation, total messenger RNA is retrotranscribed into cDNA that is used as a template for standard PCR or for the more sensitive real-time qPCR. In principle, RNA could be extracted from whole embryos or explants; however, the embryo contains traces of mRNAs of almost all transcripts; this generates a background "noise" that could mask the expansion of the marker of interest in specific domains. In contrast, animal cap explants have very low levels of several tissue-specific transcripts and thus represent a "clean background" to monitor gene inductions. For PCR primers, please refer to De Robertis Lab website (http://www.hhmi.ucla.edu/derobertis/index.html).

4.4 Reporter Gene Assays

This is an ideal first choice to assess Smad transcriptional activity. The reporter construct is injected at the 2–4 cell stage either in the marginal zone (in the region that will form the mesodermal belt), or in the animal pole, depending if when one needs to test the effect of a candidate TGF-β regulator on Smad signaling triggered by endogenous or exogenous TGF-β ligands, respectively.

Two controls should be included in reporter assays: (a) The signal from Smad reporter construct must be compared to an empty vector containing only the basal promoter, or, better, a promoter with mutated Smad-binding elements. The pGL4 plasmid (from Promega) has been engineered to reduce the number of binding sites for transcription factors, reducing the probability of aspecific transcription and background signal; (b) Co-injection of mRNA coding for β-galactosidase (50 pg) and/or pCMV-Renilla DNA (5 pg) as normalizers for injection efficiency.

5 How to Study Smad2/3 Activation

5.1 Phenotypic Analysis in Whole Embryo

5.1.1 Gain-of-Function of Smad2/3 Signaling

Where to inject: mRNA is injected into 2–4 blastomeres in the animal pole or in the Ventral Marginal Zone (VMZ, Fig. 3a–d).

Positive controls as reference: Smad2/3, Nodal ligands (Xnrs, Vg1), constitutively activated receptors type I (such as ACVR1B, TGF-βRI) [25].

Expected results:

1. Generation of ectopic secondary axis. These axes are never complete (i.e., they do not form a head), ranging from truncated double axis (i.e., with trunk and tail) to double tail.
2. At stage 10.5 (when the dorsal lip blastopore is evident), overactivation of Smad2/3 signaling in animal cap results in the formation of ectopic "bottle cells" that tend to invaginate and fall into the blastocoele [26]. Induction of bottle cells resembles the formation of an ectopic blastopore lip. These cells appear pigmented, apically constricted with the cell body elongating

inward. The formation of these structures is typical of very strong Smad2/3 activation; indeed, other mesoderm inducers (such as FGF activation) do not form ectopic bottle cells.

Molecular markers to be used in in situ hybridization: expansion of pan-mesoderm markers Xbra, Veg-T, Eomes at stage 11.

Notes

1. Overexpression of TGF-β1/2/3 mRNA has no or little effect in Xenopus embryos because lack of expression of TGF-β Receptor type I.
2. Induction of bottle cells or excessive induction of endoderm is incompatible with gastrulation movements and arrests development.

5.1.2 Loss-of-Function of Smad2/3 Signaling

Injection sites: injection in each blastomere at the 4-cell stage in the marginal/vegetal zone (Fig. 3e–f).

Positive controls as reference: Smad4 morpholinos, Ectodermin mRNA [27], dominant-negative Smad4, p53 morpholinos [24], truncated Activin receptor type II (tAR) [28], Cerberus-Short (CerS) [29].

Expected results: inhibition of Smad2/3 activity in the whole meso-endodermal tissue may cause a range of phenotypes, depending on

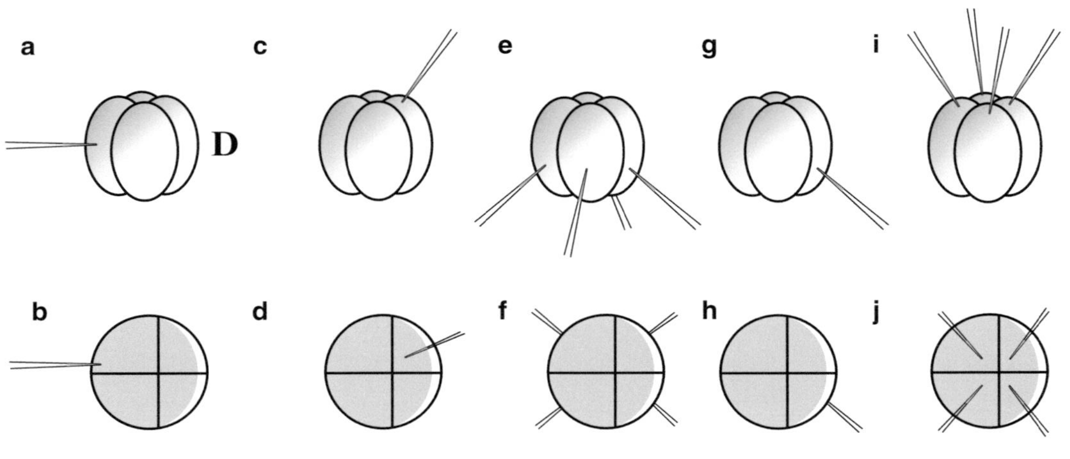

Fig. 3 (**a–j**) Simplified drawings of Xenopus embryos depicting classical injection sites for studying Smad activity. The same injection scheme is showed from the lateral view (*top*) and from the animal pole view (*bottom*). Dorsal (D) and ventral poles are indicated. In Xenopus embryos, following sperm entry into the oocyte, the cortical cytoplasm rotates relatively to the central cytoplasm. This has two consequences that can be exploited to recognize specific blastomeres: (i) the pigmentation of the animal pole is slightly shifted towards the future ventral side and (ii) the ventral blastomeres appear larger. When localizing injections in particular regions, coinjection of a tracer is recommended. A suitable reagent for this is GFP mRNA (300 pg). GFP protein starts being detectable around stages 9–10 under a fluorescent microscope. At this stage, injected cells can be visualized and only embryos with the correct spatial distribution selected for analysis. Note that GFP labeling is stable and can be followed until neurula

the degree of inhibition: a shorter embryo, incomplete blastopore closure, reduction of all anterior structures (cement gland and eyes) [11].

Molecular markers to be used in in situ hybridization experiments: mesoderm and Spemann's Organizer markers should disappear or decrease. At stage 11 good markers are Xbra, Eomes, zygotic Veg-T, Mix.2, Xlim1, Pintallavis, Lefty1/Antivin, Xnr1, Chd, XnotI, Gsc, ADMP. At neurula (stages 14–20): MyoD.

Note

Left/Right patterning is regulated by Xnr1 and Lefty/Antivin at the neurula stage and can be nicely visualized by the left-specific expression of the *Pitx2c* marker. Disturbance of Nodal signaling (e.g., through injection of mRNA encoding for the Nodal antagonist Caronte) may result in either randomization of the normal left-sided expression, bilateral or no expression of *Pitx2c*. Later, at stage 45, this may be visualized morphologically in the reverse coiling of the gut [30].

5.2 Animal Caps Elongation Assay

A key function of Smad2/3 activity in Xenopus embryos is to induce gastrulation movements. This property has been extensively used in animal cap explants to induce, or inhibit, Smad activity [23, 31]. For this assay, animal caps are maintained in polyHEMA-coated tissue culture plates, in a medium (0.3× MMR or 1× LCMR) and cultivated until tadpole. Elongation is a great qualitative readout but it may be difficult to quantify; note that it would occur only when muscle is induced while stronger activations of Smads, leading to endoderm formation, will not elongate the explants.

5.3 PCR Analysis of Animal cap Explants

Animal caps are explanted at the late blastula stage (around stage 9) and cultivated in 0.3× MMR. Groups of ten animal caps should be explanted, and each group harvested at defined time points (depending on the natural timing of expression of the marker to be assayed).

Injection sites: injection at the 2–4 cell stage in the animal pole (Fig. 3c, d, i, j).

Positive controls as reference: as detailed above.

Molecular markers: at stage 11.5: Xbra, Eomes, Veg-T, Snail, Xpo, Xlim1, Gsc, Mix.2, Mixer, Sox17, GATA4. At stage 38 (tadpole): Muscle Actin, MyoD, HoxB9.

Note

Treatment with Activin ligands of explanted animal caps is often useful as it allow several hours for morpholino-mediated knockdown to be effective. For this, caps should be explanted at blastula and left either untreated or treated with soluble Activin protein (R&D, 15 ng/mL in 0.3× MMR) in polyHema-treated tissue culture dishes.

5.4 PCR Analysis of Marginal Zone Explants

When assaying for gain- or loss-of-function of TGF-β, also Ventral or Dorsal Marginal Zone explants might be informative (VMZ or DMZ respectively). These tissues must be explanted at the onset of gastrulation as these endogenously express markers for ventral or dorsal mesoderm. If not perturbed, VMZ will express MyoD, Muscle Actin, GATA1, β-globin (when harvested at stage 38). Alternatively, blunting Smad signaling in DMZ may impair expression of Organizer markers at stage 11 (Chd, Gsc).

5.5 Luciferase Reporter Analysis

The following reporters could be used as alternatives to endogenous markers to measure quantitatively the effects of an injected mRNAs, miRNAs or morpholinos on Nodal/Smad responsiveness (either endogenously present or triggered ectopically by co-injected Xnr1 mRNA or by treatment with Activin protein of the animal caps). It is sufficient to inject luciferase reporters in a single blastomere in marginal zone to exploit endogenous Smad signaling, or in animal cap together with activating mRNAs (Fig. 3a–d). Reporters for Smad2/3/4 activity to be used in Xenopus embryos are the following:

1. Mix.2/-290-lux: this has been the first promoter reported to respond as immediate early target of Activin. FoxH1 (FAST1) binds to Smad2/4 complex and contacts ARE (Activin Responsive Element) sequences in the promoter. FoxH1 presence is indispensable for Smad2 signaling during early embryogenesis [32]. This reporter also contains p53-binding elements that are critical for its full activation [24].
2. ARE3-lux: this is obtained by multimerization of ARE elements from Mix.2/-290-lux [33].
3. Gsc-lux: this reporter contains the Goosecoid natural promoter and contains two domains, a Distal Element (DE) and a Proximal Element (PE) regulated by Smad2/3 and β-catenin, respectively [34].

Note

pCAGA12-lux, the widely used and sensitive TGF-β reporter in mammalian cells, does not work in frog embryos.

6 How to Study Smad1/5 Activation

6.1 Phenotype Analysis in Whole Embryo

6.1.1 Gain-of-Function of Smad 1/5 Signaling

Injection sites: mRNA injection in individual blastomeres at the 4-cell stage in animal cap or Marginal Zone (Fig. 3e, f, i, j).

Positive controls as reference: Smad1/5 mRNA, BMP ligands (for example: BMP2, 4, 7), constitutively activated BMP Receptor type I (as caBR [35]).

Expected results: A range of phenotypes, from reduced head to completely ventralized embryos.

Molecular markers to be used in in situ hybridization: upregulated at stage 12: BMP4 and 7, Szl, BAMBI, XVent1/2, Msx1/2. Upregulated at tadpole stage: Siszled (Szl), β-globin, epidermal keratin. Downregulated at stage 12: Myf5, Chd, MyoD. Downregulated at tadpole: XAG1 (cement gland), Pax6 (eye field), Six3, Rx1, Otx2, Bf1, Sox2, Sox3, NCAM.

6.1.2 Loss-of-Function of Smad1/5 Signaling

Injection sites: single or radial injection at the 4-cell stage in the Marginal Zone (Fig. 3a, b, e, f respectively).

Injected reagents as reference: dominant-negative BMP2/4 Receptor type I (tBR) [17], dominant-negative BMP2 [35], Chordin (low doses).

Expected results:

1. Incomplete duplicated axis (that is, an embryo showing a second trunk but lacking secondary head).
2. When attenuation of BMP signaling is obtained by radial injections in all blastomeres, the embryo displays increased dorsal structures (and neural expansion), enlargement of the head and anterior structures (bigger eyes and cement gland). The secondary axis should be positive for Sox2, MyoD and negative for Six3 or Pax6. Generation of complete secondary axes, with head duplication, is typical of activation of maternal Wnt signaling or of dual inhibition of Nodal and zygotic Wnt signaling.

6.2 PCR Analysis of Animal cap Explants

Endogenously expressed BMP4 and BMP7 basally activate Smad1/5 in animal cap explants. As such, cultured animal caps will start expressing epidermal markers starting from stage 12, in absence of other exogenous signals. This property has been exploited during the last 20 years to study neural induction by soluble BMP antagonists (Chordin, Noggin). Since animal cap endogenously produce BMPs, this experimental set up is more suitable for loss-of-function of BMP/Smad1/5 components, rather than gain-of-function.

Injection sites: single or radial injections in animal cap (Fig. 3c, d, i, j respectively).

Positive controls as reference: Chordin, Noggin or dominant-negative BMP receptor mRNAs.

Molecular markers: at stage 12, upregulation of Otx2, Sox2; downregulation of XVent1/2. At stage 22, upregulation of Otx2, Sox2, NCAM; downregulation of Epidermal Keratin, β-globin (for additional markers refer to [36]).

6.3 Luciferase Reporter Analysis

The following luciferase reporters are effectively modulated by gain or loss-of-function when injected in Xenopus embryos:

1. XVent2-lux: this contains the endogenous promoter of the ventral BMP target gene XVent2. As for Mix.2 promoter, it contains additional sequences bound by the Smad1/Smad4 cofactor OAZ [37].
2. Id1-BRE-lux: a synthetic sensor generated by fusing the two copies of Smad1 binding elements of mammalian Id1 promoter [38]. This is nicely working also in Xenopus and is more sensible than XVent2-lux reporter.

Note

Injection in Ventral Marginal Zone results in higher basal activation and it is thus suggested for loss-of-function assays.

References

1. Bauer DV, Huang S, Moody SA (1994) The cleavage stage origin of Spemann's Organizer: analysis of the movements of blastomere clones before and during gastrulation in Xenopus. Development 120(5):1179–1189
2. Thuret R, Papalopulu N (2012) Following the fate of neural progenitors by homotopic/homochronic grafts in Xenopus embryos. Methods Mol Biol 916:203–215
3. Hill CS (2001) TGF-beta signalling pathways in early Xenopus development. Curr Opin Genet Dev 11(5):533–540
4. Nieuwkoop PD (1973) The organization center of the amphibian embryo: its origin, spatial organization, and morphogenetic action. Adv Morphog 10:1–39
5. Massague J (2000) How cells read TGF-beta signals. Nat Rev Mol Cell Biol 1(3):169–178
6. Wilson PA, Lagna G, Suzuki A, Hemmati-Brivanlou A (1997) Concentration-dependent patterning of the Xenopus ectoderm by BMP4 and its signal transducer Smad1. Development 124(16):3177–3184
7. Gurdon JB, Harger P, Mitchell A, Lemaire P (1994) Activin signalling and response to a morphogen gradient. Nature 371(6497):487–492
8. Zhang J, Houston DW, King ML, Payne C, Wylie C, Heasman J (1998) The role of maternal VegT in establishing the primary germ layers in Xenopus embryos. Cell 94(4):515–524
9. Kofron M, Demel T, Xanthos J, Lohr J, Sun B, Sive H et al (1999) Mesoderm induction in Xenopus is a zygotic event regulated by maternal VegT via TGFbeta growth factors. Development 126(24):5759–5770
10. Harland R, Gerhart J (1997) Formation and function of Spemann's organizer. Annu Rev Cell Dev Biol 13:611–667
11. Inui M, Montagner M, Ben-Zvi D, Martello G, Soligo S, Manfrin A et al (2012) Self-regulation of the head-inducing properties of the Spemann organizer. Proc Natl Acad Sci U S A 109(38):15354–15359
12. Reversade B, De Robertis EM (2005) Regulation of ADMP and BMP2/4/7 at opposite embryonic poles generates a self-regulating morphogenetic field. Cell 123(6):1147–1160
13. Plouhinec JL, Zakin L, De Robertis EM (2011) Systems control of BMP morphogen flow in vertebrate embryos. Curr Opin Genet Dev 21(6):696–703
14. Ishibashi S, Kroll KL, Amaya E (2008) A method for generating transgenic frog embryos. Methods Mol Biol 461:447–466
15. Gurdon JB, Lane CD, Woodland HR, Marbaix G (1971) Use of frog eggs and oocytes for the study of messenger RNA and its translation in living cells. Nature 233(5316):177–182
16. Rupp RA, Snider L, Weintraub H (1994) Xenopus embryos regulate the nuclear localization of XMyoD. Genes Dev 8(11):1311–1323
17. Suzuki A, Thies RS, Yamaji N, Song JJ, Wozney JM, Murakami K et al (1994) A truncated bone morphogenetic protein receptor affects dorsal-ventral patterning in the early Xenopus embryo. Proc Natl Acad Sci U S A 91(22):10255–10259
18. Sive HL, Grainger RM, Harland RM (2000) Early development of Xenopus laevis: a laboratory manual. Cold Spring Harbor Laboratory Press, Cold Spring Harbor, NY

19. Liu XJ (2006) Xenopus protocols: cell biology and signal transduction. Humana Press, Totowa, NJ
20. Sharpe PT, Mason I (2008) Molecular embryology: methods and protocols, 2nd edn. Humana Press, New York, NY
21. Hoppler S, Vize PD (2012) Xenopus protocols: post-genomic approaches, 2nd edn. Humana Press, New York, NY
22. Pelegri FJ (2011) Vertebrate embryogenesis: embryological, cellular, and genetic methods. Humana Press, New York, NY
23. Baker JC, Harland RM (1996) A novel mesoderm inducer, Madr2, functions in the activin signal transduction pathway. Genes Dev 10(15):1880–1889
24. Cordenonsi M, Dupont S, Maretto S, Insinga A, Imbriano C, Piccolo S (2003) Links between tumor suppressors: p53 is required for TGF-beta gene responses by cooperating with Smads. Cell 113(3):301–314
25. Armes NA, Smith JC (1997) The ALK-2 and ALK-4 activin receptors transduce distinct mesoderm-inducing signals during early Xenopus development but do not co-operate to establish thresholds. Development 124(19): 3797–3804
26. Kurth T, Hausen P (2000) Bottle cell formation in relation to mesodermal patterning in the Xenopus embryo. Mech Dev 97(1-2): 117–131
27. Dupont S, Zacchigna L, Cordenonsi M, Soligo S, Adorno M, Rugge M et al (2005) Germ-layer specification and control of cell growth by Ectodermin, a Smad4 ubiquitin ligase. Cell 121(1):87–99
28. Hemmati-Brivanlou A, Melton DA (1992) A truncated activin receptor inhibits mesoderm induction and formation of axial structures in Xenopus embryos. Nature 359(6396):609–614
29. Piccolo S, Agius E, Leyns L, Bhattacharyya S, Grunz H, Bouwmeester T et al (1999) The head inducer Cerberus is a multifunctional antagonist of Nodal, BMP and Wnt signals. Nature 397(6721):707–710
30. Cheng AM, Thisse B, Thisse C, Wright CV (2000) The lefty-related factor Xatv acts as a feedback inhibitor of nodal signaling in mesoderm induction and L-R axis development in xenopus. Development 127(5):1049–1061
31. Hawley SH, Wunnenberg-Stapleton K, Hashimoto C, Laurent MN, Watabe T, Blumberg BW et al (1995) Disruption of BMP signals in embryonic Xenopus ectoderm leads to direct neural induction. Genes Dev 9(23):2923–2935
32. Yeo CY, Chen X, Whitman M (1999) The role of FAST-1 and Smads in transcriptional regulation by activin during early Xenopus embryogenesis. J Biol Chem 274(37):26584–26590
33. Dennler S, Itoh S, Vivien D, ten Dijke P, Huet S, Gauthier JM (1998) Direct binding of Smad3 and Smad4 to critical TGF beta-inducible elements in the promoter of human plasminogen activator inhibitor-type 1 gene. EMBO J 17(11):3091–3100
34. Watabe T, Kim S, Candia A, Rothbacher U, Hashimoto C, Inoue K et al (1995) Molecular mechanisms of Spemann's organizer formation: conserved growth factor synergy between Xenopus and mouse. Genes Dev 9(24):3038–3050
35. Suzuki A, Kaneko E, Ueno N, Hemmati-Brivanlou A (1997) Regulation of epidermal induction by BMP2 and BMP7 signaling. Dev Biol 189(1):112–122
36. Kuroda H, Wessely O, De Robertis EM (2004) Neural induction in Xenopus: requirement for ectodermal and endomesodermal signals via Chordin, Noggin, beta-Catenin, and Cerberus. PLoS Biol 2(5):E92
37. Hata A, Seoane J, Lagna G, Montalvo E, Hemmati-Brivanlou A, Massague J (2000) OAZ uses distinct DNA- and protein-binding zinc fingers in separate BMP-Smad and Olf signaling pathways. Cell 100(2):229–240
38. Korchynskyi O, ten Dijke P (2002) Identification and functional characterization of distinct critically important bone morphogenetic protein-specific response elements in the Id1 promoter. J Biol Chem 277(7):4883–4891

Chapter 16

Animal Cap Assay for TGF-β Signaling

Chenbei Chang

Abstract

TGF-β signals regulate a variety of processes during early vertebrate development, from stem cell maintenance and differentiation to tissue patterning and organogenesis. Detailed understanding of how this signaling pathway operates and what genes control activities of the signaling components of the pathway is therefore important for us to comprehend temporal- and tissue-specific TGF-β functions in vertebrate embryogenesis. Xenopus model system has been employed extensively in research on TGF-β signals, and much insight about TGF-β signaling mechanisms has been gained from these studies. Besides using whole embryos, explants from the ectodermal region of Xenopus, also known as animal caps, are used widely in investigations of the activities of an array of signal transducers as well as regulators of the pathway. This chapter introduces methods for dissection of animal caps and analyses of TGF-β signaling effects on animal caps.

Key words Animal cap assay, Cell dissociation, RT-PCR, Activin, BMP, Noggin

1 Introduction

TGF-β signals play key roles in every germ layer and at every stage of early vertebrate development and thus are regulated stringently during embryogenesis [1]. In Xenopus, the two branches of the TGF-β signals participate in distinct cell fate determination. Activin/nodal-like growth factors signal in a graded fashion to specify endoderm and mesoderm along the vegetal-animal axis. Within the mesoderm, activin/nodal also acts in a temporally and spatially controlled manner to influence the formation of the dorsal (e.g., notochord and anterior somite), the lateral (e.g., heart, kidney), and the ventral (e.g., posterior somite) types of the mesoderm [2, 3]. In contrast, BMP-like growth factors partake in patterning of ventral tissues in all three germ layers. BMP activities inhibit the formation of the neural tissues in the ectoderm and are required for ventral and posterior mesoderm development during axis specification [4]. Because of the easily identifiable TGF-β signal-associated phenotypes and changes in marker expression,

Xin-Hua Feng et al. (eds.), *TGF-β Signaling: Methods and Protocols*, Methods in Molecular Biology, vol. 1344,
DOI 10.1007/978-1-4939-2966-5_16,

Xenopus has been used as a favorite animal model to interrogate transduction, regulation, mechanisms, and functions of the TGF-β signals during early vertebrate development. Many novel regulators of the pathway were first characterized in this system, including both secreted factors, such as noggin [5], chordin [6], and Cerberus [7], and the cytoplasmic and nuclear proteins, such as Smurf [8] and OAZ [9]. The ease to co-express different genes in Xenopus and observe the embryonic phenotypes in less than 2 days further facilitates research on epistatic relationship between various molecules, providing basis for mechanistic analyses of the signaling system. Xenopus also produces a great quantity of large-sized embryos, making biochemical studies on protein purification, interaction, modification, localization, and enzymatic activities suitable and relatively easy [10, 11]. Thus the Xenopus system is not only good for examination of in vivo requirement of TGF-β signaling in vertebrate embryogenesis but also advantageous in dissection of detailed signaling mechanisms.

In addition to whole Xenopus embryos, explants from different regions of the embryos are often isolated and used to study the effects of altered TGF-β signals on tissue development. One particularly powerful explant system is the ectodermal explants, also called animal caps. Cells in the animal caps are pluripotent and can differentiate into various cell types in accordance with the signals they are exposed to. The differentiation mimics the endogenous fates of the cells that receive similar signals, so that in the presence of activin/nodal-like growth factors, animal caps develop along the mesodermal and endodermal pathways; whereas the presence of BMP ligands induces ventral mesodermal gene expression in the caps. Inhibition of BMP signals also leads to cell differentiation toward the neural lineage. As animal caps can be dissected easily and are amenable to biochemical, cell biological, as well as functional assays, they serve as an excellent system for investigation of novel regulators of the TGF-β signals. In this chapter, I describe the methods used to cut animal caps and to analyze the effects of varied levels of TGF-β signals on morphology and marker expression of the animal caps. Basic procedures dealing with handling of adult Xenopus, including maintenance of Xenopus colonies, obtaining testes from male frogs, and priming female frogs for egg production, are not described here. Researchers can find information about these manipulations in [12].

2 Materials

2.1 Tools Needed to Make Animal Caps (See Fig. 1)

1. Forceps: two pairs of sharp watchmaker's forceps, such as Dumont #5 Forceps. Some people prefer using forceps with tips bent toward each other, but it is not necessary. The important thing is to have the tips of the forceps meet exactly and evenly when they close.

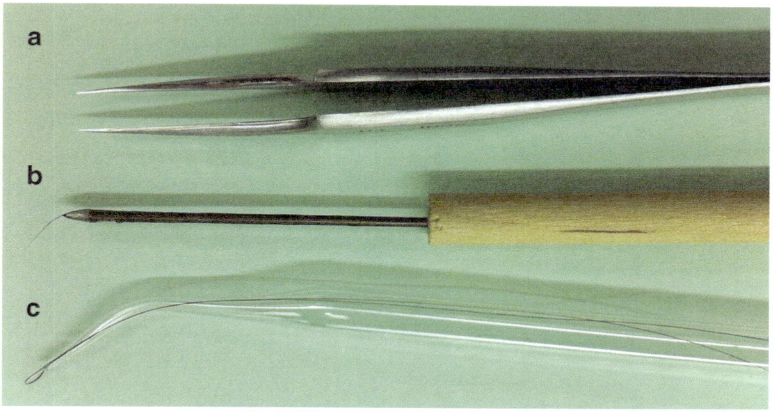

Fig. 1 Crucial tools needed to dissect animal caps: (**a**) sharp watchmaker's forceps; (**b**) an eyebrow hair knife; (**c**) a hair loop

2. Hair knife: use eyebrow or eyelash hairs. Attach the hair to the tip of a colony picker or a pulled glass pipet scored and broken at the tip. Arrange the hair so that it has an angle of about 120° to the colony picker/pipet. Glue the hair to position with nail polisher or molten wax (*see* **Note 1**).
3. Hair loop: pull a disposable glass pipet over the fire, score and break the end of the pipet to make a relatively small diameter tip. Insert half of the length of a hair (preferably twice as long as the pipet) into the tip of the pipet with the forceps, then bend the hair and insert the other end into the pipet to make a loop. The diameter of the loop should be about 1–2 mm. Secure the position of the hair loop by sealing the pipet with nail polisher or molten wax.
4. Transfer pipets: disposable plastic pipets are used to transfer embryos between dishes, and disposable glass pipets are used to transfer dissected explants.
5. Agarose-coated dishes: coat 60 mm and 4-well tissue culture dishes with 1 % agarose in sterile water to prevent cells and explants from adhering to the surface of the dishes.

2.2 Solutions and Culture Media

1. Cysteine solution: 2 % (w/v) cysteine in water or 1/3× MMR (see below), adjust pH to 8.0 with NaOH. Prepare fresh on the day of injection for dejellying embryos.
2. Ficoll: 2–3 % Ficoll (w/v) in 1/3× MMR, for temporary culture of embryos (2–4 h) following RNA injection.
3. MMR (Marc's Modified Ringers): 0.1 M NaCl, 2 mM KCl, 1 mM $MgSO_4$, 2 mM $CaCl_2$, 5 mM HEPES, pH 7.8, 0.1 mM EDTA.

4. MBS (Modified Barth's saline): 88 mM NaCl, 1 mM KCl, 0.7 mM $CaCl_2$, 1 mM $MgSO_4$, 5 mM HEPES, pH 7.8, 2.5 mM $NaHCO_3$ (*see* **Notes 2** and **3**).
5. CMFB (Calcium Magnesium-Free Buffer): 88 mM NaCl, 1 mM KCl, 2.4 mM $NaHCO_3$, 5 mM HEPES, pH 7.8.
6. Solutions for semi-quantitative RT-PCR: Solutions: Lysis buffer (20 mM Tris–Cl, pH 7.8, 1 % SDS, 100 mM NaCl, 10 mM EDTA); 10× DNase buffer (400 mM Tris, pH 7.9, 100 mM NaCl, 60 mM $MgCl_2$, 1 mM $CaCl_2$); DNase I mix (10 μL for each sample, containing 5.6 μL dH_2O, 2.5 μL 10× DNase buffer, 1.25 μL 20 mM DTT, 0.5 μL RNAsin, 0.2 μL RNAse-free DNase. Make enough mix for all the samples right before the DNase treatment step and aliquot 10 μL into each reaction); RT mix (9 μL for each sample, containing 4 μL 5× MMLV reverse transcription buffer, 2.3 μL dH_2O, 1 μL 20 mM DTT, 1 μL 10 mM dNTP mix, 0.5 μL RNAsin, and 0.2 μL MMLV Reverse Transcriptase. Make enough for all the samples and aliquot 9 μL into each reaction); 5× TBE (450 mM Tris-borate, 10 mM EDTA).

3 Methods

3.1 Dissection of Animal Caps

Animal caps isolated early from Xenopus embryos are naïve tissues devoid of influence from signals emitted from their neighbors. They self-differentiate into atypical epidermis. However, cells in animal caps remain pluripotent for a long time and can respond to growth factor signals or inhibitors of growth factors to differentiate into cell types of all germ layers, including mesodermal, endodermal, neural, or neural crest cells. Growth factors and soluble inhibitors can be supplied as proteins in culture media after isolation of animal caps; or alternatively, RNAs encoding these factors as well as their downstream cytoplasmic and nuclear signal transducers and regulators can be injected into cleavage stage embryos. Animal caps from the injected embryos can be dissected then and the signaling and the regulatory properties of the injected molecules be analyzed. As mentioned above, animal cap assays have been used extensively in research on both activin/nodal/TGF-β and BMP/GDF signals, including studies on signaling components and regulators at the extracellular (e.g., ref. [13]), membrane (e.g., ref. [14]), cytoplasm (e.g., ref. [15]), and nuclear (e.g., ref. [16]) levels. In the following, the method to dissect animal caps from Xenopus embryos is described.

Experimental Procedures

1. Perform in vitro fertilization of Xenopus embryos by collecting freshly laid eggs (through gentle squeeze of female frogs)

in a culture dish and mixing them with a small piece of testis (isolated from euthanized male frogs and stored in 4 °C refrigerator in MMR). Unlike unfertilized eggs which have pigmented hemisphere (animal hemisphere) pointing in all directions, fertilized eggs all turn their animal hemisphere upward in about 20 min.

2. Dejelly the fertilized embryos by immersing the embryos in 2 % cysteine solution and gently swirling the dish for about 4 min. During this period, fragmented jelly pieces become visible in the dish, and the embryos sever their connection with the jelly coat and start to pack at the bottom of the dish. Wash extensively (5 times or more) with 1/3× MMR to get rid of cysteine and store the embryos in 1/3× MMR (or 0.1× MBS) or in Ficoll solution if injection is involved (*see* **Note 4**).
3. Inject 2-cell stage embryos into the animal regions of both cells with RNAs encoding signaling or regulatory components of the TGF-β pathway. RNAs are made using in vitro RNA synthesis kit (mMessage mMachine RNA synthesis kit, Ambion), following the manufacturer's instruction (*see* **Note 5**).
4. Culture the injected embryos in Ficoll solution for 4–5 h at room temperature until they reach mid- to late blastula stages (*see* **Note 6**).
5. Remove the thin, transparent vitelline membrane surrounding the embryos before cutting animal caps. This is done by transferring embryos with disposable plastic pipet to an agarose-coated 60-mm tissue culture dish containing 1/3× MMR. Take hold of the vitelline membrane with one pair of forceps in a region close to the marginal zone (the area surrounding the junction between the pigmented and the nonpigmented hemispheres) or the vegetal pole (the nonpigmented region). Use the second pair of forceps to grab the vitelline membrane in a nearby region and move the forceps to the opposite directions to tear the membrane apart. The embryos are now released in the medium and are ready for dissection (*see* **Note 7**).
6. Place the embryos with the animal hemisphere up. Insert the eyebrow hair knife into the animal region of the embryos, so that the tip of the hair knife comes out from the other side of the animal region. Flick the hair knife upward while holding the embryos with the hair loop to make a cut. Alternatively, use the hair loop or the forceps to rub against the hair knife to make a cut. Insert the hair knife again at a 90° angle and make another cut. Repeat the process until a square piece of the animal tissue is isolated from the embryos (Fig. 2). The animal caps should be even in thickness and devoid of yolky larger cells which are mesodermal or endodermal in origin (*see* **Note 8**). Normally 5–10 caps are needed for each sample condition.

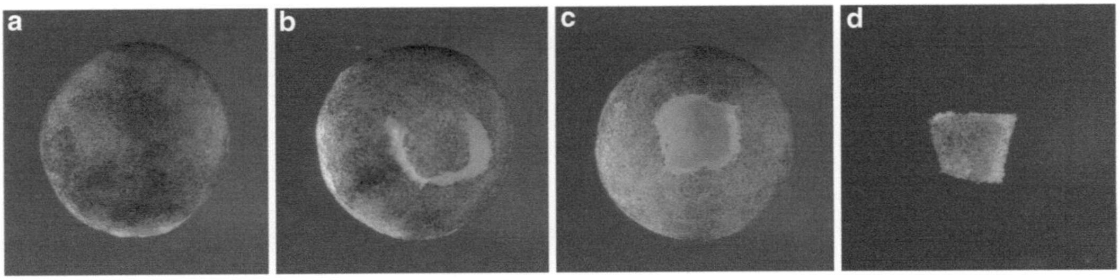

Fig. 2 Dissection of the animal caps: (**a**) an intact embryo released from the vitelline membrane; (**b**) an embryo cut with the hair knife in the animal region; (**c**) the remaining embryo with the animal cap removed; (**d**) an isolated animal cap

7. Transfer the animal caps with a disposable glass pipet to an agarose-coated 4-well tissue culture dish and incubate them in 1/2× MMR or 1/2× MBS until the desired stages (*see* **Note 9**). Soluble TGF-β ligands, antagonists, or chemical inhibitors can be added to the culture medium during animal cap incubation (*see* **Note 10**). Untreated animal caps from sibling control embryos should be included to control for the absence of TGF-β signaling components or their regulators. Intact uninjected embryos should also be cultured at the same time for staging of embryonic development.

3.2 Animal Cap Dissociation Assay

Although animal caps isolated at the blastula stages are not subjected to signals from their neighboring tissues, cells in the animal caps nonetheless signal to each other. Disruption of animal cap-autonomous BMP signals has been shown to convert the animal caps to differentiate from the epidermal to the neural fates. Dissociation of animal caps into single, individual cells interferes with cell-cell signaling and has been used to study the "default" fate of animal cells [17]. Dissociated cells, which are more uniformly exposed to the growth factors than the animal cap cells, have also been used to examine cell response to gradient TGF-β inducing signals as well as the requirement for cell contact in tissue development [18]. In this section, the method to dissociate animal caps is discussed.

Experimental Procedure

1. Dissect animal caps as described above. Five to ten animal caps are normally used in cell dissociation assays.
2. Transfer animal caps to an agarose-coated 60-mm tissue culture dish that contains calcium- and magnesium-free buffer CMFB. Place the caps with the outer pigmented layer up. Leave the caps undisturbed in CMFB for 15–20 min.

3. The cells in the inner layers of the animal caps (facing blastocoel) are dissociated more readily than the cells in the outer pigmented layer. They separate from each other and the outer cell sheet after 20 min. Carefully peel back the outer pigmented layer with the eyebrow knife and discard it. Transfer the dissociated inner cap cells into agarose-coated 4-well dishes containing CMFB. Pipet the solution up and down to disperse the cells evenly in the well.
4. Dissociate the animal cap cells in CMFB for 3–5 h (*see* **Note 11**). Growth factors, such as activin or BMPs, can be added in CMFB during the incubation period. Gently shake the dish every 30 min to disperse replicated cells and maintain cell dissociation.

5.1. At the end of the dissociation period, gently swirl the dish in a circular fashion to concentrate the cells in the center. Replace CMFB with the MBS buffer. Repeat buffer changes three times to prevent dilution of MBS by remaining CMFB. Swirl the dish to concentrate the cells after the last buffer change and leave the cells undisturbed for 1–2 h. Cells should reaggregate efficiently and form a round ball.

5.2. An alternative method for cell reaggregation is to collect cells into an Eppendorf tube which has just been coated with 1 % BSA (Bovine Serum Albumin) in water (simply wash the Eppendorf tube with 1 % BSA solution once). Spin down the cells at the low speed of 1000 rpm for 1 min with a microcentrifuge. Remove the CMFB buffer and disperse the cells in 1 mL of MBS. Spin down the cells again. Replace the solution with 1 mL of fresh MBS once. Spin down the cells at 1000 rpm for 1 min. Leave the cell pellet in the Eppendorf tube undisturbed for 1–2 h. Cells will reaggregate and form a round ball.

6. Collect reaggregated cells at the desired stages for analyses for marker expression in the presence or absence of added factors.

3.3 Assay for TGF-β Signaling in Animal Caps

During early Xenopus development, TGF-β signals regulate formation and patterning of all three germ layers. The activities of TGF-β signals are recapitulated in animal caps, so that activin/nodal-like ligands induce ventral mesodermal, dorsal mesodermal, and endodermal tissues in a dose-dependent manner (e.g., *see* Fig. 3), whereas BMP-like growth factors participate in ventral mesoderm formation (Fig. 4). Endogenous BMP signals are also involved in epidermal specification, and inhibition of this pathway leads to neural induction in animal caps (Fig. 4). Some of the activities of the TGF-β signals are manifested in changes in morphology of the animal caps; but in most cases, additional steps are required to assay for gene expression patterns in these explants. In this section, methods used to analyze TGF-β responses in animal caps are described, with the main focus on the semi-quantitative reverse transcription-PCR (RT-PCR) assay.

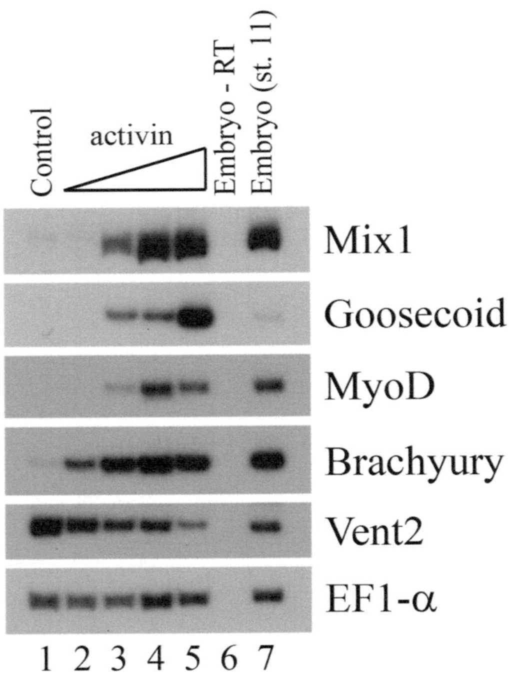

Fig. 3 Dose-dependent induction of mesodermal and endoderm genes by activin in animal caps. Embryos were injected with 1, 2, 4, and 10 pg activin RNA at the 2-cell stage. Animal caps were dissected at the blastula stage (stage 9) and incubated to the mid-gastrula stage (stage 11). RT-PCR was performed to assay for gene expression in the caps. The marker genes used are Vent2, ventral ectodermal and mesodermal marker; Brachyury, pan-mesodermal marker; MyoD, ventral-lateral mesodermal marker; Goosecoid, dorsal anterior (head) mesodermal marker; Mix1, endodermal marker. As the doses of activin increased, the ventral markers were downregulated, whereas the dorsal mesodermal and the endodermal markers were upregulated

3.3.1 Morphological Assessment of TGF-β Activities in Animal Caps

Animal caps cultured in saline solutions heal quickly and form round balls. At late tailbud to tadpole stages, surface cells differentiate into multiciliated epidermal cells so that animal caps begin to spin and "glide" on agarose-coated dishes. When induced with activin/nodal-like growth factors, animal caps form mesodermal tissues. At low doses of the ligands when the ventral and the lateral mesoderm are induced, the caps lose the round shape at the neurula to tailbud stages and show rugged edges. At high doses when the dorsal mesoderm is induced, the caps display elongated morphology, reflecting the convergent extension movements of axial and paraxial mesoderm (Fig. 5). At even higher doses when the head mesoderm and the endoderm are induced, the caps no longer elongate but tend to collapse and spread on substrates on tissue culture dishes (if the dishes are not coated with agarose).

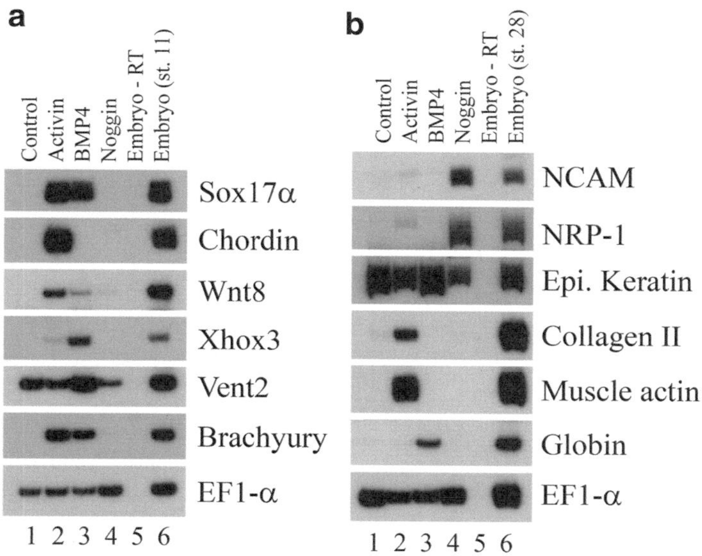

Fig. 4 Gene expression in animal caps in response to TGF-β signal activation or inhibition. Embryos were injected with 4 pg activin, 20 pg BMP4-flag, and 40 pg Noggin RNAs at the 2-cell stage. Animal caps were dissected at the blastula stage (stage 9) and incubated to the mid-gastrula (stage 11, panel (**a**)) or the tailbud (stage 28, panel (**b**)) stages. RT-PCR was performed to assay for gene expression in these caps. The marker genes used are Brachyury, pan-mesodermal marker; Vent2, ventral ectodermal and mesodermal marker; Xhox3, ventral mesodermal marker; Wnt8, ventral-lateral mesodermal marker; Chordin, dorsal mesodermal marker; and Sox17α, endodermal marker, at the gastrula stage; and Globin, ventral mesodermal (blood) marker, Muscle actin, paraxial mesodermal marker; Collagen II, axial mesodermal marker; Epidermal Keratin, epidermal marker; and NRP-1 and NCAM, neural markers. While activin induced dorsal mesodermal markers, BMP4 was capable of only inducing ventral types of the mesoderm. Both growth factors induced endoderm formation in the animal caps. Noggin, a BMP inhibitor, did not induce mesoderm or endoderm in the caps, but it interfered with the epidermal development and induced neural formation

This is due to the migratory rather than intercalating behaviors of the head mesoderm. Although the morphologies of the caps correlate with the types of the mesoderm induced by activin/nodal, it is by no means accurate. Often the animal caps contain mesodermal tissues from both the dorsal and the ventral regions, and the morphology is dominated by the amount of dorsal or ventral mesodermal cells present. Direct analysis of marker expression in these caps is required (see below). Animal caps induced by the activated BMP signals remain round and do not have special appearance, but they do not glide on the bottom of the culture dish at the tadpole stages. Inhibition of the BMP pathway, which induces cement gland and neural tissues, often results in animal

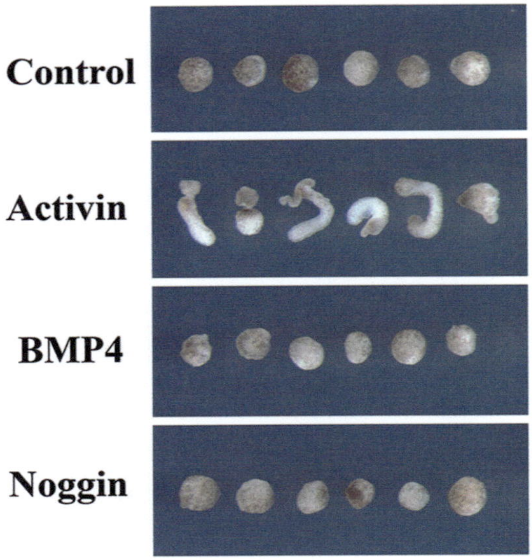

Fig. 5 Morphology of animal caps in response to TGF-β signal activation or inhibition. Animal caps isolated from Xenopus embryos normally round up to form a tight ball. Induction of dorsal mesoderm by activin leads to cap elongation, whereas induction of ventral mesoderm by BMP4 does not alter the cap morphology in an obvious way. Inhibition of BMP signal by noggin often results in formation of cement gland in the animal caps, which stick to the hair knife or hair loop when they are in contact

caps with dark patches of cells. These cells have differentiated along the cement gland lineage and are very sticky. They tend to adhere to the hair knife, hair loop, or the transfer pipets and need to be handled carefully during cap collection.

3.3.2 Semi-quantitative RT-PCR Assay

Assay for marker expression by RT-PCR has the advantage of assessing expression levels of multiple genes at the same time and is therefore a widely used method to analyze TGF-β signal transduction and regulation in animal caps. Caps can be collected at gastrula, neurula, tailbud, or tadpole stages depending on particular issues to be addressed and markers to be analyzed. The protocol for RT-PCR assay is described in the following (*see* **Note 12**).

Experimental Procedures

1. RNA extraction from animal caps: Homogenize samples (5–10 caps or 1 embryo) in 250 μL of Lysis buffer containing 250 μg/mL Proteinase K by pipetting up and down. Incubate at 42 °C for 30 min. Extract the sample with 250 μL of phenol/chloroform/isoamyl alcohol once. Transfer the clear supernatant to a new RNase-free tube. Add 25 μL of 10 M NH_4OAc, 1 μL of 10 mg/mL glycogen, and 650 μL of EtOH. Precipitate RNA in −70 °C freezer for more than 15 min.

2. DNase treatment: Spin down RNA in the cold room and wash the pellet once with cold 70 % EtOH in DEPC dH_2O. Resuspend the RNA pellet in 15 μL pf DEPC dH_2O. Add 10 μL of DNase I mix. Incubate at 37 °C for 30 min. Bring the reaction volume to 100 μL with DEPC dH_2O and extract once with an equal volume (100 μL) of phenol/chloroform/isoamyl alcohol. Transfer the supernatant to a new RNase-free tube. Precipitate RNA with 10 μL of 10 M NH_4OAc, 1 μL of 10 mg/mL glycogen, and 275 μL of EtOH. Leave in −70 °C freezer for more than 15 min.
3. Reverse transcription: Spin down RNA in the cold room and wash the pellet once with cold 70 % EtOH. Resuspend the pellet in 10 μL of DEPC dH_2O. Add 1 μL of 0.1 mg/ml random hexamer primers and 9 μL of RT mix. Incubate at 42 °C for 30 min.
4. PCR: Take 1–2 μL cDNA from the RT reaction, add 1 μL of 0.1 mg/mL primer pairs for the genes assayed, 5 μL of 5× Taq polymerase buffer, 1 μL of 10 mM dNTP mix, 0.1 μL [α-^{32}P]-dATP, 0.2 μL Taq DNA polymerase, and dH_2O to make the final volume of 25 μL. PCR for 20–30 cycles.
5. Gel electrophoresis: Make 5 % polyacrylamide gel (19:1) in 0.5× TBE. Load 5 μL of PCR reaction on the gel and run PAGE. Transfer the gel to 3 MM filter paper and dry the gel in the gel dryer. Expose the gel to the X-ray film and develop autoradiography to examine the PCR products.

3.3.3 In Situ Hybridization and Immunohistochemistry

Responses to TGF-β signals in animal caps can also be analyzed via examination of marker expression by in situ hybridization (for RNA expression) or immunohistochemistry (for protein expression). Although levels of gene expression cannot be accurately assessed in these assays, regionalized expression patterns of distinct genes can be observed directly in animal caps, especially when double staining of different markers is performed. The protocols for these experiments are similar to those used in analyses of gene expression patterns in whole embryos and have been described in detail in ref. [12].

4 Notes

1. People have eyebrow and eyelash hairs with different strength, length, or curvatures. Soft eyebrow hairs tend to bend easily and cannot cut Xenopus tissues efficiently, whereas coarse hairs may damage tissues easily. It is thus advisable to shop around and collect hairs from different people for comparison. Baby hair is often the best in terms of resilience and fineness. Instead of hair knifes, some people prefer using tungsten needles

or fine glass needles. These are normally used for dissection of later stage embryos, but can be used to cut animal cap too.

2. To make MBS, prepare 0.1 M $CaCl_2$ and 10× MBS salt solution without $CaCl_2$ separately and autoclave each solution. To make final MBS, mix 100 mL of 10× MBS salt solution and 7 mL of 0.1 M $CaCl_2$. Adjust the volume to 1 L.
3. Several other solutions can also be used to culture animal caps, such as 3/4× NAM (Normal Amphibian Medium). The full strength NAM is consisted of: 110 mM NaCl, 2 mM KCl, 1 mM $Ca(NO_3)_2$, 1 mM $MgSO_4$, 0.1 mM EDTA, 1 mM $NaHCO_3$, and 2 mM sodium phosphate pH 7.4. Regardless of the solutions used, always add antibiotics to the solution to prevent bacterial infection.
4. Embryos are wrapped in a thick, protective layer of extracellular matrix called jelly coat. Removal of the jelly materials (dejelly) is required for all embryonic manipulations, such as injection or dissection. Be sure that dejelly is complete, as remaining jelly coat surrounding the embryos makes it difficult for injection (embryos tend to roll around at the touch of the injection needle, making needle penetration difficult). However, be careful not to over-dejelly the embryos, as prolonged treatment with cysteine damages the embryos.
5. The doses of RNAs required to achieve induction or inhibition by a particular molecule differ depending on the molecules involved, and sometimes the constructs used by different labs have different efficiency in their activities. One may need to do a titration experiment to obtain the optimal dose for a newly acquired construct.
6. The length of the incubation period after embryo injection varies depending on the temperature used, so that the embryos need less or more time to reach the blastula stages when they are cultured at higher or lower temperatures, respectively. The range of the tolerable temperatures for embryos is 13–25 °C, but we normally keep the embryos between 15 and 22 °C.
7. At the blastula to gastrula stages, a space is created between cells at the animal hemisphere and the vitelline membrane. It is therefore easiest to remove the vitelline membrane by taking hold of it from the animal side. However, without enough practice, one can damage the animal region easily this way. Proceeding from the marginal zone or the vegetal regions is thus recommended here, as damage in these regions will not affect the integrity of the animal caps.
8. Animal caps can be dissected in several other ways. Some people prefer using two pairs of sharp forceps to cut out the animal caps directly from the embryos. The embryos can also

be placed with the vegetal side up and opened through the vegetal tissue (endodermal mass) to reveal the animal tissues on the other side of the blastocoel. A square piece of animal cap can then be trimmed out with the hair knife. In addition, the embryos can be placed on their side with the blastocoel and the animal tissues on one side. Animal caps are then dissected out with a single cut using the eyebrow knife. Regardless of the method used, an important point to keep in mind is to check for potential contamination of mesodermal/endodermal cells in the animal caps. These cells are larger and contain more yolky substance. Contamination with mesodermal or endodermal cells can obscure research results and conclusions and needs to be avoided. Simply trim off the tissue containing the mesodermal cells or gently scrape off loose endodermal/mesodermal cells adhering to the caps with the eyebrow knife.

9. When transfer animal caps, make sure that there is enough medium in the transfer pipet before and after taking the caps into the pipet. Avoid contact between animal caps and the surface of the solution, as surface tension destroys the animal caps (they explode).

10. Animal caps heal quickly and curl up to form a ball. This can prevent the access of soluble factors to the inner cells of the animal caps. In the case that soluble factors are used, caps should be cut quickly and transferred immediately to the conditioned culture medium. One can also divide the embryos into small batches, with each containing only a small number of the embryos (e.g., [3]). This way, dissected caps can be transferred in batch to prevent them from healing and to facilitate access of all cells to the added soluble factors.

11. For neural induction to occur in the absence of any exogenous factors, animal cells need to be dissociated for at least 3 h. Cells dissociated for more than 5 h are unhealthy and do not reaggregate efficiently. They remain as a loose group and may have undergone apoptosis. Gene expression assay indicates that these cells do not express any differentiated markers.

12. Instead of using semi-quantitative PCR, many groups now use quantitative PCR (qRT-PCR) to assay for gene expression. qRT-PCR helps to guarantee that the expression is in the lineage range of amplification and can give more precise information on expression levels of various genes.

Acknowledgement

This work is supported by NIH grant R01GM098566.

References

1. Wu MY, Hill CS (2009) TGF-β superfamily signaling in embryonic development and homeostasis. Dev Cell 16:329–343
2. Faure S, Lee MA, Keller T, ten Dijke P, Whitman M (2000) Endogenous patterns of TGFβ superfamily signaling during early Xenopus development. Development 127:2917–2931
3. Lee MA, Heasman J, Whitman M (2001) Timing of endogenous activin-like signals and regional specification of the Xenopus embryo. Development 128:2939–2952
4. De Robertis EM, Kuroda H (2004) Dorsal-ventral patterning and neural induction in Xenopus embryos. Annu Rev Cell Dev Biol 20:285–308
5. Smith WC, Harland RM (1992) Expression cloning of noggin, anew dorsalizing factor localized to the Spemann organizer in Xenopus embryos. Cell 70:829–840
6. Sasai Y, Lu B, Steinbeisser H, Geissert D, Gont LK, de Robertis EM (1994) Xenopus chordin: a novel dorsalizing factor activated by organizer-specific homeobox genes. Cell 79:779–790
7. Bouwmeester T, Kim S, Sasai Y, Lu B, de Robertis EM (1996) Cerberus is head-inducing secreted factor expressed in the anterior endoderm of Spemann's organizer. Nature 382:595–601
8. Zhu H, Kavsak P, Abdollah S, Wrana JL, Thomsen GH (1999) A SMAD ubiquitin ligase targets the BMP pathway and affects embryonic pattern formation. Nature 400: 687–693
9. Hata A, Seoane J, Lagna G, Montalvo E, Hemmati-Brivanlou A, Massague J (2000) OAZ uses distinct DNA- and protein-binding zinc fingers in separate BMP-Smad and Olf signaling pathways. Cell 100:229–240
10. Zimmerman LB, de Jesus-Escobar JM, Harland RM (1996) The Spemann organizer signal noggin binds and inactivates bone morphogenetic protein 4. Cell 86:599–606
11. Piccolo S, Sasai Y, Lu B, de Robertis EM (1996) Dorsoventral patterning in Xenopus: inhibition of ventral signals by direct binding of chordin to BMP-4. Cell 86:589–598
12. Sive HL, Grainger RM, Harland RM (2000) Early development of Xenopus laevis: a laboratory manual. Cold Spring Harbor Laboratory Press, Cold Spring Harbor, NY
13. Chang C, Holtzman DA, Chau S, Chickering T, Woolf EA, Holmgren LM, Bodorova J, Gearing DP, Holmes WE, Hemmati-Brivanlou A (2001) Twisted gastrulation can function as a BMP antagonist. Nature 410:483–487
14. Harms PW, Chang C (2003) Tomoregulin-1 (TMEFF1) inhibits nodal signaling through direct binding to the nodal coreceptor Cripto. Genes Dev 17:2624–2629
15. Zhang Y, Chang C, Gehling DJ, Hemmati-Brivanlou A, Derynck R (2001) Regulation of Smad degradation and activity by Smurf2, an E3 ubiquitin ligase. Proc Natl Acad Sci U S A 98:974–979
16. Alliston T, Ko TC, Cao Y, Liang YY, Feng XH, Chang C, Derynck R (2005) Repression of bone morphogenetic protein and activin-inducible transcription by Evi-1. J Biol Chem 280:24227–24237
17. Wilson PA, Hemmati-Brivanlou A (1995) Induction of epidermis and inhibition of neural fate by Bmp-4. Nature 376:331–333
18. Green JB, Smith JC, Gerhart JC (1994) Slow emergence of a multithreshold response to activin requires cell-contact-dependent sharpening but not prepattern. Development 120:2271–2278

Chapter 17

Detection of Smad Signaling in Zebrafish Embryos

Xingfeng Liu, Qiang Wang, and Anming Meng

Abstract

Nodal and BMPs play critical roles in germ layer induction and patterning in early zebrafish embryos. Smad2/3 and Smad1/5/8 are intracellular effectors of Nodal and BMPs, respectively. These Smads regulate, in cooperation with other factors, transcription of hundreds of target genes in the nucleus. The activity and stability of Smads are regulated by phosphorylation modifications. To better understand the regulatory network of Smads-mediated signaling and its biological implications, it is necessary to monitor the signaling activity in an in vivo model system. In this chapter, we describe the methods used in zebrafish embryos for dissecting Smads signaling, including TGF-β/Nodal- and BMP-responsive luciferase reporter assays, Western blotting for Smads, co-immunoprecipitation for Smads and their interacting proteins, chromatin-immunoprecipitation for identification of Smad2-binding sites, and immunostaining for detection of active Smad1/5/8.

Key words TGF-β, BMP, Nodal, Smad, Signal transduction, Zebrafish, Embryo

1 Introduction

Zebrafish has been successfully used as a vertebrate model to investigate implication of components of several important signaling pathways in embryonic development because of high reproductivity, external development of embryos, and powerful genetic tools. Transforming growth factor-β superfamily members such as Nodal and bone morphogenetic proteins (BMPs) play important roles in germ layer induction and patterning during early development of zebrafish embryos [1, 2]. The deficiency of Nodal signals in zebrafish embryos causes loss of most mesodermal and endodermal tissues [3, 4], demonstrating an essential role in mesoderm and endoderm induction. Nodal signals are also crucial for the formation of the embryonic shield, the fish dorsal organizer that is indispensable for the development of dorsal tissues [5, 6]. Like in *Xenopus* [7], Bmp signaling in zebrafish embryos is required for the epidermal fate of ventral ectodermal cells and the specification of the ventral mesoderm [8–10].

Xin-Hua Feng et al. (eds.), *TGF-β Signaling: Methods and Protocols*, Methods in Molecular Biology, vol. 1344,
DOI 10.1007/978-1-4939-2966-5_17, © Springer Science+Business Media New York 2016

TGF-β ligands initiate signal transduction by forming heterotetrameric complexes with two type II receptors and two type I receptors, leading to activation of the type I receptors [11]. The activated type I receptors then recruit and activate intracellular effectors R-Smads, including Smad1, Smad2, Smad3, Smad5, and Smad8, the receptor-regulated Smads, by phosphorylating two serine residues in their C-terminal SXS motif. Usually, Smad1, Smad5, and Smad8 mediate BMP signals whereas Smad2 and Smad3 are mediators of Nodal/TGF-β signals. The activated R-Smads form complexes with Smad4 in the cytoplasm, which then translocate into the nucleus and regulate transcription of many target genes in cooperation with other transcription factors (e.g., see [12]). It has been found that the activated R-Smads can be inactivated by dephosphorylation of phospho-SXS motif [13, 14] or phosphorylation of the linker region [15].

Although developmental functions of TGF-β signaling have been elucidated, its spatial and temporal regulation during development had not yet been fully understood. There must be many new players in the TGF-β signaling pathway. One can take the advantages of the zebrafish model to investigate their functions in the TGF-β signal transduction and in embryonic development.

2 Materials

2.1 Zebrafish Embryos

Any zebrafish strains such as AB strain and Tubingen strain can be used. Embryos are incubated in Holtfreter's solution at 28.5 °C in an incubator.

2.2 Reporter Plasmids

1. *BRE-Luc*, in which the expression of the firefly luciferase is driven by the BMP-specific responsive elements (BRE) [16].
2. *ID1-Luc*, in which the firefly luciferase expression is under the control of the promoter of the BMP target gene *Inhibitor of Differentiation 1* (ID1) [17].
3. *ARE$_3$-Luc*, in which the firefly luciferase expression is simulated by Nodal/TGF-β signals due to the presence of three copies of the Activin responsive element (ARE) found in the Mix.2 promoter [18].
4. *CAGA$_{12}$-Luc*, in which the firefly luciferase expression can be induced by Nodal/TGF-β signals due to the presence of 12 tandem repeats of the Smad binding site (CAGA) in the promoter [16].
5. *Renilla*, in which the Renilla luciferase is constitutively expressed. Its expression level is used as an internal control to normalize the expression level of the firefly luciferase reporters.

2.3 Antibodies

The anti-Smad2/3 rabbit polyclonal antibody (Cat. #3102, Cell Signaling Technology) can be used to detect zebrafish Smad2 by Western blotting; anti-phospho-Smad2(Ser465/467) rabbit polyclonal antibody (Cat. #3101, Cell Signaling Technology) can react with activated zebrafish Smad2, in which Ser466 and Ser468 residues in the C-terminal SXS motif are phosphorylated; anti-phospho-Smad2(Ser245/250/255) rabbit polyclonal antibody (Cat. #3104, Cell Signaling Technology) can recognize zebrafish Smad2 with phospho-Ser246/251/256 residues in the linker region; anti-phospho-Smad1(Ser463/465)/Smad5(ser463/465)/Smad8(Ser426/428) rabbit polyclonal antibody (Cat. #9511, Cell Signaling Technology) can recognize the C-terminal phospho-SXS motif in zebrafish Smad1, Smad5, or Smad8. Anti-rabbit IgG-Peroxidase secondary antibody is the product of GE Healthcare (Cat. #NA934). Fluorescein-labeled Goat Anti-Rabbit IgG secondary antibodies (Jackson Immuno Research) are used for immunostaining.

2.4 Reagents

2.4.1 Reagents for Reporter Assay Reagents

1. Holtfreter's solution: 60 mM NaCl, 2.4 mM $NaHCO_3$, 0.9 mM $CaCl_2$, 0.7 mM KCl, pH 7.0. It should be made freshly.
2. Passive lysis buffer: 50 mM Tris–HCl, pH 7.4, 1 mM DTT, 0.01 % Triton X-100.
3. LAR buffer: 20 mM Tricine pH 7.8, 1.07 mM $(MgCO_3)_4{\cdot}Mg(OH)_2{\cdot}5H_2O$, 2.67 mM $MgSO_4$, 0.1 mM EDTA.
4. Luciferin reagent: 33.3 mM DTT, 270 μM $CoANa_2$ (Cat. #C3144, Sigma-Aldrich), 530 μM ATP (Cat. #A2383, Sigma-Aldrich), 470 μM d-luciferin (Cat. #1603, Promega) in the LAR buffer. The prepared stocks are 1 M DDT in 0.01 M NaAc-Acetic Acid, pH 5.2, stored at −20 °C; 13 mM $CoANa_2$ in water, stored at −20 °C; 5 M ATP in water pH 7.0, stored at −80 °C; 10 mM Luciferin in water, stored at −20 °C.
5. Coelenterazine reagent: 0.09 μM coelenterazine (Coelenterazine, Cat. #C3355, Sigma), 100 mM NaCl, 2.5 mM Tris–HCl, pH 7.5. The stock is 0.09 mM coelenterazine in 20 mM HCl/MeOH.

2.4.2 Reagents for Western Blotting

1. TNE lysis buffer: 10 mM Tris–HCl, pH 7.4, 150 mM NaCl, 5 mM EDTA, 1 % Triton X-100. When use, one complete Protease Inhibitor Cocktail Tablet per 50 mL (Cat. #11 697 498 001, Roche) is added. In detection of phosphorylated Smads, 100 mM NaF and 1 mM orthovanadate (Na_3VO_4) are added to the TNE lysis buffer to inhibit phosphatase activities.
2. 6× Sample loading buffer: 60 % Glycerol, 12 % SDS, 300 mM Tris–HCl, pH 6.8, 300 mM DTT, 0.01 % Bromophenol Blue.

3. 10× Running buffer: 250 mM Tris base, 200 mM glycine, 1 % SDS.
4. 1× Transfer buffer: 25 mM Tris base, 20 mM glycine, 10–20 % Methanol.
5. 10× TBS: 0.5 M Tris–HCl, pH 7.6, 1.5 M NaCl. When use, the stock is diluted to 1× with addition of 0.1 % Tween-20.
6. 1× TBST: 50 mM Tris–HCl, pH 7.6, 150 mM NaCl, 0.1 % Tween-20.
7. Antibody dilution solution: 1× TBS, 3 % BSA, 0.05 % Tween-20.
8. ECL solution: mix Solution I and Solution II with equal volume immediately before use. Solution I: 200 mM Tris–HCl, pH 8.5, 2 mM Luminol (3-aminophthalhydrazide, Cat. #A8511, Sigma-Aldrich), 0.4 mM *p*-coumaric acid (Cat. #C9008, Sigma-Aldrich), sterilized and stored at 4 °C in a light-proof bottle. Solution II: dissolve 60 μL of 30 % H_2O_2 in 100 mL of H_2O, stored at 4 °C in a light-proof bottle.

2.4.3 Reagents for Chromatin Co-immunoprecipitation

1. Phosphate buffered saline (PBS): 137 mM NaCl, 2.7 mM KCl, 10 mM Na_2HPO_4, 2 mM KH_2PO_4, pH 7.4. Pre-made PBS powder (Gibco, Cat. #21600-044) or solution is commercially available.
2. Cell lysis buffer: 10 mM Tris–HCl, pH 8.0, 10 mM NaCl, 0.5 % NP-40.
3. Nuclei lysis buffer: 50 mM Tris–HCl, pH 8.0, 10 mM EDTA, 1 % SDS
4. IP dilution buffer: 20 mM Tris–HCl, pH 8.0, 2 mM EDTA, 150 mM NaCl, 0.01 % SDS.
5. ChIP wash solution I: 20 mM Tris–HCl, pH 8.0, 2 mM EDTA, 1 % Triton X-100, 150 mM NaCl, 0.1 % SDS.
6. ChIP wash solution II: 20 mM Tris–HCl, pH 8.0, 2 mM EDTA, 1 % Triton X-100, 500 mM NaCl, 0.1 % SDS.
7. ChIP wash solution III: 10 mM Tris–HCl, pH 8.0, 1 mM EDTA, 0.25 M LiCl, 1 % NP-40, 1 % deoxycholate (sodium salt).
8. TE buffer: 10 mM Tris–HCl, pH 8.0, 1 mM EDTA.
9. Elution buffer: 25 mM Tris–HCl, pH 8.0, 10 mM EDTA, 0.5 % SDS.

2.4.4 Immunostaining

1. 4 % Paraformaldehyde: dissolve a required amount of paraformaldehyde (Cat. #158127, Sigma-Aldrich) in PBS.
2. PBST: 0.1 % Tween-20 in PBS.
3. PBSTX: 0.3 % Triton X-100 in PBS.

4. Blocking solution: 3 % BSA, 10 % fetal calf serum, and 1 % dimethylsulfoxide in PBST.
5. Anti-quenching agent: 2 % propyl gallate (Cat. #P3130, Sigma-Aldrich) in 70 % glycerol/PBS.
6. DAPI: dissolve DAPI (Cat. #D8417, Sigma-Aldrich) in sterile water to a concentration of 1 μg/mL.

3 Methods

3.1 Microinjection of Embryos

Plasmid DNAs, mRNAs, or morpholinos are dissolved in nuclease-free water (or in 0.1 mM KCl) and injected into one-cell stage zebrafish embryos. The injected volume is about 1 nL per embryo. DNAs are usually injected into the cytoplasm of the embryos, while mRNAs and morpholinos can be injected into either the cytoplasm or the yolk. The detail of microinjection has been described previously [19].

3.2 Reporter Assays

1. 80 pg of corresponding luciferase reporter DNA and 10 pg of *Renilla* DNA are co-injected into individual one-cell embryos. For the ARE_3-luc reporter, 30 pg of FAST2 plasmid DNA is also co-injected. When necessary, mRNAs or morpholinos may be co-injected. For each assay, more than 80 embryos need to be injected. Embryos are incubated in Holtfreter's solution at 28.5 °C.
2. Collect 40 live embryos at desired stages (from shield to bud stages) and remove their chorion by protease treatment.
3. Transfer the embryos into 200 μL of Holtfreter's solution in a 1.5-mL Eppendorf tube. Crash the yolk by gently sipping and blowing the embryos with a 200-μL tip. Allow the embryonic tissues to naturally settle down for a few seconds and then remove the aqueous layer.
4. Add 1 mL of Holtfreter's solution and gently mix. Discard the aqueous layer following centrifugation for 2 min at 100 × *g*.
5. Add 80 μL of passive lysis buffer (2 μL per embryo) and allow the lysis for 5–10 min. The tubes should be rocked or vibrated to facilitate cell lysis.
6. Centrifugate the lysate for about 2 min at 16,000 × *g* and collect the supernatant to a new 1.5-mL Eppendorf tube.
7. Transfer six aliquots, each 10 μL, of the lysate individually to a 96-well microplate. Place the microplate into a Centro LB 960 Microplate Luminometer (Berthold Technologies) that is connected to two injectors, one for the luciferin reagent and the other for the coelenterazine reagent.

8. Read the light units in the wells, three for luciferin (firefly luciferase) and three for coelenterazine (*Renilla* luciferase).
9. Calculate the relative luciferase activity as firefly luciferase light units/*Renilla* luciferase light units.

3.3 Western Blotting

1. Collect embryos at desired stages. The chorion and yolk are removed as described in **steps 2–4** of Subheading 3.2.
2. Cool the cell pellet and add precooled TNE lysis buffer in a ratio of 1–2 μL per embryo (before the bud stage). Allow the lysis to proceed at 4 °C for 10 min. The tubes should be rocked or vibrated to facilitate the lysis.
3. Centrifugate the lysate at 16,000 × *g* for 10 min at 4 °C and transfer the supernatant to a new Eppendorf tube.
4. Add 6× sample loading buffer and mix. Place the tube in the boiling water for 3–5 min and spin shortly. Then, put it on ice or at room temperature to cool down.
5. Prepare an appropriate percentage SDS-PAGE gel (8.5 cm × 8.5 cm). Load 10–20 μL of sample (equivalent to 10–20 embryos) onto a gel well.
6. Run the gel at appropriate voltages for about 3 h. The voltage is usually less than 100 V when the samples are still in the stacking gel and it is raised to 150 V when the samples have migrated into the separation gel.
7. Cut a piece of nitrocellulose membrane (GE Healthcare) to a size matching the gel and rinse it in 1× transfer buffer for 1–2 min.
8. Take off the gel and place it on 2–4 pieces of filter paper in the gel holder cassette of the Mini Trans-Blot cell (Bio-Rad). Place the rinsed membrane on the gel and avoid any air bubbles between them. Put another 2–4 pieces of filter paper on the top of the membrane.
9. Transfer the proteins in the gel to the membrane at an appropriate voltage (30–100 V) for 1–2 h in the Mini Trans-Blot cell. The proteins migrate into the membrane from the negative electrode to the positive electrode.
10. Take off the membrane and place it in 7–10 mL of block buffer (3 % BSA or 5 % skimmed milk in TBST solution) in a small plastic box. Incubate for 1 h at room temperature to block the unoccupied protein.
11. Discard the block buffer and wash the membrane with TBST solution twice.
12. Discard TBST solution and add 1–6 mL of diluted primary antibody. The primary antibody is usually diluted in a ratio of 1:1000 in antibody dilution solution.

13. Following incubation overnight at 4 °C, collect the primary antibody solution, which can be reused subsequently.
14. Wash the membrane with TBST three times, 10 min each.
15. Add 6 mL of diluted secondary antibody (1:5000 to 1:10,000) and incubate for 1 h at room temperature.
16. Remove the secondary antibody. Wash the membrane with TBST three times, 10 min each.
17. Incubate the membrane in 0.2 mL (0.2–0.3 mL per membrane) of ECL solution with gentle agitation for 1–2 min at room temperature.
18. Take out the membrane and place it on tissue towels. Wrap it with plastic wrap while it is still damp.
19. Place the membrane against an X-ray film in an autoradiography cassette in a dark room. Allow exposure at room temperature for a few seconds to 30 min and then develop the film.

3.4 Co-immunoprecipitation with Anti-Smad2/3 Antibody

1. Collect 300 or more embryos at desired stages. The chorion and yolk are removed as described in **steps 2–4** of Subheading 3.2.
2. Cool the cell pellet and add precooled TNE lysis buffer in a ratio of 2–4 μL per embryo (before the bud stage). Slowly rotate the tube in a rocker at 4 °C for 10–20 min to allow complete lysis.
3. Centrifugate the lysate at 16,000 × *g* for 10 min at 4 °C and transfer the supernatant to a new Eppendorf tube.
4. Take out 30 μL of the lysate as input for detecting individual proteins later on. Add 3 μL of anti-Smad2/3 antibody to the remaining lysate and incubate for 1–2 h at 4 °C.
5. Add protein A/G Sepharose Bead (Cat. #sc-2003, Santa Cruz Biotechnology) in a ratio of 1:20. Incubate for 4–6 h at 4 °C.
6. Centrifugate the mixture at 1000 × *g* for 3 min at 4 °C.
7. Discard the supernatant. Wash the pellet three times with 1 mL of TNE lysis buffer. For each wash, gently rotate the tube in a rocker for 3 min at 4 °C, followed by centrifugation at 1000 × *g* for 3 min at 4 °C.
8. After the final wash, add 30–50 μL of 2× sample loading buffer.
9. Boil the sample for 5 min in a water bath and load an aliquot of 25–30 μL onto a SDS-PAGE gel.
10. Run the gel as described for Western blotting. The Smad2-interacting protein in the blot is detected by Western blotting using the specific antibody essentially as described in Subheading 3.3.

3.5 Chromatin Co-immuno-precipitation (ChIP) with Anti-Smad2/3 Antibody

1. Collect approximately 2000 embryos at desired stages (from shield to bud stages). Remove the chorion as described above.
2. Transfer the embryos to a 15-mL conical tube and add Holtfreter's solution to 9.5 mL.
3. Add 500 μL of 37 % formaldehyde solution and gently rotate for 15 min at room temperature to crosslink protein-DNA complexes.
4. Add 500 μL of 2.5 M glycine solution and mix gently. Incubate for 5 min at room temperature to quench the cross-linking reaction.
5. Discard the supernatant after the natural sedimentation of the embryos. Wash the embryos with 10 mL of precooled PBS three times. The embryos in PBS may be snap-frozen for use in the future.
6. Remove PBS and add 2 mL of cell lysis buffer. Pipette the embryos with a 200-μL tip and vibrate for 15 min at 4 °C.
7. Centrifugate the mixture at 1300 ×*g* for 5 min at 4 °C. Discard the supernatant and add 1 mL of nuclei lysis buffer. Following mixing with a pipette, transfer the lysate to a 1.5-mL Eppendorf tube and vibrate for 10 min at 4 °C.
8. Divide the lysate into two parts, 500 μL each, and put each into a 5-mL tube. Add 3 mL of precooled IP dilution buffer and mix well.
9. Ultrasonicate DNA in the lysate in an ultrasonic homogenizer (SCIENTZ-IID) for 20 min (set to 180 W, work 5-s work and 7-s rest per cycle). After ultrasonication, DNA fragments of 300–1000 bp in length should be predominant, which can be monitored by electrophoresis in an agarose gel.
10. Split the lysate and transfer each into a 1.5-mL Eppendorf tube for easy centrifugation. Centrifugate at 20,000 ×*g* for 10 min at 4 °C.
11. Collect the DNA-containing supernatants from all of the tubes and transfer them to a new 15-mL tube. Add 3 mL of IP dilution buffer and 400 μL of 10 % Triton X-100 and mix well. Now, the lysate is approximately 10 mL and can be snap-frozen for use later.
12. Pre-equilibrate Protein A Sepharose beads (Cat. #101090, Life Technologies Co.) by incubating 100 μL of Protein A Sepharose beads with 1 mL of PBS and 0.5 % BSA for 10 min at 4 °C, followed by centrifugation at 100 ×*g* for 3 min. Remove the supernatant and rinse the beads two more times. After the final equilibration, the beads are suspended in 100 μL of PBS and 0.5 % BSA.

13. To pre-clean the lysate, split the lysate into two 15-mL conical tubes, 5 mL each. Add 100 μL of pre-equilibrated Protein A Sepharose beads to each tube and mix by slowly rotating at 4 °C for 30 min.
14. Centrifugate at 100 × *g* for 1 min at 4 °C. Transfer the supernatant in each tube to a new 15-mL conical tube. Take out 100 μL of the cleaned lysate and store at −20 °C for future use as input controls.
15. Add 50 μL of anti-Smad2/3 antibody to one of the lysate-containing tubes and 50 μL of IgG to the other tube for control. Mix by slowly rotating in a rocker overnight at 4 °C.
16. Add 100 μL of pre-equilibrated Protein A Sepharose beads and gently mix by rotating for 6 h at 4 °C.
17. Centrifugate at 100 × *g* for 1 min at 4 °C and transfer the beads to new 1.5-mL tubes.
18. Wash the beads in each tube sequentially with 1 mL ChIP wash solution I, II, III and TE. Each wash lasts 10 min at 4 °C, followed by centrifugation at 100 × *g* for 1 min at 4 °C.
19. After the final wash, add 200 μL of elution buffer to the beads and incubate at 65 °C for 15 min with a brief vortexing every 2 min.
20. Centrifugate at 100 × *g* at room temperature at 4 °C and collect the supernatant (eluent) into a new 1.5-mL Eppendorf tube.
21. Add another 200 μL of elution buffer to the beads and incubate at room temperature with slow rotation for 10 min. Following centrifugation at 100 × *g*, collect the eluent and combine it with the previous eluent.
22. Incubate 400 μL of DNA eluent at 65 °C overnight. The previously left 100 μL input control should also be treated side by side after the addition of 300 μL of elution buffer.
23. Add RNase A to a final concentration of 0.2 μg/μL and incubate for 1 h at 37 °C.
24. Add protease K to a final concentration of 0.2 μg/μL and incubate for 2 h at 55 °C.
25. Clean DNA by phenol:chloroform and chloroform extractions: Add an equal volume of phenol:chloroform (1:1) mixture and mix well. Centrifugate at 16,000 × *g* for 10 min at room temperature (RT), transfer the supernatant to a fresh 1.5-mL tube.
26. Add an equal volume of chloroform and mix well. Centrifugate at 16,000 × *g* for 10 min at RT, transfer the supernatant to a fresh 1.5-mL tube.

27. To the final supernatant, add 20 μg glycogen and 25 μL of 5 M NaCl (final concentration of 0.3 M). After mixing, add 900 μL of ethanol and store the mixture at −20 °C for 30 min.
28. Centrifugate at 16,000×*g* at 4 °C for 10 min. Discard the supernatant and wash the DNA pellet with 1 mL 75 % ethanol.
29. Centrifugate at 16,000×*g* at 4 °C for 10 min. Discard the supernatant and dissolve DNA in 20–70 μL TE buffer or sterilized water.
30. Take out 2 μL of DNA solution to check by PCR the existence of DNA fragments containing the binding sites of Smad2 target genes (e.g., *sqt* and *lim1*). Usually, the pair of specific primers surrounding the Smad2-binding sites are designed, allowing amplification of a fragment of 150–250 bp in length. PCR amplification cycle number is 30–35, depending on the template concentration and primers. Run an agarose gel to separate and visualize the amplified fragment. The fragment in the gel can be quantified after digital imaging using the software ImageJ. The isolated DNA can also be used for other purposes.

3.6 Immunostaining with Anti-phospho-Smad1/5/8 Antibody

1. Collect 20–30 embryos at desired stages and remove their chorion as described in **step 2** of Subheading 3.2.
2. Fix the embryos in a 1.5-mL Eppendorf tube with 4 % paraformaldehyde at 4 °C overnight.
3. Remove the paraformaldehyde and dehydrate the embryos sequentially in graded PBST/methanol series (3:1, 1:1, 1:3), 5 min each. Store the embryos in 100 % methanol at −20 °C for more than 30 min.
4. Rehydrate the embryos sequentially in graded methanol/PBST series (3:1, 1:1, 1:3), 5 min each.
5. Wash the embryos twice with PBST, 5 min each.
6. Replace PBST with acetone and store at −20 °C for more than 7 min.
7. Wash the embryos twice with PBST, 5 min each.
8. Wash the embryos with PBSTX for 5 min.
9. Incubate the embryos in 100 μL of blocking solution in a 1.5-mL Eppendorf tube for 1 h at room temperature.
10. Add 0.5 μL anti-phospho-Smad1/5/8 antibody (1:100–300 dilution) and incubate overnight at 4 °C.
11. Take out the solution, which can be stored at 4 °C and reused. Wash the embryos with 1 mL of PBST for 5 min and repeat wash twice.
12. To the embryos, add 100 μL of blocking solution and anti-rabbit secondary antibody (1:200 dilution). Incubate for 1 h at room temperature.

13. Wash with 1 mL of PBST for 5 min and repeat wash once.
14. Incubate with 100 μL of DAPI (1 μg/mL in PBST) for 5 min at room temperature.
15. Wash with 1 mL PBST for 5 min and repeat wash once.
16. Discard PBST. Add about 100 μL of anti-quenching agent and store at 4 °C overnight.
17. Mount the embryos with coverslips and observe under a confocal microscope.

4 Notes

1. The methods described in this chapter can generally be used to analyze other proteins in zebrafish embryos. However, the amount of samples needs to be adjusted according to the abundance of the target proteins.
2. Currently, a few antibodies are suitable for immunostaining zebrafish embryos. Among Smads, only anti-phospho-Smad1/5/8 is successfully used for immunostaining.
3. Reporter assays involve reporter DNA injection. High doses of exogenous DNA, e.g., above 100 pg per embryo, usually cause embryonic development. For reproducibility, it is necessary to well clean DNA, inject DNA into the cytoplasm at the one-cell stage, and harvest embryos at comparable stages among different batches.
4. For experiments requiring cell lysis, the sample can be robustly vortexed at the end of lysis to ensure complete lysis.
5. Antibodies used for Western blotting and immunostaining can be reused. The second-time use of the antibodies usually gives better results because some impurities in the antibody solution could have been absorbed during the first-time use.
6. Some antibodies may not well recognize antigens (target proteins) in embryos. In this case, after rehydration and wash with PBST, embryos can be boiled in 1 mM EDTA, pH 8.0 solution for 10 min and then cool down slowly to repair the antigens. We found that this antigen repair treatment improved immunostaining effect for many antibodies.

Acknowledgement

This work was supported by grants from Major Science Programs of China (#2011CB943800).

References

1. Tian T, Meng AM (2006) Nodal signals pattern vertebrate embryos. Cell Mol Life Sci 63(6): 672–685
2. Langdon YG, Mullins MC (2011) Maternal and zygotic control of zebrafish dorsoventral axial patterning. Annu Rev Genet 45:357–377
3. Feldman B, Gates MA, Egan ES, Dougan ST, Rennebeck G, Sirotkin HI et al (1998) Zebrafish organizer development and germ-layer formation require nodal-related signals. Nature 395(6698):181–185
4. Gritsman K, Zhang J, Cheng S, Heckscher E, Talbot WS, Schier AF (1999) The EGF-CFC protein one-eyed pinhead is essential for nodal signaling. Cell 97(1):121–132
5. Gritsman K, Talbot WS, Schier AF (2000) Nodal signaling patterns the organizer. Development 127(5):921–932
6. Dougan ST, Warga RM, Kane DA, Schier AF, Talbot WS (2003) The role of the zebrafish nodal-related genes squint and cyclops in patterning of mesendoderm. Development 130(9):1837–1851
7. De Robertis EM, Kuroda H (2004) Dorsal-ventral patterning and neural induction in Xenopus embryos. Annu Rev Cell Dev Biol 20:285–308, PMCID: 2280069
8. Kishimoto Y, Lee KH, Zon L, Hammerschmidt M, Schulte-Merker S (1997) The molecular nature of zebrafish swirl: BMP2 function is essential during early dorsoventral patterning. Development 124(22):4457–4466
9. Dick A, Hild M, Bauer H, Imai Y, Maifeld H, Schier AF et al (2000) Essential role of Bmp7 (snailhouse) and its prodomain in dorsoventral patterning of the zebrafish embryo. Development 127(2):343–354
10. Schmid B, Furthauer M, Connors SA, Trout J, Thisse B, Thisse C et al (2000) Equivalent genetic roles for bmp7/snailhouse and bmp2b/swirl in dorsoventral pattern formation. Development 127(5):957–967
11. Schmierer B, Hill CS (2007) TGFbeta-SMAD signal transduction: molecular specificity and functional flexibility. Nat Rev Mol Cell Biol 8(12):970–982
12. Liu Z, Lin X, Cai Z, Zhang Z, Han C, Jia S et al (2011) Global identification of SMAD2 target genes reveals a role for multiple co-regulatory factors in zebrafish early gastrulas. J Biol Chem 286(32):28520–28532, PMCID: 3151094
13. Lin X, Duan X, Liang YY, Su Y, Wrighton KH, Long J et al (2006) PPM1A functions as a Smad phosphatase to terminate TGFbeta signaling. Cell 125(5):915–928
14. Duan X, Liang YY, Feng XH, Lin X (2006) Protein serine/threonine phosphatase PPM1A dephosphorylates Smad1 in the bone morphogenetic protein signaling pathway. J Biol Chem 281(48):36526–36532
15. Wrighton KH, Lin X, Feng XH (2009) Phospho-control of TGF-beta superfamily signaling. Cell Res 19(1):8–20, PMCID: 2929013
16. Dennler S, Itoh S, Vivien D, ten Dijke P, Huet S, Gauthier JM (1998) Direct binding of Smad3 and Smad4 to critical TGF beta-inducible elements in the promoter of human plasminogen activator inhibitor-type 1 gene. EMBO J 17(11):3091–3100
17. Korchynskyi O, ten Dijke P (2002) Identification and functional characterization of distinct critically important bone morphogenetic protein-specific response elements in the Id1 promoter. J Biol Chem 277(7): 4883–4891
18. Huang HC, Murtaugh LC, Vize PD, Whitman M (1995) Identification of a potential regulator of early transcriptional responses to mesoderm inducers in the frog embryo. EMBO J 14(23):5965–5973
19. Meng A, Jessen JR, Lin S (1999) Transgenesis. Methods Cell Biol 60:133–148

Chapter 18

Role of TGF-β Signaling in Coupling Bone Remodeling

Janet L. Crane, Lingling Xian, and Xu Cao

Abstract

TGF-β signaling plays a key role in the temporal and spatial regulation of bone remodeling. During osteoclast bone resorption, TGF-β is released from the bone matrix and activated. Active TGF-β recruits mesenchymal stem cells to the bone resorption pit through the SMAD signaling pathway. Mesenchymal stem cells undergo osteoblast differentiation and deposit new bone filling in the resorption pit and maintaining the structural integrity of the skeleton. Thus, TGF-β signaling plays a key role in the coupling process and disruptions to the TGF-β signaling pathway lead to loss of skeletal integrity. This chapter describes methods on how to quantitate bone matrix TGF-β and assess its role in mesenchymal stem cell migration both in vitro and in vivo.

Key words TGF-β1, SMAD, Mesenchymal stem cells, Migration, TRAP, Immunohistochemistry co-staining

1 Introduction

In the adult skeleton, bone is continuously being formed and resorbed to maintain mineral homeostasis and the structural integrity of the skeleton. Bone resorption and formation do not occur along the bone surface at random. Rather, it occurs at specific anatomical sites and follows a well-defined sequence of events known as the bone remodeling cycle [1]. The bone remodeling process is accomplished by the precise coordination of the activities of two cell types: osteoblasts, which deposit the calcified bone matrix, and osteoclasts, which resorb bone [2]. Cell differentiation into osteoblasts for subsequent bone formation, and ultimately skeletal remodeling are processes governed by various growth factors and their intracellular signals [3, 4]. The bone morphogenetic protein (BMP) family of proteins, through a cascade of downstream signals, coordinates the coupling process and thereby regulates many aspects of skeletal development and remodeling, including osteoblast differentiation.

Xin-Hua Feng et al. (eds.), *TGF-β Signaling: Methods and Protocols*, Methods in Molecular Biology, vol. 1344,
DOI 10.1007/978-1-4939-2966-5_18,

Coupling of bone resorption and formation is believed to occur through release of factors from the bone matrix during osteoclastic bone resorption that then directs migration of bone mesenchymal stem cells (BMSCs) to the bone resorptive surfaces [5–9]. Transforming growth factor-β1 (TGF-β1) has been identified as one such factor [10]. TGF-β1 is synthesized as a large precursor molecule, which is cleaved into active TGF-β1 and latency-associated protein (LAP). LAP remains noncovalently linked to active TGF-β1, masking the receptor-binding domains of TGF-β1 and rendering it inactive [11, 12]. TGF-β1 is then secreted and deposited in the bone matrix as an inactive, latent complex [13, 14] and is one of the most abundant cytokines in the bone matrix (200 μg/kg) [15–17]. During osteoclastic bone resorption, TGF-β1 is released from the extracellular matrix and activated by osteoclasts by cleavage of LAP [18]. Active TGF-β1 then recruits MSCs to the bone resorptive sites [10], where the MSCs then undergo osteoblastic differentiation to form new bone.

Signaling by TGF-β1 in BMSCs occurs through Smad-family signal-transduction proteins. TGF-βs bind to two major types of membrane-bound serine/threonine kinase receptors, the type-I and type-II receptors [19, 20], where the type-II receptor then transphosphorylates the type-I receptor. Various phosphorylated type-I receptors, in turn, phosphorylate different receptor-regulated Smads (R-Smads = Smad 1, 2,3, 5, and 8) [21–24]. R-Smads then rapidly dissociate from the receptor to form complexes with common-partner Smad, (Co-Smad = Smad4) and migrate into the nucleus to regulate transcription of target genes [21, 23–26]. Smad1 and Smad5 have been shown to be the major signaling molecules for inducing differentiation of osteoblasts in vitro, while Smad2 and Smad3 have been shown to be involved in cell migration [10, 27, 28]. Smad8 and Smad4 exhibit modest synergy in mesenchymal cells [29]. In contrast to R-Smads and Co-Smad, Inhibitory-Smads (Smad6 and Smad7) stably bind to the type-1 receptors and compete with R-Smads for activation, resulting in inhibition of signaling [29, 30].

In the methods below, we first describe how to quantitate TGF-β1 concentrations in the bone matrix. We then describe the transwell migration assay and wound scratch assay as two in vitro methods that may be used to assess cell migration at a simulated bone remodeling site. Finally, we describe how to visualize the bone remodeling site in vivo with double immunohistochemical staining with tartrate resistance acid phosphatase (TRAP) and downstream TGF-β signaling with phosphorylated Smad2/3. Manipulations to the TGF-β signaling pathway may be done by adding recombinant TGF-β1 or inhibiting the signaling pathway with antibodies, inhibitors, or siRNA specific to components of the TGF-β signaling pathway.

2 Materials

2.1 Reagents

1. Extraction solution (4 M Guanidine-HCl, 0.05 M EDTA, 30 mM Tris, 1 mg/mL bovine serum albumin, pH = 7.4).
2. Protease inhibitors (1 mM phenylmethyl-sulfonyl fluoride, 5 mM benzamidine-HCl, 0.1 M E-aminocaproic-acid, 2 μg/mL leupeptin).
3. 10× PBS (phosphate buffered saline): dilute to 1× PBS with dH_2O.
4. 1 % Triton in 1× PBS.
5. Ethanol: 10, 70, 80, 95, and 100 %.
6. 4 % Paraformaldehyde.
7. Xylene.
8. 1× TBST (Tris-Buffered Saline with 20-Tween).
9. Violet Crystal (Sigma-Aldrich).
10. Endogenous enzyme blocking buffer: 0.3 % H_2O_2 in 1× PBS.
11. Serum blocking buffer: 3 % BSA in 1× PBS.
12. Antibody Dilution Buffer (Dako).
13. DAB Chromogen (Dako).
14. Citrate Buffer: heat-induced epitope retrieval buffer.
15. Hematoxylin.
16. Collagen type I (BD Biosciences).
17. M-CSF.
18. RANKL.
19. BMSCs (Scal-1(+), CD29(+), CD11b(−), and CD45(−)): can be extracted and sorted from mice or purchased from Texas A&M Health Science Center College of Medicine Institute.
20. BMSCs culture medium: α-MEM with 20 % Fetal Bovine Serum.
21. Lipofectamine.
22. Human natural TGF-β1 protein (Cat# T1654, Sigma-Aldrich).
23. TβRI inhibitor SB 505124 (Sigma-Aldrich).
24. TGF-β1 neutralizing antibody (R&D Systems).
25. SiRNA-GFP SMAD2, SMAD3, and SMAD4 (Santa Cruz Biotechnology Inc.).
26. p-Smad2/3 antibody (Santa Cruz Biotechnology Inc., diluted 1:50).

2.2 Equipment

1. Biopulverizer (BioSpec Products, Inc.).
2. Sonicator—Single chamber ultrasonic cleaning system.

3. Transwell migration chambers: diameter 6.5 mm, pore size 8 μm (Corning Inc.).
4. Cell culture incubator: 37 °C and 5 % CO_2.

2.3 Other Materials

1. Fresh bovine diaphysis bone may be purchased from a local supermarket.
2. 24-Well plates, 6-well plates.
3. Quik-Change Site-Directed Mutagenesis Kit.
4. TRAP staining kit (Sigma-Aldrich).
5. Mini dialysis Slide-A-Lyzer MINI Dialysis Devices, 3.5 K molecular weight cut-off (MWCO).
6. Coverslip mount buffer.
7. Protein G-Sepharose (Amersham Biosciences).
8. Motic Image Advanced 3.1 software.

3 Methods

3.1 Detection of TGF-β1 in Mouse or Rat Bone Matrix and Bone Marrow

1. From freshly sacrificed animal, remove bones of interest and dissect free from muscle and fat (*see* **Note 1**).
2. Cut bone at growth plate.
3. Suspend bone in a 1.5 mL Eppendorf tube, proximal side down (Fig. 1).

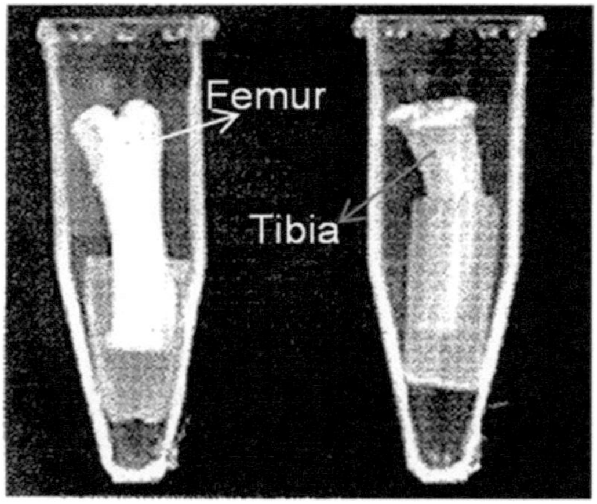

Fig. 1 Remove bone marrow from long bone for matrix protein extraction. A cut is made through the growth plates of the long bone, removing the subchondral bone and exposing the bone marrow cavity. Place femur (*left*) or tibia (*right*) proximal side down in a small tube, such as a 250 μL PCR tube that has the bottom and lid cut off to create a reservoir for collection of bone marrow in a larger (1.5) mL Eppendorf tube

4. Microcentrifuge for 15 min at 5000 rpm at 4 °C to collect bone marrow (*see* **Note 2**). Allow bone marrow to clot by storing at 4 °C for 2 h. Centrifuge for 20 min at 2000 × *g* to collect bone marrow serum.
5. Transfer bone to new Eppendorf tube. Microcentrifuge for 15 min at 17,000 × *g* at 4 °C to remove the residual cells. Remove bone from the tube and examine closely to make certain there is no tissue (*see* **Note 3**).
6. Immerse cleaned femurs in liquid nitrogen and mechanically crush into small fragments of several millimeters in diameter (*see* **Note 4**). Wash the fragments repeatedly in cold distilled water until the washings are free of blood. Defat fragments by washing in cold isopropylether.
7. Collect the bone fragments by picking up with forceps. Grind the defatted bone fragments into smaller particles using a biopulverizer.
8. Weigh bone powder on an electronic scale; for one mouse femur the expected yield is 20–40 mg. Place in a mini dialysis unit that has a regenerated cellulose membrane with 3.5 K Molecular Weight Cut-Off.
9. Add 0.5 mL of extraction solution (4 M Guanidine-HCl, 0.05 M EDTA, 30 mM Tris, 1 mg/mL BSA, pH = 7.4) with protease inhibitors (1 mM phenylmethyl-sulfonyl fluoride, 5 mM benzamidine-HCl, 0.1 M E-aminocaproic-acid, 2 μg/mL leupeptin) to the mini dialysis unit to decalcify bone tissue and extract proteins from bone matrix.
10. Cover the mini dialysis tube with a cap and place in a float such that the unit's membrane is in contact with the extraction solution in a beaker (300 mL for 15 mini dialysis units). Carry out the extraction procedure at 4 °C on a stir plate with a low-speed setting for 72 h.
11. Following the extraction process, re-dialyze the bone sample in 1× PBS for 72 h at 4 °C on a stir plate at a low-speed setting.
12. The extract may be stored at −80 °C until assayed for growth factor activity.
13. The amount of active and/or total TGF-β1 in the bone matrix or bone marrow can be determined with the DuoSet ELISA Development kit (R&D Systems) according to the manufacturer's instructions (*see* **Note 5**).

3.2 In Vitro Methods for Cell Migration

3.2.1 Transwell Cell Migration Assay

Preparation of Bone Slices

1. Strip adherent tissue and bone marrow from fresh bovine diaphyseal bone with scalpel.
2. Using 10 % ethanol as a cooling solution, cut bone into small blocks with a diamond wafering blade.

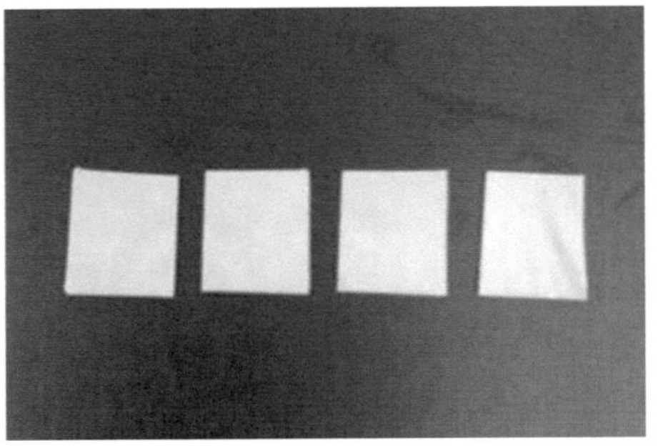

Fig. 2 Bone slices used for making resorption conditioned medium. Bovine cortical bone slices (*white*) are cut to 1 mm × 10 mm × 15 mm to be used for making osteoclastic resorption conditioned medium

3. While on ice, rinse bone blocks with distilled water 3×.
4. Sonicate bone blocks using a single chamber ultrasonic cleaning system at 40 kHz in 1 % triton in 1× PBS for 30 min 2× to lyse the cells on the bone surface (*see* **Note 6**).
5. Rinse bone blocks with distilled water 3× and sonicate again in distilled water for 15 min.
6. Cut bone blocks with a wafering blade into 1-mm-thick square slices (Fig. 2) to 10 mm × 15 mm size, keeping in cold 1× PBS.
7. Select slices with the same shape and scale and rinse them in 50 mL of distilled water 3×. Sonicate in distilled water for 20 min 3×.
8. Sterilize bone slices in 70 % ethanol for 5 min 1×, then 15 min 2×.
9. Wash in 1× PBS for 5 min 3×.
10. Store bone slices in sterilized 1× PBS at 4 °C.

Preparation of Osteoclastic Bone Resorption Conditioned Medium

1. In 4–8-week-old mice, isolate whole bone marrow by flushing bone marrow cavity with serum-free media (*see* **Note 7**).
2. Seed cells on 10-cm plate suspending in 10 % FBS-α MEM × 24 h.
3. Collect cells unattached to plate.
4. Generation of conditioned and control media: Plate 1×10^5 cells per well on 24-well tissue culture plates in 10 % FBS-α-MEM (Figs. 3 and 4).
5. For macrophages/osteoclastic precursor cells (Fig. 3a), add macrophage colony-stimulating factor (M-CSF) 44 ng/mL when seeding the cells (*see* **Note 8**).

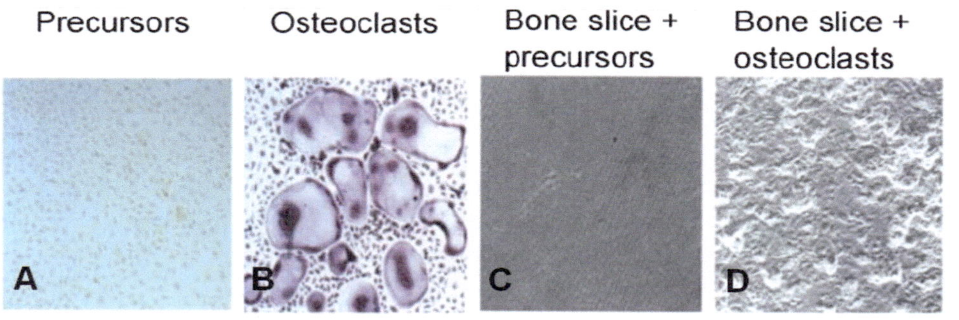

Fig. 3 Preparation of osteoclastic bone resorption conditioned medium. (**a**) Whole bone marrow cells from 4 to 8 week-old mice are cultured in the presence of M-CSF (44 ng/mL) to generate osteoclastic precursors. (**b**) Osteoclast formation can be induced by both M-CSF (44 ng/mL) and RANKL (100 ng/mL), as indicated by TRAP staining (indicated by *red* color). (**c**) Scanning electronic microscopy image of bone slice cultured with osteoclastic precursors. (**d**) Scanning electronic microscopy image of bone slice cultured with osteoclasts

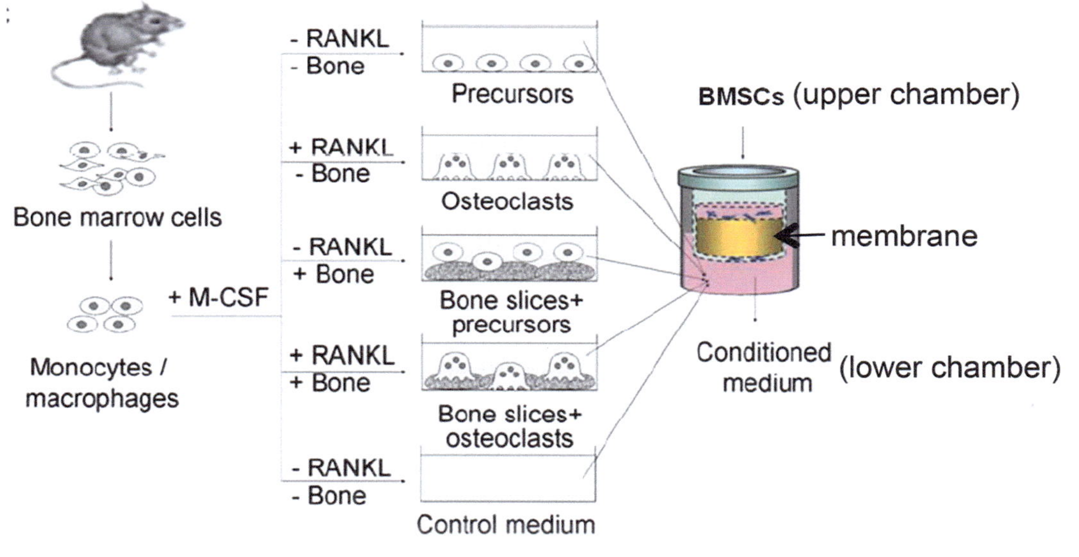

Fig. 4 Preparation of transwell migration assay. Whole bone marrow cells are cultured and nonadherent cells are collected, representing monocytes/macrophages. Five cultured mediums are generated: (1) precursors from monocytes cultured in M-CSF, (2) osteoclasts from monocytes cultured with M-CSF and RANKL, (3) bone slices + precursors from monocytes cultured in M-CSF plus bone slice placed in control medium; (4) bone slices + osteoclasts from monocytes cultured in M-CSF and RANKL plus bone slice placed in control medium; and (5) control medium without monocytes, but M-CSF added. For assembly of the transwell, a membrane separates the upper and lower chambers. Conditioned medium is added to the lower chamber, and BMSCs are seeded in upper chamber with 10 % α-MEM medium

6. For osteoclasts, culture the cells with both M-CSF (44 ng/mL) and RANKL (100 ng/mL) when seeding the cells (*see* **Note 9**). Differentiation into osteoclasts may be confirmed by TRAP staining after 8–10 days (Fig. 3b).
7. Add bone slices to culture plates of plated cells with M-CSF alone and M-CSF + RANKL to generate media for bone slice + precursors and bone slice + osteoclasts, respectively (Fig. 3c, d) (*see* **Note 10**).
8. Place culture plates in cell culture incubator at 37 °C and 5 % CO_2. Change the media and replace the additives every 2 days.
9. Harvest the conditioned media by pipetting media into a conical tube after 8–10 days of culture. Centrifuge media at $5000 \times g$ for 2 min to pellet out the cells.
10. If not used immediately, store at −80 °C in small aliquots to avoid repeated freeze/thaw cycles.

Transwell In Vitro Migration Assay of BMSCs

1. Perform cell migration assays using Transwell migration chambers. Coat both sides of the 8 μm pore membrane with 10 μg/mL of collagen I in 1× PBS overnight at 4 °C.
2. Treat BMSCs with 10 μg/mL mitomycin C for 2 h to prevent cell proliferation.
3. Rinse BMSCs with 1× PBS and starve in low serum media (0.5 mL; 0.5–0.1 % serum in α-MEM) overnight.
4. Plate BMSCs with 10 % α-MEM medium in the upper chambers and undiluted conditioned media from the osteoclastic bone resorption system to the lower chambers of the Transwell (Fig. 4).
5. Incubate for 5 h at 37 °C.
6. Fix cells for at least 10 min with 4 % paraformaldehyde.
7. Remove non-migrated cells on the upper membrane with a cotton swab.
8. Stain migrated cells with 0.05 % Violet Crystal for 30 min, rinse with tap water 2×. Count cells under microscope (*see* **Note 11**).

Manipulation of TGF-β Signaling Pathway

1. To create an artificial TGF-β1 microenvironment, TGF-β1 may be added to the cell culture media at varying concentrations (0.5–50 ng/mL) (*see* **Note 12**).
2. Neutralizing antibodies such as TGF-β1 Ab may be added to the conditioned media or the conditioned media (BRCM) may be depleted of active TGF-β1 by three cycles of immunoprecipitation using a monoclonal antibody specific for TGF-β1 adsorbed to protein G-Sepharose.

3. In cell culture, the downstream SMAD proteins may be disrupted by the use of SiRNA with GFP. SMAD2, SMAD3, and SMAD4 may be knocked down in BMSCs according to the manufacturer's protocol.

3.2.2 Wound Scratch Assay

1. Culture BMSCs in 24-well plates coated with 50 μg/mL collagen type I to reach 90 % confluency.
2. Treat the cells with 10 μg/mL mitomycin C for 2 h to prevent cell proliferation.
3. Rinse cells with 1× PBS and starve in low serum media (0.5 mL; 0.5–0.1 % serum in α-MEM) overnight.
4. On the day of the assay:
5. Draw a line with a marker on the bottom of the dish (Fig. 5).
6. Using a sterile 200 mL pipet tip, scratch two separate lines through the cells, moving perpendicularly through the maker line drawn in the previous step.
7. Wash cells three times with serum-free medium (*see* **Note 13**).
8. Culture cells in 0.5 mL of media containing additives. (For additives, refer to Subheading 3.2.1, **step 4**—Preparation of Osteoclastic Bone Resorption Conditioned Medium).
9. Take a picture under the microscope at 20× magnification using phase contrast at time 0, 5, and 8 h (*see* **Note 14**).
10. The migration area or number of migration cells is counted by Motic Image Advanced 3.1 software.

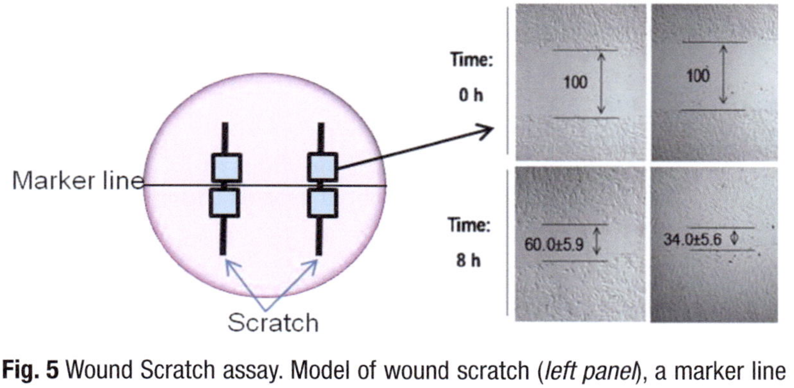

Fig. 5 Wound Scratch assay. Model of wound scratch (*left panel*), a marker line is drawn on the surface of the dish or plate with a pen. Two scratch wounds are made, and four fields (*blue square*) close to the marker line are chosen to be observed. (*Right*) Represented images of one field of scratch wound healing at 0 and 8 h visualized under a light microscope at 200× magnification

3.3 In Vivo Methods for Cell Migration: Double Immuno-histochemical Staining with TRAP and pSmad2/3

Preparation of Bone Slices

1. From freshly sacrificed animal, remove bones of interest and dissect free from muscle and fat.
2. Fix in 10 % buffered formalin for 24–48 h.
3. Decalcify tissue in 10 % ethylenediamine tetraacetic acid (EDTA), pH 7.4 at 4 °C on a shaker for 21 days, changing EDTA solution every 2–3 days (*see* **Note 15**).
4. Embed tissues in paraffin or O.C.T. compound (Sakura Finetek USA Inc.) and cut 4-μm-thick sagittally oriented sections. Adhere sections to slides by incubating at 65 °C for 30 min (*see* **Note 16**).
5. Store slides at 4 °C until used for staining.

Deparaffinization/Rehydration

1. Heat the paraffin embedded bone slides at 56 °C in a humidified incubator overnight, and then on hot plate for 2 more hours (*see* **Note 17**).
2. Wash slides in xylene for 5 min 3×.
3. Wash slides in 100 % ethanol for 5 min 2×.
4. Wash slides in 95 % ethanol for 5 min 2×.
5. Rinse slides in dH_2O for 5 min 2×.

TRAP Staining (See Note 18)

1. Preheat 300 mL of 1× TBST to 65 °C in incubator.
2. Prepare TRAP staining buffer: 200 μL of Diazotized Fast Garnet GBC solution (mix 100 μL of Fast Garnet GBC base solution and 100 μL of sodium nitrite solution, stand for 2 min), 100 μL of Naphthol AS-BI phosphate, 400 μL of Acetate solution, 200 μL of Tartrate solution, 9 mL of ddH_2O, mix and heat to 65 °C in the incubator (15–20 min).
3. Rinse slides in 1× TBST for 5 min 2× at room temperature.
4. Immerse the slides in 1× TBST preheated to 65 °C for 20 min in the incubator to allow the slides to reach 65 °C.
5. Lie slides flat in a humidified tissue box, wiping or suctioning off excess TBST. Draw a circle around each section with a liquid repeller slide marker pen.
6. TRAP staining: In an incubator heated to 65–68 °C, drop the TRAP staining buffer on the preheated slides, and stain for 40 s to 2 min.
7. Rinse the slides with tap water and check the staining under the microscope (*see* **Note 19**).

Immunohistochemical Staining of pSmad2/3

1. To unmask antigen, bring slides to a boil in 10 mM sodium citrate buffer pH 6.0, and Maintain at a sub-boiling temperature for 10–20 min.
2. Cool slides on bench top for 30 min, remaining submerged in sodium citrate buffer.
3. To block endogenous enzymes, incubate in a glass trough with 0.3 % H_2O_2 in 1× PBS for 20 min.
4. Block specimen with serum blocking buffer (3 % BSA in 1× PBS) for 30 min.
5. While blocking, prepare primary antibody by diluting p-Smad 2/3 antibody 1:50 in Antibody Dilution Buffer.
6. Aspirate blocking solution, apply diluted primary antibody. Incubate overnight at 4 °C in a humidified chamber.
7. Rinse in 1× TBST for 5 min 3×.
8. Incubate with secondary antibody for 60 min at room temperature.
9. Incubate sections in 50 mL DAB solution for 45 s.
10. Rinse with dH_2O.
11. Counterstain if required with hematoxylin for 2 min at room temperature.
12. Rinse in running tap water for 5 min.
13. Dehydrate Tissue (*see* **Note 20**) 100 % Xylene for 5 min 3×.
14. Mounting the slides. Place one drop of permount buffer and apply coverslip to slides. Allow to dry at room temperature.

4 Notes

1. Be careful not to allow the specimen to dry out. To keep tissue from drying, place tissue in 1× PBS in small culture dish.
2. Centrifuge at low speed to prevent cell lysis.
3. To remove any remaining exterior tissue, scrape the exterior with a scalpel and wipe with kimwipes. Also, make certain there is no marrow interior. Bones must be cleaned of blood cells and fat. If not clean, specimens may be centrifuged again or the bone marrow cavity may be flushed with 1× PBS or serum-free media using a 25–27G needle and syringe.
4. Specimens may be crushed by cutting bone into small pieces using a scalpel or alternatively may be grinded with a mortar and pestle.
5. As the antibody used in the kit recognizes only active TGF-β1, the measurement of total TGF-β1 can be obtained by activation

of bone matrix latent TGF-β1. To activate latent TGF-β1, acidify bone matrix extract in conditioned media with 1 N HCl for 10 min, then neutralized with 1.2 N NaOH/0.5 M HEPES.

6. The volume of 1 % Triton in 1× PBS needed depends on the size of the bone block. Use enough to cover the bone, usually 100–150 mL. For the second sonication, triton is not essential.
7. Use a 10 mL syringe and 25–27G needle, placing needle in bone marrow cavity and flushing with serum-free media.
8. M-CSF is a lineage-specific growth factor that is responsible for the proliferation and differentiation of committed myeloid progenitors into cells of the macrophage/monocyte lineage.
9. Under these conditions, osteoclasts begin to form and resorb bone at 6–7 days.
10. Resorption of the bovine cortical bone slices may be determined by either light microscopy or scanning electronic microscopy between 6 and 7 days. Resorption will be seen in bone slices in M-CSF + RANKL cultured cells, but not in the M-CSF alone cultured cells.
11. At least four fields at 200× magnification should be counted, either as directly visualized with a microscope or after taking a photomicrographic image. Experiments should be performed in triplicate to validate results.
12. Before use, the TGF-β1 product should be activated by treating with 4 mM HCl containing 1 mg/mL BSA for 10 min, then neutralizing with 1.2 N NaOH/0.5 M HEPES.
13. Wash the cells very gently to prevent the loss of cells as confluency or near confluency is needed for this assay to work.
14. To help orient the picture for measurements, make certain the line just appears in each picture, taking a picture just above and just below the marker line. After each picture, replace the old media with the fresh media/additives as prepared in **step 4** (refer to Subheading 3.2.1, **step 4** - Preparation of Osteoclastic Bone Resorption Conditioned Medium).
15. To decalcify tissue more quickly, cut bone at the growth plate or mid-diaphyseal shaft and/or change EDTA more frequently. Intact whole bone and younger animals may require longer periods of decalcification.
16. Keep the incubator humidified by placing a container of tap water in the incubator.
17. If using a frozen section, do not heat overnight. Simply, dry the slides at 40 °C for 1 h. Additionally, if using frozen sections, skip to **step 5**.
18. When doing double staining with TRAP and immunohistochemistry of pSmad, always do the TRAP staining first, then immunohistochemistry.

19. If further staining is to be performed, do not let the slides dry out. Keep the slides wet by coating sections with a small amount of tap water while visualizing staining efficacy. If no additional staining is to be done, then keeping the samples hydrated is less critical.
20. Do not dehydrate in ethanol after TRAP staining, as the ethanol will fade the red color of the TRAP staining. If TRAP staining is not performed, then immerse slides in 95 % ethanol for 5 min 1×, then 100 % ethanol for 5 min, prior to Xylene step.

References

1. Hill PA (1998) Bone remodelling. Br J Orthod 25:101–107
2. Zaidi M (2007) Skeletal remodeling in health and disease. Nat Med 13:791–801
3. Tuan RS (2003) Cellular signaling in developmental chondrogenesis: N-cadherin, Wnts, and BMP-2. J Bone Joint Surg Am 85-A(Suppl 2):137–141
4. Karsenty G, Wagner EF (2002) Reaching a genetic and molecular understanding of skeletal development. Dev Cell 2:389–406
5. Oreffo RO, Mundy GR, Seyedin SM, Bonewald LF (1989) Activation of the bone-derived latent TGF beta complex by isolated osteoclasts. Biochem Biophys Res Commun 158:817–823
6. Mundy GR (1994) Peptides and growth regulatory factors in bone. Rheum Dis Clin North Am 20:577–588
7. Martin TJ, Allan EH, Fukumoto S (1993) The plasminogen activator and inhibitor system in bone remodelling. Growth Regul 3:209–214
8. Pfeilschifter J, Wolf O, Naumann A et al (1990) Chemotactic response of osteoblastlike cells to transforming growth factor beta. J Bone Miner Res 5:825–830
9. Hill PA, Tumber A, Meikle MC (1997) Multiple extracellular signals promote osteoblast survival and apoptosis. Endocrinology 138:3849–3858
10. Tang Y, Wu X, Lei W et al (2009) TGF-beta1-induced migration of bone mesenchymal stem cells couples bone resorption with formation. Nat Med 15:757–765
11. Dallas SL, Park-Snyder S, Miyazono K et al (1994) Characterization and autoregulation of latent transforming growth factor beta (TGF beta) complexes in osteoblast-like cell lines. Production of a latent complex lacking the latent TGF beta-binding protein. J Biol Chem 269:6815–6821
12. Gentry LE, Lioubin MN, Purchio AF, Marquardt H (1988) Molecular events in the processing of recombinant type 1 pre-pro-transforming growth factor beta to the mature polypeptide. Mol Cell Biol 8:4162–4168
13. Pfeilschifter J, Bonewald L, Mundy GR (1990) Characterization of the latent transforming growth factor beta complex in bone. J Bone Miner Res 5:49–58
14. Pedrozo HA, Schwartz Z, Robinson M et al (1999) Potential mechanisms for the plasmin-mediated release and activation of latent transforming growth factor-beta1 from the extracellular matrix of growth plate chondrocytes. Endocrinology 140:5806–5816
15. Bismar H, Kloppinger T, Schuster EM et al (1999) Transforming growth factor beta (TGF-beta) levels in the conditioned media of human bone cells: relationship to donor age, bone volume, and concentration of TGF-beta in human bone matrix in vivo. Bone 24:565–569
16. Roberts AB, Frolik CA, Anzano MA, Sporn MB (1983) Transforming growth factors from neoplastic and nonneoplastic tissues. Fed Proc 42:2621–2626
17. Seyedin SM, Thomas TC, Thompson AY et al (1985) Purification and characterization of two cartilage-inducing factors from bovine demineralized bone. Proc Natl Acad Sci U S A 82:2267–2271
18. Dallas SL, Rosser JL, Mundy GR, Bonewald LF (2002) Proteolysis of latent transforming growth factor-beta (TGF-beta)-binding protein-1 by osteoclasts. A cellular mechanism for release of TGF-beta from bone matrix. J Biol Chem 277:21352–21360
19. Zwijsen A, Verschueren K, Huylebroeck D (2003) New intracellular components of bone morphogenetic protein/Smad signaling cascades. FEBS Lett 546:133–139
20. Massague J (1998) TGF-beta signal transduction. Annu Rev Biochem 67:753–791
21. Attisano L, Wrana JL (2002) Signal transduction by the TGF-beta superfamily. Science 296:1646–1647

22. Kretzschmar M, Massague J (1998) SMADs: mediators and regulators of TGF-beta signaling. Curr Opin Genet Dev 8:103–111
23. Heldin CH, Miyazono K, ten Dijke P (1997) TGF-beta signalling from cell membrane to nucleus through SMAD proteins. Nature 390:465–471
24. Miyazono K (2000) Positive and negative regulation of TGF-beta signaling. J Cell Sci 113(Pt 7): 1101–1109
25. Derynck R (1994) TGF-beta-receptor-mediated signaling. Trends Biochem Sci 19: 548–553
26. Cohen PT (2002) Protein phosphatase 1 – targeted in many directions. J Cell Sci 115: 241–256
27. Fujii M, Takeda K, Imamura T et al (1999) Roles of bone morphogenetic protein type I receptors and Smad proteins in osteoblast and chondroblast differentiation. Mol Biol Cell 10:3801–3813
28. Ten DP, Korchynskyi O, Valdimarsdottir G, Goumans MJ (2003) Controlling cell fate by bone morphogenetic protein receptors. Mol Cell Endocrinol 211:105–113
29. Hanyu A, Ishidou Y, Ebisawa T et al (2001) The N domain of Smad7 is essential for specific inhibition of transforming growth factor-beta signaling. J Cell Biol 155:1017–1027
30. Imamura T, Takase M, Nishihara A et al (1997) Smad6 inhibits signalling by the TGF-beta superfamily. Nature 389:622–626

Chapter 19

Studying the Functions of TGF-β Signaling in the Ovary

Chao Yu, Jian-Jie Zhou, and Heng-Yu Fan

Abstract

In mammals, ovulation is a multistep physiological process that includes preovulatory follicle growth, oocyte meiotic maturation, cumulus-oocyte complex (COC) expansion, follicle rupture, and luteinization. TGF-β signaling pathway has multiple functions in mammalian ovary, as its complexity in ovarian function has been demonstrated by mouse models with knockouts of TGF-β receptors and SMADs. We describe the protocol that we use to study functions of TGF-β signaling pathway in follicle development and ovulation. Because total knockout of TGF-β pathway components often causes embryonic lethality, which prevents further investigation of these genes in ovarian functions, people have generated ovarian cell type-specific knockout mouse strains for TGF-β signaling pathway genes. These mouse models are also described.

Key words Ovary, Follicle, Oocyte, Ovulation, Female fertility

1 Introduction

In mammals, ovulation is a multistep physiological process that includes preovulatory follicle growth, oocyte meiotic maturation, cumulus-oocyte complex (COC) expansion, follicle rupture, and luteinization. The pituitary-secreted gonadotropins, follicle stimulating hormone (FSH), and luteinizing hormone (LH) are major regulators of these events. During each estrus cycle, FSH facilitates the rapid growth of preantral and early antral follicles to enter the preovulatory stage. Then, a mid-estrous LH surge triggers the initiation of ovulation and the terminal differentiation of GCs into luteal cells [1].

In addition to gonadotropins, local ovarian signaling factors also play crucial roles during specific steps of ovulation. For example, EGF-like factors are intrafollicular mediators of LH effects, including triggering oocyte germinal vesicle breakdown (GVBD) and COC expansion [2, 3]. Granulosa and cumulus cell-produced prostaglandin E2 (PGE2) is important for organizing the COC matrix and in positive feedback regulation of EGF-like factors [4].

Xin-Hua Feng et al. (eds.), *TGF-β Signaling: Methods and Protocols*, Methods in Molecular Biology, vol. 1344,
DOI 10.1007/978-1-4939-2966-5_19, © Springer Science+Business Media New York 2016

It is now well accepted that the oocyte is not only a passenger carried by the follicle, but is also an active regulator of follicle growth and ovulation. Oocyte-derived factors, including growth and differentiation factor 9 (GDF9) and bone morphogenic protein 15 (BMP15), promote COC expansion and the expressions of several key target genes involved in ovulation regulation by LH (*Has2*, *Ptgs2*, *Tnfaip6*, and *Ptx3*) [5]. GDF9 and BMP15 are both signaling molecules of the TGF-β super family. They bind to their respective membrane receptors and trigger the serine/threonine protein kinase activity of these receptors [6]. As a result, several members of the SMAD family of transcription factors become phosphorylated; SMAD1/5/8 in response to BMPs and SMAD2/3 in response to other TGF-β family ligands. Subsequently, phosphorylated SMAD1/5/8 and SMAD2/3 form heterodimers with SMAD4 and enter the nucleus, which regulates the expression of their target genes. This is the canonical TGF-β signaling pathway [7].

The TGF-β signaling pathway has multiple functions in the mammalian ovary. For example, in the ovaries of new born Gdf9 KO mice, oocytes cannot recruit the surrounding ovarian somatic cells to form follicles, which indicate that this paracrine factor is essential for follicle formation [8]. Knockout of inhibin and activin, TGF-β family ligands that are secreted by GCs, causes multiple ovarian defects by disrupting the feedback regulation between the pituitary and ovary [9]. In both murine and human ovaries, the TGF-β family ligand AMH is an intraovarian growth factor that regulates primordial follicle recruitment and the FSH sensitivity of growing follicles in an inhibitory manner [10].

The complexity of the TGF-β signaling pathway in ovarian functions has been further demonstrated by mouse models with knockouts of TGF-β receptors and SMADs. Conditional knockout of BMP receptor 1A/1B (BMPR1A/B) [11] or SMAD1/5/8 [12] in the GCs of developing follicles results in the oncogenic transformation of these cells. Deletion of both SMAD2 and 3 dramatically reduces female fertility, which is associated with disrupted multiple ovarian processes, including follicular development, ovulation, and COC expansion [13]. Most importantly, GC-specific depletion of SMAD4, the central component of the canonical TGF-β signaling pathway, causes premature luteinization of GCs followed by ovulation failure [14].

2 Materials

2.1 Cell Culture Reagents

Minimum Essential Medium (MEM), Dulbecco's Modified Eagle's Medium/F12 (DMEM/F12), calcium and magnesium free phosphate buffered saline (PBS), fetal bovine serum (FBS), penicillin/streptomycin sulfate, and 0.25 % Trypsin/EDTA were purchased from Invitrogen. Bovine serum albumin (BSA), mineral oil, milrinone, and M2 and M16 media are purchased from Sigma.

2.2 Chemicals

Recombinant amphiregulin, GDF9, BMP15, and activin are purchased from R&D systems. TGF-β receptor I inhibitor SB431542 and pregnant mare serum gonadotropin (PMSG) are purchased from Calbiochem. EGF, prostaglandin E2, forskolin, and PMA are purchased from Sigma. Human chorionic gonadotropin (hCG) is purchased from American Pharmaceutical Partners (Schaumburg, IL).

2.3 Solutions

Acidic M2 medium: M2 medium containing 1% saturated hydrochloric acid.

Immunofluorescent staining wash buffer: PBS with 0.1 % Tween and 0.01 % Triton X-100.

Immunofluorescent staining blocking buffer: wash buffer with 1% BSA.

Tail buffer: 5 mL of 1 M Tris–HCl, pH 8.0, 7.5 mL of 5 M NaCl, 25 mL of 0.5 M EDTA, 10 % SDS, and adjusted to 500 mL with dH_2O.

2.4 Cre-expressing Mouse Strains

Amhr2-Cre: The expression of CRE recombinase in this mouse strain is under the control of the anti-Müllerian hormone type II receptor (*Amhr2*) promoter. CRE activity is found in the ovary as early as embryonic day 17.5. Throughout the postnatal ovary, CRE activity is found in GCs of all secondary and small antral follicles. However, low CRE activity is also found in some theca cells [14]. In addition, as expected, the muscular layer of the uterus is CRE positive, as Amhr2 is expressed in the mesenchyme of the Müllerian duct that gives rise to the uterine musculature.

Cyp19-Cre: The expression of Cre DNA recombinase in Cyp19-Cre mice is under the control of GC-specific Cyp19a1 promoter II. CRE activity is primarily detected during the preantral/early antral stage of follicle development.

Gdf9-Cre: Oocyte-specific deletion of ovarian genes using Cre/loxP technology provides an excellent tool for determining their physiological roles in the TGF-β signaling pathway during folliculogenesis, oogenesis, and pre-implantation embryonic development. A transgenic mouse line that expressed improved Cre recombinase (*iCre*) driven by the mouse growth differentiation factor-9 (GDF-9) promoter was used. In this mouse strain, Cre recombinase is only expressed in postnatal ovaries. Within the ovary, Cre recombinase is exclusively expressed in the oocytes of primordial follicles and follicles at later developmental stages. The expression of iCre in GDF-9-iCre mice is earlier than Cre expression in Zp3-Cre mice (see below).

Zp3-Cre: The expression of Cre recombinase in this mouse strain is under the control of the *Zp3* promoter. The *Zp3* gene encodes the mouse zona pellucida protein ZP3. In female *Zp3-Cre* transgenic mice, Cre activity is first detected in activated oocytes of primary follicles, but not in dormant oocytes of primordial follicles.

Therefore, *Zp3-Cre* is used to efficiently delete target genes in activated oocytes without affecting the ovarian oocyte storage pool (primordial follicles). Both *Gdf9-Cre* and *Zp3-Cre* are not expressed in male germ cells.

3 Methods

3.1 Mouse Granulosa Cell Harvesting and Culture

1. Intraperitoneally inject pubertal female mice (21-days-old) with 4 IU of PMSG to stimulate granulosa cell proliferation and follicle maturation. Humanly sacrifice mice 24 h after PMSG injection (*see* **Notes 1** and **2**).
2. Humanely sacrifice mice and isolate their ovaries. Place ovaries on a piece of clean wet paper tower, and carefully remove the fat and connective tissues and the oviduct adjacent to the ovaries. Then, place the ovaries in a sterile 3.5 mm cell culture dish (5–6 ovaries per dish) containing 2 mL of DMEM/F12 medium.
3. Release undifferentiated GCs from antral follicles by repeatedly puncturing the ovary with a 26.5 gauge needle. Remove the ovarian tissue debris from the dish after thorough puncturing.
4. Place the cell suspension in a sterile 15 mL centrifuge tube, and centrifuge at $700 \times g$ for 5 min.
5. Carefully remove the supernatant and resuspend the cells in 10 mL of DMEM/F12. Repeat **step 4** twice.
6. After a final wash, resuspend at a density of 1×10^6 cells/mL in DMEM/F12 medium (Invitrogen) supplemented with 5 % fetal bovine serum (FBS; Invitrogen), 100 U/mL of penicillin, and 100 μg/mL of streptomycin, and culture in 24-well culture dishes. After overnight incubation, cells are washed and cultured in serum-free DMEM/F12 medium before any further treatments.

3.2 In Vitro Hormone Treatment

3.2.1 Evaluating the Role of the TGF-β Signaling Pathway in FSH Actions

1. Pre-treat cultured granulosa cells with the ALK inhibitor SB431542 (10 μM) to block the TGF-β pathway, or with recombinant TGF-β ligands (GDF9, BMP15, activin, or AMH) to activate the TGF-β pathway.
2. Treat the cells with FSH (100 ng/mL) by directly adding this hormone to the medium.
3. FSH-induced phosphorylation of the cAMP/PKA pathway, the RAS/MAPK pathway, and the PI3K/AKT pathway can be determined by Western blot analysis. Harvest cells 30–60 min after FSH treatment by adding SDS protein loading buffer directly to the wells of a cell culture dish (*see* **Note 3**).
4. To determine transcriptional regulation of ovarian genes by FSH and TGF-β, harvest cells at 4, 12, and 24 h after FSH treatment by adding Trizol Reagent (300 μL) directly to each

well of a cell culture dish. The following genes are usually analyzed by real-time RT-PCR: *Cyp19a1*, which encodes aromatase, the key enzyme of estradiol biosynthesis that is upregulated by FSH and TGF-β signals; *Fshr*, which encodes the FSH receptor that is upregulated by FSH in a positive feedback manner; *Lhcgr*, which encodes the LH receptor that is upregulated by FSH, but is downregulated by TGF-β ligands.

3.2.2 Evaluating the Role of the TGF-β Signaling Pathway in LH Actions

1. Pre-treat cultured granulosa cells with the ALK inhibitor SB431542, or with recombinant TGF-β ligands, as described in **step 1** of Subheading 3.2.1.
2. To induce in vitro luteinization, treat the cells with LH (500 ng/mL) by directly adding this hormone to the medium. Alternatively, treat cells with forskolin (10 μM) plus PMA (20 nM) to mimic LH actions and induce more robust luteinization in cultured granulosa cells.
3. To determine the transcriptional regulation of ovulation-related genes by FSH and TGF-β, harvest the cells 2–4 h after LH treatment. The following genes are usually analyzed by real-time RT-PCR: *Areg, Ereg,* and *Btc*, which encode the EGF-like factors amphiregulin, epiregulin, and beta-cellulin, respectively, which are intrafollicular mediators of LH actions and whose expressions are known to be downregulated by TGF-β signals; *Has2, Ptgs2, Ptx3*, and *Tnfaip6*, which are essential for cumulus expansion and ovulation and are known to be significantly upregulated by LH in granulosa cells and cumulus cells in a TGF-β-dependent manner; *Lhcgr*, which encodes the LH receptor that is upregulated by FSH, but is downregulated by TGF-β ligands.
4. To determine the transcriptional regulation of luteinization-related genes by FSH and TGF-β, harvest the cells 24 h after LH treatment. The following genes are usually analyzed by real-time RT-PCR: *Star* and *Cyp11a1*, which encode steroidogenic acute regulatory protein and cholesterol side-chain cleavage enzyme, respectively, which are essential for progesterone biosynthesis; *Lhcgr*, which encodes the LH receptor; and *Sfrp4*, which encodes secreted frizzled-related protein 4. The expressions of these genes are known to be strongly induced by LH, but are repressed by TGF-β ligands. Primer sequences are listed in Table 1.

RNA Sample Preparation and Reverse Transcription

1. Intraperitoneally inject female mice with 5 IU of pregnant mare serum gonadotropin (PMSG) to stimulate preovulatory follicle development.
2. After 44–48 h, humanely sacrifice mice by cervical dislocation, dissect out the ovaries, and remove fat and connective tissues.

Table 1
Primers used for RT-PCR

Gene name	Primer sequence
Cyp19a1	Forward: 5′-TGTTGTGGACTTGGTCATGC-3′ Reverse: 5′-GGCTGAAAGTACCTGTAGGGAA-3′
Fshr	Forward: 5′-ACCATGGCTTAGAAAATCTGAAG-3′ Reverse: 5′-GATCCCCAGGCTGAGTCATA-3′
Lhcgr	Forward: 5′-TGCCCTCCAAAGAAAAATTC-3′ Reverse: 5′-AATCGTAATCCCAGCCACTG-3′
Areg	Forward: 5′-TTTATCTTCACACATCTCTTTATGTACAG-3′ Reverse: 5′-GAGGCTTCGACAAGAAAACG-3′
Ereg	Forward: 5′-GTCCCCTGAGGTCACTCTCTC-3′ Reverse: 5′-ACACTGGTCTGCGATGTGAG-3′
Btc	Forward: 5′-CTCCCTCCTGCATCTGTGAG-3′ Reverse: 5′-CTCTTGAATATCTTCACTTATGGGAG-3′
Has2	Forward: 5′-GTTGGAGGTGTTGGAGGAGA-3′ Reverse: 5′-ATTCCCAGAGGACCGCTTAT-3′
Ptgs2	Forward: 5′-TGTACAAGCAGTGGCAAAGG-3′ Reverse: 5′-CCCCAAAGATAGCATCTGGA-3′
Ptx3	Forward: 5′-GTGGGTGGAAAGGAGAACAA-3′ Reverse: 5′-GGCCAATCTGTAGGAGTCCA-3′
Tnfaip6	Forward: 5′-TTCCATGTCTGTGCTGCTGGATGG-3′ Reverse: 5′-AGCCTGGATCATGTTCAAGGTCAAA-3′
Star	Forward: 5′-CACACATTTTGGGGAGATGC-3′ Reverse: 5′-GAACTCTATCTGGGTCTGCGATA-3′
Cyp11a1	Forward: 5′-GTCTTACACAGACGCATCAAGC-3′ Reverse: 5′-ACACTGGTGTGGAACATCTGG-3′
Sfrp4	Forward: 5′-ATGCTCCGCTCCATCCTGGTG-3′ Reverse: 5′-TGGCCAGGATGGCGTTCTCC-3′

3. Place the ovaries in M2 medium and prick follicles using sharp needles under a stereoscope.
4. Collect fully grown germinal vesicle (GV)-stage oocytes under a stereoscope with a mouth-controlled micropipette.
5. Extract total RNA from 100 to 150 GV-stage oocytes using an RNeasy Micro kit (Qiagen) according to the manufacturer's instructions.

6. Add 1× first-strand buffer, 10 mM DDT, 1 mM dNTP, 25 nM oligo-dT, 20 U of RNaseOUT, and 25 U of SuperScriptII.
7. Reverse transcribe total RNA for 60 min at 42 °C and then 10 min at 95 °C.

Identify samples by PCR.

3.3 Mouse Cumulus-Oocyte Complex (COC) Culture and In Vitro COC Expansion Assay

The TGF-β family members GDF9 and BMP15 are oocyte-secreted factors that regulate cumulus cell differentiation and ovulation. Therefore, in vitro COC expansion experiments are typically used to study the role of TGF-β signaling in ovulation.

1. Intraperitoneally inject mice (3-weeks-old) with 5 IU of pregnant mare serum gonadotropin (PMSG) to stimulate preovulatory follicle development.
2. After 44–48 h, humanely sacrifice mice by cervical dislocation, dissect out the ovaries, and remove fat and connective tissues.
3. Cumulus-oocyte complexes (COCs) are released from antral follicles by puncturing with a 26.5 gauge needle and placed in COC medium (MEM w/ES, 0.25 mM sodium pyruvate, 3 mM L-GLUTAMINE, 1 mg/mL of BSA, 100 U/mL of penicillin, 100 μg/mL of streptomycin) with 1 % FBS.
4. Collect nonexpanded cumulus-oocyte complexes (COCs) with a micropipette of a suitable diameter, transfer the cumulus-oocyte complexes (COCs) in 50 μL of COC medium with 1 % FBS, and cover with mineral oil in a nunclon plate.
5. COC expansion can be induced by adding FSH (100 ng/mL), amphiregulin (250 ng/mL), epidermal growth factor (EGF, 10 ng/mL), forskolin (10 μM), or prostaglandin E2 (PGE2, 500 nM) to the medium. Culture COCs overnight in a 5 % CO_2 incubator at 37 °C.
6. Check the expansion status of COCs with a microscope the next day.

3.4 Mouse Denuded Oocyte Harvesting and Culture

1. Intraperitoneally inject female mice with 5 IU of pregnant mare serum gonadotropin (PMSG) to stimulate preovulatory follicle development.
2. After 44–48 h, humanely sacrifice mice by cervical dislocation, dissect out the ovaries, and remove fat and connective tissues.
3. Place the ovaries in M2 medium with 2.5 mM milrinone and prick follicles using sharp needles under a stereoscope.
4. Frequently agitate and pipette the tissues.
5. Collect fully grown germinal vesicle (GV) oocytes under a stereoscope with a mouth-controlled micropipette of a suitable diameter.

6. Wash GV oocytes three times by transferring the oocytes into fresh M2 medium without milrinone.
7. Transfer fully grown GV oocytes to a mini-drop of M16 medium covered with mineral oil by micropipette.
8. Culture germinal vesicle (GV)-stage oocytes in a 5 % CO_2 incubator at 37 °C.

3.5 Mouse Superovulation Procedures

1. Intraperitoneally inject female mice with 5 IU of pregnant mare serum gonadotropin (PMSG).
2. After 48 h, inject female mice with 5 IU of human chorionic gonadotrophin (hCG).
3. After 16 h, humanely sacrifice mice by cervical dislocation, dissect out the oviducts, and remove fat and connective tissues.
4. Place oviducts in M2 medium, blow out oocytes from oviducts using a mouth-controlled micropipette under a stereoscope.
5. Record the numbers of oocytes harvested from oviducts.

3.6 Immunofluorescent and Confocal Microscopy Imaging

1. Humanely sacrifice mice by cervical dislocation, dissect out the ovaries, and remove fat and connective tissues.
2. Place ovaries in M2 medium and prick follicles using sharp needles under a stereoscope.
3. Agitate and pipette the tissues, and collect oocytes under the stereoscope with a mouth-controlled micropipette of an appropriate diameter.
4. Remove the Zona Pellucida of oocytes with acidic M2 medium, 2–3 s each time.
5. Wash oocytes in fresh M2 medium.
6. Transfer oocytes to 4 % paraformaldehyde in PBS, and fix oocytes for at least 30 min at room temperature.
7. Permeabilize oocytes with a membrane-permeabilizing solution (PBS plus 0.3 % Triton X-100) for 30 min at room temperature in a humid chamber.
8. Incubate oocytes with a blocking buffer (1 % BSA) for 1 h at room temperature.
9. Incubate oocytes with a primary antibody (diluted in blocking buffer) overnight at 4 °C or for 1 h at room temperature.
10. Wash oocytes three times (5 min each) in wash solution (PBS containing 0.1 % Tween 20).
11. Label oocytes with a secondary antibody (diluted 1:200 in wash solution) for 1 h at room temperature.
12. Counterstain oocytes with 4′,6-diamidino-2-phenylindole (DAPI; 5 μg/mL in wash solution) for 5–10 min.
13. Wash oocytes three times (5 min each) in wash solution.

14. Drop approximately 2–3 μL of 80 % glycerol with a pipette tip on a glass slide and carefully transfer oocytes to the slide with a minimal amount of wash solution.
15. Place four small columns of Vaseline around the glycerol drop with a pipette tip, and mount a cover slip on the slide.
16. Examine the pattern and localization of the fluorescent signals in oocytes using a confocal microscope.
17. Take oocyte images for each 0.5–1 μm thickness using the *Z*-stack function of the confocal microscope.

3.7 Genotyping

1. Cut approximately 0.5 cm from the end of a mouse tail, place in a centrifuge tube, and label it.
2. Add 400 μL of tail buffer (contains protease K; 1:200) into the tube, and digest the tail overnight at 65 °C.
3. Add 400 μL of a mixture of phenol, chloroform, and isoamyl alcohol (25:24:1), vortex, and let stand for 5 min at room temperature.
4. Centrifuge at 15,000 × *g* for 10 min using a microfuge.
5. Transfer 350 μL of the supernatant to a clean centrifuge tube.
6. Add 1 mL of 100 % ethanol into the tube, and gently tip the tube upside down a few times.
7. Centrifuge at 12,000 rpm for 5 min using a microfuge.
8. Pour out the supernatant, add 1 mL of 70 % ethanol into the tube, and shake.
9. Centrifuge at 12,000 rpm for 5 min using a microfuge.
10. Pour out the 70 % ethanol, and completely remove any remaining ethanol with a pipette.
11. Dry DNA at room temperature, and then add 300–400 μL of TE buffer to dissolve the DNA (65 °C, 30 min).
12. PCR reaction.
13. Apply the PCR products to agarose gel electrophoresis using an appropriate gel concentration to determine mouse genotypes.

    PCR reaction mixtures for genotyping include the following:

    Template: 1 μL.

    10× PCR buffer: 2 μL.

    dNTP (10 μM): 0.5 μL.

    Forward primer: 0.5 μL.

    Reverse primer: 0.5 μL.

    DNA polymerase: 0.2 μL.

    Add ddH_2O up to 20 μL.

4 Notes

1. To harvest undifferentiated granulosa cells and study ovarian responses to exogenous gonadotropins, PD21 female mice should be used to avoid the complexity of ovarian functions associated with estrous cycles and endogenous gonadotropin surges. In addition, the ovaries of adult mice contain the corpus lutea of different developmental stages, which will affect the purity of harvested undifferentiated granulosa cells.
2. Based on our experience, 24 h after PMSG injection is the best time point for harvesting undifferentiated granulosa cells. If the mouse granulosa cells are isolated at later time points, such as 44–48 h after PMSG injection, they usually spontaneously luteinize in culture, which precludes the analyses of TGF-β functions in undifferentiated granulosa cells.
3. It is not necessary to measure the protein concentrations in each sample, as only small amounts of protein are present in this experiment. Typically, 200 μL of SDS-loading buffer is added to each well of a 24-well plate to lyse cells, and 25–30 μL of the lysate is loaded into each lane for Western blotting. These lysates are sticky and difficult to load on SDS-PAGE gels. Repeated freezing and thawing can reduce the lysates' stickiness.

Acknowledgements

The research described was supported by the National Basic Research Program of China (2011CB944504, 2012CB944403), the National Natural Science Foundation of China (81172473), the Natural Science Foundation of Zhejiang Province (R2100145), and Basic Scientific Research Funding of Zhejiang University (2011QN81001).

References

1. Fan HY, Richards JS (2010) Minireview: physiological and pathological actions of RAS in the ovary. Mol Endocrinol 24(2):286–298, PMCID: 2817603
2. Fan HY, Liu Z, Shimada M, Sterneck E, Johnson PF, Hedrick SM et al (2009) MAPK3/1 (ERK1/2) in ovarian granulosa cells are essential for female fertility. Science 324(5929):938–941
3. Park JY, Su YQ, Ariga M, Law E, Jin SL, Conti M (2004) EGF-like growth factors as mediators of LH action in the ovulatory follicle. Science 303(5658):682–684
4. Shimada M, Hernandez-Gonzalez I, Gonzalez-Robayna I, Richards JS (2006) Paracrine and autocrine regulation of epidermal growth factor-like factors in cumulus oocyte complexes and granulosa cells: key roles for prostaglandin synthase 2 and progesterone receptor. Mol Endocrinol 20(6):1352–1365
5. Diaz FJ, Wigglesworth K, Eppig JJ (2007) Oocytes determine cumulus cell lineage in mouse ovarian follicles. J Cell Sci 120(Pt 8): 1330–1340
6. Pangas SA, Matzuk MM (2005) The art and artifact of GDF9 activity: cumulus expansion

and the cumulus expansion-enabling factor. Biol Reprod 73(4):582–585

7. Pangas SA (2012) Bone morphogenetic protein signaling transcription factor (SMAD) function in granulosa cells. Mol Cell Endocrinol 356(1-2):40–47, PMCID: 3203253
8. Dong J, Albertini DF, Nishimori K, Kumar TR, Lu N, Matzuk MM (1996) Growth differentiation factor-9 is required during early ovarian folliculogenesis. Nature 383(6600):531–535
9. Pangas SA, Woodruff TK (2000) Activin signal transduction pathways. Trends Endocrinol Metab 11(8):309–314
10. van Houten EL, Themmen AP, Visser JA (2010) Anti-Mullerian hormone (AMH): regulator and marker of ovarian function. Ann Endocrinol (Paris) 71(3):191–197
11. Edson MA, Nalam RL, Clementi C, Franco HL, Demayo FJ, Lyons KM et al (2010) Granulosa cell-expressed BMPR1A and BMPR1B have unique functions in regulating fertility but act redundantly to suppress ovarian tumor development. Mol Endocrinol 24(6):1251–1266, PMCID: 2875809
12. Pangas SA, Li X, Umans L, Zwijsen A, Huylebroeck D, Gutierrez C et al (2008) Conditional deletion of Smad1 and Smad5 in somatic cells of male and female gonads leads to metastatic tumor development in mice. Mol Cell Biol 28(1):248–257, PMCID: 2223289
13. Li Q, Pangas SA, Jorgez CJ, Graff JM, Weinstein M, Matzuk MM (2008) Redundant roles of SMAD2 and SMAD3 in ovarian granulosa cells in vivo. Mol Cell Biol 28(23):7001–7011, PMCID: 2593383
14. Pangas SA, Li X, Robertson EJ, Matzuk MM (2006) Premature luteinization and cumulus cell defects in ovarian-specific Smad4 knockout mice. Mol Endocrinol 20(6):1406–1422

Chapter 20

Quantitative Real-Time PCR Analysis of MicroRNAs and Their Precursors Regulated by TGF-β Signaling

Hara Kang and Akiko Hata

Abstract

The signaling pathway of TGF-β and its family member BMP has been implicated in vascular development and maintenance of homeostasis by modulating expression of small noncoding microRNAs (miRNAs). MiRNAs repress target genes, which play a critical role in regulating vascular smooth muscle cell (VSMC) growth, phenotype, and function. To understand the mechanisms by which specific miRNAs control the TGF-β and BMP signaling pathway in VSMC, it is essential to quantitate levels of specific miRNAs and their precursors whose expression are controlled by TGF-β/BMP signaling. Here, we describe a real-time quantization method for accurate and sensitive detection of miRNAs and their precursors, such as primary transcripts of miRNAs (pri-miRNAs) and precursor miRNAs (pre-miRNAs). This method requires two steps; synthesis of single-stranded complementary DNAs (cDNAs) from total RNA samples and quantization of specific pri-, pre-, or mature miRNAs by quantitative polymerase chain reaction (PCR) using a real-time PCR machine.

Key words MicroRNA, TGF-β, BMP, Quantitative real-time PCR

1 Introduction

The transforming growth factor-β (TGF-β) family of growth factors is evolutionarily conserved. These growth factors are critical modulators of various biological processes such as cell fate determination, cell differentiation, proliferation, and migration in embryos and adults from insects to mammals [1], including vascular development and the maintenance of homeostasis. The TGF-β family of growth factors, such as TGF-β and BMP, has been shown to promote differentiation of VSMC by upregulating the expression of contractile genes and inhibiting VSMC growth and migration [2, 3]. In recent studies, several miRNAs were found to be expressed in the vascular cells and deregulation of these miRNAs is involved in vascular diseases [4, 5]. Moreover, Smad proteins, as signal transducers of the TGF-β family of growth factors, have been demonstrated to regulate generation of subset of miRNAs at

Xin-Hua Feng et al. (eds.), *TGF-β Signaling: Methods and Protocols*, Methods in Molecular Biology, vol. 1344,
DOI 10.1007/978-1-4939-2966-5_20,

the transcriptional and/or posttranscriptional level [6, 7]. In this chapter we summarize the mechanism of regulation of miRNA biogenesis by the TGF-β family of growth factors and techniques to quantitate the levels of miRNAs and their precursors upon stimulation with the TGF-β family of growth factors.

1.1 Biogenesis of miRNAs

miRNAs are transcribed by RNA polymerase II as long primary transcripts known as pri-miRNAs which contain a hairpin structure. Although mature form of miRNAs is ~22 nucleotides (nt) in length, pri-miRNAs can be as long as a few kilo nt long, and are cleaved sequentially by two ribonuclease enzymes; Drosha and Dicer [8–11]. After the first processing by Drosha in the nucleus, pre-miRNAs, which are ~60–100 nt hairpins, are exported from the nucleus to the cytoplasm by Exportin 5. In the cytoplasm, the pre-miRNAs associate with Dicer, which cleaves the pre-miRNA into an approximately 22 nt miRNA. The miRNA duplex is then loaded into Argonaute (Ago) proteins, which select one strand and present it to the RNA-induced silencing complex (RISC) for recognizing target mRNAs. The mature miRNAs regulate expression of protein coding genes by promoting degradation of mRNAs or repressing their translation [12] (Fig. 1a). Although miRNAs represent a relatively abundant class of RNAs, their expression levels are found to vary greatly in different cell types, tissues, and developmental stages. To elucidate a critical role of miRNA during normal physiological and developmental processes, as well as pathogenesis of various disorders, sensitive, efficient, and accurate quantitation of miRNA levels is essential.

1.2 Control of miRNA Biogenesis by Smads and Quantitative Analysis of Pri-, Pre-, and Mature miRNAs

One of the best-characterized steps of miRNA biogenesis is at the point of pri-to-pre-miRNA processing by the Drosha microprocessor complex. It was discovered that when pulmonary artery smooth muscle cells (PASMCs) are stimulated with ligands of the TGF-β family, such as TGF-β and bone morphogenetic protein 4 (BMP4), pri-miR-21 levels are unchanged while mature miR-21 expression increases significantly [3, 6]. This result suggested a transcription-independent mechanism of miR-21 regulation, which was confirmed by the induction of miR-21 even in the presence of the transcriptional inhibitor α-amanitin [3, 6]. Biochemical analysis showed that the increase in mature miR-21 is caused by faster turnover in the biogenesis of miR-21 mediated by the nuclear translocation of Smad proteins [3, 6].

Smads represent the primary signal-transduction molecules of the TGF-β family of growth factors. Receptor activation by TGF-β family ligands induces phosphorylation of several Smad proteins, collectively called the Receptor-specific Smads (R-Smads) [1]. Phosphorylated R-Smads form a complex with the common-Smad (co-Smad), Smad4, and translocate to the nucleus where they act as transcriptional regulators to promote or inhibit gene expression [1].

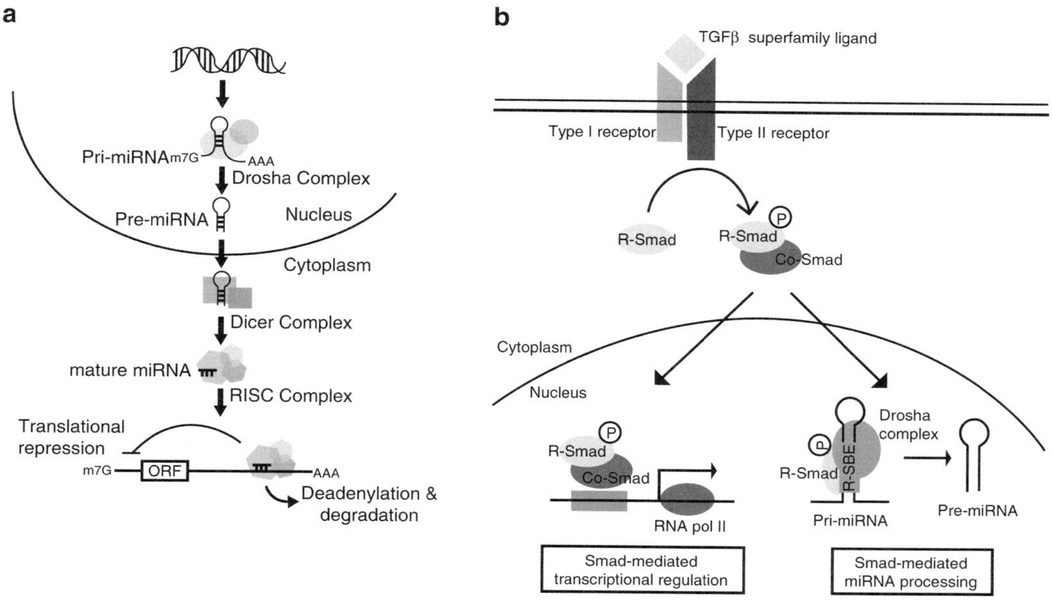

Fig. 1 Schematic diagram of miRNA biogenesis and its regulation by the TGF-β signaling pathway. (**a**) Pri-miRNAs are transcribed and processed into hairpin-structured pre-miRNAs by Drosha complex. The pre-miRNAs are exported from the nucleus to the cytoplasm and are cleaved by the Dicer complex into ~22 nt miRNAs. The mature miRNA is loaded into the RISC complex and mediates posttranscriptional repression of target mRNA by translational repression and/or deadenylation and degradation. (**b**) Activation of receptor Smads by TGF-β family ligands leads to their translocation into the nucleus. In the nucleus, R-Smads bind, independent of the transcriptionally necessary co-Smad, to a conserved sequence in pri-miRNAs which they recruit to the Drosha microprocessor complex and facilitate pri- to pre-miRNA maturation

R-Smad/co-Smad hetero-dimerization is required for the transcriptional regulation mediated by TGF-β signaling [1]. Surprisingly, Smad4 is dispensable for TGF-β mediated induction of miR-21 [3, 6]. Furthermore, stimulation with TGF-β ligands induced an association between Smads and the RNA helicase p68, indicating a direct recruitment of R-Smads to the Drosha microprocessor complex [3, 6].

More recent miRNA expression profiling experiments have revealed that stimulation of PASMCs with either TGF-β or the related BMP4 induces the expression of approximately 20 miRNAs [6]. Following stimulation, many of these miRNAs are bound directly by Smad proteins through a 5-base RNA sequence motif (R-SBE) that closely resembles the Smad DNA binding element (SBE; 5′-CAGAC-3′) in a region overlapping the encoded mature miRNA [6]. Thus, R-Smads associate with pri-miRNA in a sequence specific manner and facilitate the Drosha microprocessor activity to enhance mature miRNA expression (Fig. 1b). Interestingly, in silico analyses have found that pre-miRNA sequences encode a higher number of transcription factor biding sites than would be predicted by chance [13]. These results suggest that the mechanism for Smad-dependent regulation of pri-miRNAs may represent a widespread, though under appreciated, strategy for controlling miRNA expression.

During the last decade, more than 30 miRNA quantitation methodologies have been developed, including northern blot [14], miRNA microarray [15], next generation sequencing [16], and quantitative real-time PCR (qPCR) [17, 18]. Of these methods, qPCR is most sensitive and easily applicable to validate the data obtained from the high-throughput approaches. Our group has found that it is difficult to detect small changes in miRNA expression upon growth factor stimulation using next generation sequence (A.H. unpublished observation). qPCR-based miRNA analysis is also convenient as it does not require the use of radioactive material. Because of the small size, detection of mature miRNAs by PCR is technically challenging. A number of specific qPCR techniques have been developed for the detection of miRNAs. Here, we present qPCR methods based upon reverse transcriptase reaction with a stem-loop primer followed by a TaqMan PCR analysis to detect changes in expression levels of miR-21 and miR-302c in the cell in response to stimulation with the TGF-β family of growth factors [7, 19] (Fig. 2).

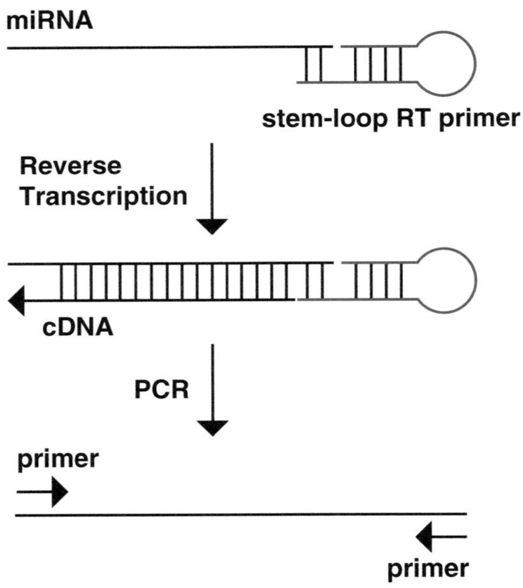

Fig. 2 Schematic description of miRNA detection by qPCR. Stem-loop RT primer binds to the 3′ end of miRNA and is reverse-transcribed. The synthesized cDNA is amplified using miRNA specific PCR primers

2 Materials

Cell culture reagents: Human pulmonary artery smooth muscle cells (PASMC) and SmGM-2 smooth muscle growth medium (CC-2581 and CC-3182) are purchased from Lonza. Dulbecco's Modified Eagle's Medium (DMEM), fetal bovine serum (FBS), penicillin/streptomycin sulfate, 0.25 % Trypsin–EDTA, and phosphate buffered saline (PBS) are purchased from Invitrogen. For cytokine stimulation, human recombinant TGF-β1 and BMP4 are purchased from R&D systems.

RNA isolation components: TRIzol and glycogen are purchased from Invitrogen. Chloroform, ethanol, and isopropanol are purchased from Sigma.

Reverse-transcription (RT) components: To synthesize cDNA of mature miRNAs, TaqMan miRNA reverse transcription kit (PN 4366596) and TaqMan MicroRNA Assays for human U6 snRNA (#001973), hsa-miR-21 (#000397), and hsa-miR-302c (#000533) including miRNA-specific RT primers are purchased from Applied Biosystems (Table 1). To synthesize cDNA of pri- or pre-miRNA, iScriptcDNA synthesis kit is purchased from Bio-Rad. For RT reactions, S1000 Thermal cycler (Bio-Rad) is used.

qPCR components: Each TaqMan MicroRNA Assay contains a mix of a miRNA-specific forward PCR primer, a specific reverse PCR primer, and miRNA-specific TaqMan probe linked to FAM dye. AmpliTaq DNA polymerase with buffer I (N8080160) and 10 mM dNTP mix are purchased from Applied Biosystems. To detect pri- and pre-miRNA levels, primers are synthesized from IDT and iQ

Table 1
miRNA, control U6 snRNA, and primer sequences used for qPCR

Name	Sequence
Has-miR-21	UAGCUUAUCAGACUGAUGUUGA
Has-miR-302c	UAAGUGCUUCCAUGUUUAGG
Human U6 snRNA	GTGCTCGCTTCGGCAGCACATATACTAAAATTGGAACGATACA GAGAAGATTAGCATGGCCCCTGCGCAAGGATGACACGCAAA TTCGTGAAGCGTTCCATATTTT
Human pri-miR-302c	TGAATCCAATTTACTTCTCCA and TCCTTTAACCTGTAACAAGC
Human pre-miR-302c	CCTTTGCTTTAACATGGGGG and CCTCCACTGAAACATGGAAG
Human pri-miR-21	TTTTGTTTTGCTTGGGAGGA and AGCAGACAGTCAGGCAGGAT
Human pre-miR-21	TGTCGGGTAGCTTATCAGAC and TGTCAGACAGCCCATCGACT
Human *GAPDH*	ACCACAGTCCATGCCATCAC and TCCACCACCCTGTTGCTGTA

SYBR Green Supermix is purchased from Bio-Rad. Primer sequences are shown in Table 1. Hard-shell thin-wall 96-well skirted PCR plates and microseal "B" adhesive seals are purchased from Bio-Rad. For qPCR reactions, the MJ Research Opticon2 continuous fluorescence detection system is used.

3 Methods

3.1 Cell Culture and Stimulation with TGF-β Family Growth Factors

Human PASMC are maintained in SmGM-2 media containing 5 % fetal calf serum. Early passage (passage 6–8) PASMC are split into 6-well plates with 120,000 cells/well. To examine changes of miRNA expression levels upon TGF-β signals, cells are cultured in DMEM with 0.2 % FBS for 24 h before growth factor stimulation. TGF-β (400 pM) or BMP4 (3 nM) is treated under starvation conditions (DMEM with 0.2 % FBS) for 24 h. Cells are cultured at 37 °C in the presence of 5 % CO_2.

3.2 RNA Isolation

1. Rinse cells with PBS and add 500 μL of TRIzol directly to the cells in a well of the 6-well plate to isolate total RNA.
2. Incubate the plate on a rotator for 5 min at room temperature and lyse the cells directly by pipetting.
3. Collect and transfer the mixture of cells and TRIzol into an eppendorf tube (E-tube). Homogenized samples can be stored at room temperature for several hours or at −80 °C for at least 1 month.
4. Add 100 μL of chloroform, shake vigorously for 15 s, and spin at 15,000 × *g* for 15 min at 4 °C. The homogenate will separate into a clear upper aqueous layer containing RNA, an interphase, and a red lower organic layer containing the DNA and proteins.
5. Collect the upper aqueous layer and transfer it into a new E-tube and add 250 μL of isopropanol and 40 μg of glycogen to the sample. Incubate for 10 min at room temperature. RNA is precipitated from the aqueous layer by the isopropanol.
6. Centrifuge the sample at 15,000 × *g* for 15 min at 4 °C.
7. Decant the supernatant and wash the pellet with 1 mL of 75 % ethanol to remove impurities.
8. Spin at 15,000 × *g* for 5 min at 4 °C to remove any remaining ethanol. Air-dry the pellet for 5–10 min.
9. Dissolve the RNA pellet in 20 μL of nuclease-free water and incubate at 55 °C for 10 min. Isolated total RNA can be frozen and stored at −80 °C.

3.3 cDNA Synthesis for Detection of Mature miRNAs

cDNA is reverse-transcribed from the isolated total RNA using specific miRNA primers from the TaqMan MicroRNA Assays and reagents from the TaqMan MicroRNA Reverse Transcription kit.

All experimental procedure should be done on ice. Vortex samples thoroughly between each step to be sure to mix.

1. Dilute the isolated total RNA to 10 ng/μL in nuclease-free water.
2. Prepare RT master mix. For each 15 μL RT reaction, RT master mix consists of 9.66 μL of nuclease-free water, 1.5 μL of 10× reverse transcription buffer, 0.19 μL of RNase inhibitor (20 U/μL), 0.15 μL of 100 mM dNTPs, and 1 μL of MultiScribe reverse transcriptase (50 U/μL).
3. Gently mix 1 μL of the diluted total RNA, 1.5 μL of specific miRNA primer with 12.5 μL of RT master mix.
4. Spin to bring solution to the bottom of the tube and incubate the tube on ice until a thermal cycler is ready.
5. To perform reverse transcription, program a thermal cycler. The reaction proceeds for 30 min at 16 °C, 30 min at 42 °C, and 5 min at 85 °C.
6. Load the reaction tube into the thermal cycler and start the reverse transcription run.

3.4 qPCR Analysis for Detection of Mature miRNAs

For qPCR, performed three replicates each of 10 μL PCR reactions (triplicate). All experimental procedure should be done on ice.

1. Prepare PCR master mix. For each 10 μL PCR reaction, PCR master mix consists of 7.2 μL of nuclease-free water, 1 μL of 10× PCR buffer, 0.2 μL of 10 mM dNTPs, and 0.1 μL of AmpliTaq DNA polymerase.
2. Gently mix the PCR master mix with 0.5 μL of the specific miRNA PCR primer from TaqMan MicroRNA Assay.
3. Add 1 μL of the synthesized cDNA from the RT reaction tube into the PCR reaction tube.
4. Mix and spin to bring the solution to the bottom of the tube and prepare the PCR reaction plate by transferring 10 μL of the complete PCR reaction mix into each well.
5. Seal the plate with an optical adhesive cover and centrifuge the plate at 220 × *g* for 2 min at 4 °C to spin down the contents and eliminate any air bubbles.
6. Load the reaction plate into the MJ Research Opticon2 continuous fluorescence detection system and start the run. The reactions are incubated at 95 °C for 10 min, followed by 40 cycles of 95 °C for 15 s and 60 °C for 1 min.

3.5 cDNA Synthesis for Detection of Pri- and Pre- miRNAs

We have found that the 5× iScript reaction mix generates some precipitation upon thawing occasionally. Mix thoroughly to resuspend the precipitate.

1. For each 20 μL RT reaction, mix 4 μL of 5× iScript reaction mix, 1 μL of iScript reverse transcriptase, and 1 μg of total RNA. Add nuclease-free water to make total volume of 20 μL.
2. Mix gently and spin to bring solution to the bottom of the tube.
3. To perform reverse transcription, incubate the reaction mix for 5 min at 25 °C, 30 min at 42 °C, and 5 min at 85 °C in the thermal cycler.

3.6 qPCR Analysis for Detection of Pri- and Pre-miRNAs

To amplify pri-miRNAs, forward and reverse primers are designed to anneal to the outside stem portion of the hairpin sequence of the miRNA precursors. On the other hand, to amplify pre-miRNAs, primers are designed to be located within the hairpin sequence of the pre-miRNA. In theory, pre-miRNA primers are able to anneal and amplify pri-miRNAs, however, due to a secondary structure of long pri-miRNAs, pre-miRNA primers do not amplify pri-miRNAs efficiently under the PCR condition described below, and selectively measure the quantity of pre-miRNAs. Pre-miRNA sequences are obtained from the miRBase website (www.mirbase.org). The extended pre-miRNA sequences to design pri-miRNA primers are obtained from Ensembl Genome Browser (www.ensembl.org). Primers are designed using Primer3Plus (http://frodo.wi.mit.edu/). Primers are designed with a maximal Tm difference between both primers of less than 2 °C and a primer length between 18 and 24 nt. An ideal Tm of 55–60 °C is selected for the primers. Following design and purchase of primers from IDT (Table 1), qPCR is carried out as follows.

1. Thaw iQ SYBR Green supermix and store on ice protected from light.
2. Prepare PCR reaction mix. For each 10 μL PCR reaction, PCR master mix consists of 5 μL of iQ SYBR Green supermix, 0.5 μL of both 10 μM forward and reverse primers, 3.5 μL of nuclease-free water, and 0.5 μL of cDNA.
3. Mix gently and spin to bring the solution to the bottom of the tube and prepare the PCR reaction plate by transferring 10 μL of the complete PCR reaction mix into each well.
4. Seal the plate with an optical adhesive cover and centrifuge the plate at 220 × *g* for 2 min at 4 °C to spin down the contents and eliminate any air bubbles.
5. Load the reaction plate into the MJ Research Opticon2 continuous fluorescence detection system and start the run. PCR cycling conditions are 94 °C for 3 min and 40 cycles of 94 °C for 15 s, 60 °C for 20 s, and 72 °C for 40 s.

3.7 Analysis of qPCR Data

Data analysis is performed using the comparative C_T method in the accompanying MJ Opticon Monitor 3 software. The threshold cycle (C_T) is defined as the fractional cycle number at which the fluorescence passes the fixed threshold. An average of three experiments, each performed in triplicate with standard errors is presented. To normalize human miR-21 or miR-302c, the constitutively expressed endogenous control, human U6 snRNA, is used. Expression levels of pri- and pre-miRNAs are normalized to the endogenous expression level of human *glyceraldehyde-3-phosphate dehydrogenase (GAPDH)*.

4 Notes

1. Use RNase ZAP spray, filter tip, and nuclease-free water throughout all the experimental procedures to avoid RNase contamination.
2. We find that the RNA is often invisible prior to centrifugation and forms a gel-like pellet on the side and bottom of the tube.
3. Do not allow the RNA to dry completely, because the RNA pellet will become difficult to solubilize. Purity of RNA samples should be examined by measuring the $OD_{260/280}$ ratio, which should be greater than 1.6.
4. RT reactions should include negative controls without addition of RNAs. All PCR reactions should include samples that are not treated with RT as negative controls.
5. Keep all TaqMan MicroRNA Assays protected from light during storage in the freezer until ready to be used. Excessive exposure to light affects the fluorescent probes and decreases the sensitivity of quantization during qPCR.
6. The RT primers which contain the stem-loop structure increase the specificity and sensitivity of the assay in comparison with regular linear RT-primers because of base stacking and spatial constraint of the stem-loop structure [17].
7. TaqMan miRNA assays are specific for mature miRNAs and discriminate between related miRNAs that differ by as little as one nucleotide. Furthermore, they are not affected by genomic DNA contamination.
8. This method uses commercially available miRNA detection kits and primer sets. It requires a unique probe and set of primers for each miRNA assay, and therefore is neither cost effective nor suitable for high-throughput analysis of miRNAs.

5 Concluding Remarks

Control of miRNA biogenesis is an integral component of cell biological activities. Deregulation of miRNA biogenesis results in aberrant expression of multiple mRNAs, which can lead to developmental defects and human diseases. Increasingly studies highlight the importance understanding transcriptional regulation as well as posttranscriptional regulation of miRNA biogenesis upon stimulation with different growth factors. We anticipate that in the future more accurate and easy methodology to quantitate miRNAs will be developed. In the meanwhile, the RT-PCR method described above provides a basis for the quantization of miRNAs and their precursors upon growth factor signaling and during pathogenesis of various human disorders.

Acknowledgement

We thank members of the Hata lab in particular Matt Blahna for critical reading of the manuscript. This work was supported by grants from the National Institute of Health: HL093154 and HL108317, the American Heart Association: 0940095N and the LeDucq foundation Transatlantic network grant to A.H. and the National Research Foundation of Korea (Basic Science Research Program; 2012R1A1A1042812) to H.K.

References

1. Massague J (2012) TGFbeta signalling in context. Nat Rev Mol Cell Biol 13:616–630
2. ten Dijke P, Arthur HM (2007) Extracellular control of TGFbeta signalling in vascular development and disease. Nat Rev Mol Cell Biol 8:857–869
3. Davis BN, Hilyard AC, Lagna G, Hata A (2008) SMAD proteins control DROSHA-mediated microRNA maturation. Nature 454:56–61
4. Hata A. Functions of microRNAs in cardiovascular biology and disease. Annu Rev Physiol 75:69–93
5. Kang H, Hata A. (2012) MicroRNA regulation of smooth muscle gene expression and phenotype. Curr Opin Hematol 19:224–231
6. Davis BN, Hilyard AC, Nguyen PH, Lagna G, Hata A. (2010) Smad proteins bind a conserved RNA sequence to promote microRNA maturation by Drosha. Mol Cell 39:373–384
7. Kang H, Davis-Dusenbery BN, Nguyen PH, Lal A, Lieberman J et al. (2012) Bone morphogenetic protein 4 promotes vascular smooth muscle contractility by activating microRNA-21 (miR-21), which down-regulates expression of family of dedicator of cytokinesis (DOCK) proteins. J Biol Chem 287:3976–3986
8. Borchert GM, Lanier W, Davidson BL (2006) RNA polymerase III transcribes human microRNAs. Nat Struct Mol Biol 13:1097–1101
9. Denli AM, Tops BB, Plasterk RH, Ketting RF, Hannon GJ (2004) Processing of primary microRNAs by the microprocessor complex. Nature 432:231–235
10. Han J, Lee Y, Yeom KH, Nam JW, Heo I et al (2006) Molecular basis for the recognition of primary microRNAs by the Drosha-DGCR8 complex. Cell 125:887–901
11. Hutvagner G, McLachlan J, Pasquinelli AE, Balint E, Tuschl T et al (2001) A cellular function for the RNA-interference enzyme Dicer in the maturation of the let-7 small temporal RNA. Science 293:834–838
12. Bartel DP (2009) MicroRNAs: target recognition and regulatory functions. Cell 136: 215–233

13. Piriyapongsa J, Jordan IK, Conley AB, Ronan T, Smalheiser NR (2011) Transcription factor binding sites are highly enriched within microRNA precursor sequences. Biol Direct 6:61
14. Valoczi A, Hornyik C, Varga N, Burgyan J, Kauppinen S et al (2004) Sensitive and specific detection of microRNAs by northern blot analysis using LNA-modified oligonucleotide probes. Nucleic Acids Res 32, e175
15. Fichtlscherer S, De Rosa S, Fox H, Schwietz T, Fischer A et al. (2010) Circulating microRNAs in patients with coronary artery disease. Circ Res 107:677–684
16. Wu Q, Lu Z, Li H, Lu J, Guo L et al (2011) Next-generation sequencing of microRNAs for breast cancer detection. J Biomed Biotechnol 2011: 597145
17. Chen C, Ridzon DA, Broomer AJ, Zhou Z, Lee DH et al (2005) Real-time quantification of microRNAs by stem-loop RT-PCR. Nucleic Acids Res 33, e179
18. Benes V, Castoldi M. (2010) Expression profiling of microRNA using real-time quantitative PCR, how to use it and what is available. Methods 50:244–249
19. Kang H, Louie J, Weisman A, Sheu-Gruttadauria J, Davis-Dusenbery BN et al. (2012) Inhibition of microRNA-302 (miR-302) by bone morphogenetic protein 4 (BMP4) facilitates the BMP signaling pathway. J Biol Chem 287:38656–38664

Chapter 21

TGF-β-Regulated MicroRNAs and Their Function in Cancer Biology

Pengyuan Yang, Yun Zhang, Geoffrey J. Markowitz, Xing Guo, and Xiao-Fan Wang

Abstract

The transforming growth factor-β (TGF-β) is known to regulate a large number of biological processes and is involved in various aspects of tumor development. Recent studies have shown that the biogenesis of miRNAs can be regulated by TGF-β signaling directly via Smad-dependent mechanisms and/or other unknown mechanisms, which may induce autoregulatory feedback loops in response to the activation of TGF-β signaling, influencing the fate of tumor cells. In this chapter, we summarize the currently described mechanisms underlying TGF-β's regulation of miRNA biogenesis, and the functional role of TGF-β-regulated miRNAs in tumor initiation, epithelial-mesenchymal transition, and tumor microenvironment modulation. Finally, we introduce methods to study TGF-β-regulated miRNAs and their functions in tumor progression and metastasis using an example of publication from our lab demonstrating the presence of a TGF-β-miR-34a-CCL22 signaling axis, which serves as a potent etiological pathway for the development of hepatocellular carcinoma venous metastases.

Key words MicroRNA, TGF-β, Cancer biology, Metastasis

1 Introduction

The cytokine TGF-β is a multifunctional factor that plays critical roles in various aspects of tumor development due to its tumor-suppressive and tumor-promoting effects. In premalignant cells, TGF-β can function as a tumor suppressor by regulating autophagy and tissue homeostasis and/or by inducing apoptosis or cell cycle arrest [1]. However, as tumor develops to a later stage, cancer cells gain the capacity to adulterate the suppressive influence of TGF-β signaling through either inactivating mutations in TGFβR2/SMAD4 or rewiring the signal transduction pathway downstream of SMADs [2]. Pathological forms of TGF-β signaling act as potent promoters of cell motility, invasion, and metastasis, and aid in tumor stem cell maintenance [3]. Multiple signal transduction pathways, using complex networks of signaling molecules, are

Xin-Hua Feng et al. (eds.), *TGF-β Signaling: Methods and Protocols*, Methods in Molecular Biology, vol. 1344,
DOI 10.1007/978-1-4939-2966-5_21, © Springer Science+Business Media New York 2016

responsible for the influence of TGF-β on various aspects of tumor biology. Therefore, further investigations are necessary to elucidate TGF-β signaling networks and outputs of these networks, and the context in which signaling occurs. Especially important among these investigations are studies to elucidate how this cytokine can be so radically switched from a tumor suppressor to a tumor promoter, which are vital to the development and informed use of potentially powerful TGF-β-targeted therapies.

MicroRNAs (miRNAs) are small noncoding RNAs of ~22 nucleotides that negatively regulate gene expression by blocking protein translation and promoting degradation of the target messenger RNA [4]. More than 60 % of protein-coding genes are potentially regulated by miRNAs [5]. As the expression pattern of the miRome is highly tissue-specific, miRNAs confer fine-tuning to protein expression in a highly cell- and context-specific and adaptable manner. Changes in the expression profiles of miRNAs have been linked to the development of various types of human diseases, including cancer [6]. MiRNAs can act as oncogenic promoters or tumor suppressors, depending on the functional nature of their specific target genes within a specific cell or tissue type.

Recent studies have demonstrated that most members of TGF-β-mediated signaling pathways, both components of the canonical signaling cascade [7, 8] and TGF-β target genes [9], are targeted by miRNAs via direct interaction. Importantly, more recent studies have shown that the biogenesis of miRNAs is also regulated by TGF-β signaling directly via Smad-dependent and/or other uncharacterized mechanisms [10–12], which may induce autoregulatory feedback loops in response to the activation of TGF-β signaling, influencing the fate of tumor cells. In this chapter we will review the recognized mechanisms underlying the regulatory effect of TGF-β on miRNA biogenesis, and the functional role of TGF-β-regulated miRNAs in tumor initiation, epithelial-mesenchymal transition, and tumor microenvironment modulation. We will also present our related research methods that provide new concepts and tools to study the functional role of TGF-β-regulated microRNAs in tumor metastasis.

1.1 TGF-β Regulates the Biogenesis of MiRNAs

The biogenesis of miRNAs begins with a long primary transcript (pri-miRNA) which bears a 7-methylguanosine cap and a poly-(A) tail, and undergoes posttranscriptional 5′- and 3′-end processing events similar to messenger RNA [13]. Pri-miRNAs contain one or more stem-loop secondary structures, which are recognized by members of the RNase III family of enzymes, including Drosha and Dicer. Drosha, associated with DGCR8 and p68, acts in the nucleus to cleave the stem-loop region of pri-miRNAs to generate a precursor miRNA (pre-miRNA), which is then exported to the cytoplasm by Exportin 5 [14, 15]. Dicer subsequently cleaves the pre-miRNA to generate a ~22-nucleotide miR/miR* complex [16].

After processing, one strand of the miRNA, corresponding to the mature miRNA, gets loaded into the miR-induced silencing complex (miRISC) and performs the functional role of miRNA through inhibiting translation of the target mRNAs.

Emerging evidence indicates regulatory functionality of TGF-β in the miRNA biogenic process. In 2009, Davis et al. reported that TGF-β and BMP signaling promotes a rapid increase in the expression of mature miR-21 through a posttranscriptional step, promoting the processing of primary transcripts of miR-21 (pri-miR-21) into precursor miR-21 (pre-miR-21) by the Drosha complex. TGF-β- and BMP-specific Smad signal transducers are recruited to pri-miR-21 in a complex with the RNA helicase p68, a component of the Drosha microprocessor complex. This processing of miR-21 is critical for control of the vascular smooth muscle cell phenotype mediated by the TGF-β and BMP signaling pathways. Further studies profiling miRNA expression in response to TGF-β or BMP4 treatment demonstrated that approximately 20 miRNAs can be induced by TGF-β signaling. Upon TGF-β stimulation, many of these miRNAs bind directly to Smad proteins utilizing a 5-base motif that closely resembles the Smad DNA binding element (SBE) in the region overlapping with the encoded mature miRNA. These studies indicated that Smads acting in TGF-β signaling pathways associate with pri-miRNA in a sequence-specific manner and facilitate the Drosha microprocessor activity to enhance mature miRNA expression.

1.2 The Roles of TGF-β-Regulated MiRNAs in Cancer Biology

1.2.1 Tumor Initiation

As described above, in normal, healthy tissues, TGF-β acts as a tumor suppressor by inhibiting cell proliferation, inducing apoptosis, and regulating autophagy. However, in malignantly transformed cells, this repressive function is often lost. Instead, cancer cells that lose the tumor-suppressive arm of TGF-β pathway accrue tumorigenic effects that directly enhance tumor growth, induce cancer cell stemness, and promote cancer cell invasion by activating the AKT, WNT, and ERK pathways. Although the mechanisms underlying the crosstalk between TGF-β and other signaling pathways are not fully understood, several studies have revealed that TGF-β-regulated miRNAs play key roles in the modulation of pathological TGF-β signaling that benefits tumor growth.

AKT activation downstream of TGF-β-induced signaling promotes tumorigenesis, but the mechanism underlying this regulation is still unclear. Natarajan and colleagues [17] found the regulation of TGF-β-activated AKT is mediated by miR-216a and miR-217, both of which target PTEN (phosphatase and tensin homologue), an inhibitor of AKT activation. These miRNAs are located within the second intron of a noncoding RNA (RP23-298H6.1-001). The RP23 promoter was activated in their studies by TGF-β and miR-192 through E-box-regulated mechanisms [18]. AKT activation by these miRNAs led to cell survival, which was similar to the effect of activation by TGF-β. Due to the diversity of PTEN functions in tumorigenesis, this miRNA-amplifying circuit may have more extended roles in promoting tumor initiation and progression.

TGF-β and its family members have been implicated in regulating both normal (embryonic and somatic) stem cells and cancer stem cells (CSCs), with recent studies emphasizing that TGF-β signaling-induced stemness of cancer cells plays a crucial role in tumorigenesis. Wang et al. observed that exposure to TGF-β increased the population of breast cancer cells that can form mammospheres in suspension, a feature endowed to a population of cells with characteristics of stem cells. Interestingly, this function of TGF-β was mediated by the miR-181 family, which was upregulated by TGF-β at the posttranscriptional level [19]. Levels of miR-181 family members were elevated in mammospheres grown in undifferentiating conditions, compared with the cells grown in two-dimensional conditions. Overexpression of miR-181a/b was sufficient to induce sphere formation in breast cancer cells via downregulating Ataxia telangiectasia mutated (ATM), a target gene of miR-181, and its substrate CHK2. This study and others demonstrate a new mechanism by which TGF-β-regulated miRNAs enhance tumorigenesis:regulation of cancer stem cell properties in response to TGF-β.

1.2.2 Epithelial-Mesenchymal Transition

The epithelial-mesenchymal transition (EMT) is a well-coordinated process that contributes to tumor invasion and dissemination owing to a junction-free, motile cell phenotype. This process has been observed in transformed epithelial progenitor cells with tumor-propagating ability [20]. TGF-β is considered to be an important inducer of EMT, and promotes EMT in combination with a set of transcription factors including the zinc-finger proteins Snail and Slug, the bHLH factor Twist, the zinc-finger/homeodomain proteins ZEB-1 and -2, and the forkhead factor FoxC3. Goodall and colleagues have reported that TGF-β-regulated miRNAs are involved in specifying the epithelial phenotype, and TGF-β induction of EMT involves alleviating the repressive effects of these miRNAs on EMT drivers [21]. In response to TGF-β treatment, five members of miR-200 family (miR-200a, miR-200b, miR-200c, miR-141, and miR-429) and miR-205 were markedly downregulated in cells that had undergone EMT. These miRNAs cooperatively inhibit expression of the E-Cadherin transcriptional repressors ZEB1 and SIP1, by which sufficiently regulate the EMT; re-repression of ZEB1 and SIP1 via these microRNAs in mesenchymal cells was also sufficient to initiate mesenchymal to epithelial transition (MET). These data suggest that the regulation of miRNAs may be an important step in the pathological process of TGF-β-induced EMT.

TGF-β-regulated miRNAs can also play multiple context-specific roles, acting as either oncogenes or tumor suppressors. MiR-29a, a miRNA which can be suppressed by TGF-β [22, 23], was found to be upregulated in mesenchymal, metastatic RasXT cells relative to epithelial EpRas cells. Overexpression of miR-29a suppressed the expression of tristetraprolin (TTP), a protein involved in the degradation of messenger RNAs with AU-rich 3′-untranslated regions, and led to EMT and metastasis in cooperation with oncogenic Ras signaling. This report by Gebeshuber et al. [24] demonstrating a promotion of EMT and resultant cancer progression contrasts with previous studies describing tumor-suppressive effects of miR-29 family members in hematopoietic and lung tumors [25, 26], emphasizing the context-specific functionality of TGF-β-regulated miRNAs.

1.2.3 Tumor Microenvironment Modulation

Once cancer cells bypass the tumor-suppressive effects of TGF-β, they benefit from this signaling pathway to proliferate and metastasize. At this stage, in addition to the roles that TGF-β plays in changing intrinsic properties of tumor cells, it also facilitates cancer progression through shaping the tumor microenvironment. Accumulating evidence suggests that TGF-β-regulated microRNAs function as important mediators in this process.

Several studies indicate the involvement of TGF-β-regulated microRNAs in tumor microenvironmental regulations through extracellular matrix (ECM) remodeling, angiogenic induction, and myofibroblast generation, which are processes also related to organ fibrosis. Although the precise mechanistic link between fibrosis and cancer progression is not well characterized, it is likely to play an oncogenic role given that organ fibrosis usually induces a microenvironment with inflammation, oxidative stress, and hypoxia, which are also conditions that can facilitate tumor development [27]. Moreover, fibrosis entails the differentiation of mesenchymal precursors to myofibroblasts, which produce matrix metalloproteases, cytokines, and chemokines to promote cancer cell proliferation, invasion, and neoangiogenesis and facilitate tumor development [28]. TGF-β signaling, and specifically TGF-β-regulated microRNAs, plays a crucial role in this fibrotic process. The miR-29 family negatively regulates expression of collagens and adhesion molecules in both fibrosis and the tumor microenvironment. TGF-β reduces expression of the miR-29a/b/c family via a SMAD3-dependent signaling mechanism in multiple epithelial cell types [29, 30], and in both renal and liver fibrosis models, TGF-β has been shown to promote expression of ECM components including collagens I and IV by downregulating miR-29 family members [31]. TGF-β-induced suppression of miR-29b/c yields upregulation of the expression of collagens, promoting the differentiation of myoblasts into myofibroblasts [28]. Conversely, another TGF-β-regulated microRNA, miR-192, has been shown to promote renal fibrosis after its SMAD3-dependent upregulation by TGF-β [32]. Finally, TGF-β also downregulates the programmed cell death 4 (PDCD4) gene through promoting miR-21, eliciting myofibroblast differentiation [28, 33]. These data demonstrate multiple mechanisms by which TGF-β-regulated miRNAs play important roles in shaping the tumor microenvironment.

The immune system plays an important role in suppressing the growth of primary and disseminated tumor cells. In this process, regulatory T (Treg) cells, which maintain immune homeostasis by suppressing the activities of T effector cells, are found to accumulate in tumors, and positively contribute to cancer progression by inhibiting immune surveillance. In a recent study, Yang et al. [34] found that in hepatocellular carcinoma patients, TGF-β downregulated miR-34a expression and thus led to an enhanced expression of the chemokine CCL22 which recruits Tregs to the tumor site. Through this mechanism, the TGF-β-miR-34a-CCL22 axis facilitates portal vein tumor thrombus (PVTT) development by creating an immune-subversive microenvironment. This axis also demonstrates another mechanism by which TGF-β-regulated miRNAs impact tumor progression: immune modulation. Here, we will use this study as an example to demonstrate methods to elucidate the functional role of TGF-β-regulated miRNAs in tumor progression and metastasis.

2 Materials

2.1 Cell Lines and Culture Reagents

1. Minimum Essential Medium (MEM) (Invitrogen-Gibco).
2. Dulbecco's Modified Eagle's Medium (DMEM) (Invitrogen-Gibco).
3. Phosphate buffered saline (Invitrogen-Gibco).
4. Calcium and Magnesium free (PBS) (Invitrogen-Gibco).
5. Fetal bovine serum (FBS) (Invitrogen-Gibco).
6. Penicillin/streptomycin sulfate (Invitrogen-Gibco).
7. Non-essential amino acids (NEAA) (Invitrogen-Gibco).
8. 0.25 % Trypsin/EDTA (Invitrogen-Gibco).
9. HepG2, Hepa1-6, and WRL68 cells (ATCC and cultured in MEM with 10 % FBS plus adding non-essential amino acids, 2 mM L-glutamine, 1 mM sodium pyruvate, and 1500 mg/L sodium bicarbonate).
10. MHCC97 cells (Dr. Zhaoyou Tang).
11. 4T1 cells were (Dr. Robert A. Weinberg).
12. MHCC97 and 4T1 cells (cultured in DMEM with 10 % FBS).

2.2 Antibodies

1. Mouse anti-human CD4-APC (Biolegend).
2. Mouse anti-human CD25-PE (Biolegend).
3. Mouse anti-human FoxP3-PB (Biolegend).
4. PB anti-mouse CD4 (eBioscience).
5. PE anti-mouse CD25 (eBioscience).
6. PE-Cy5 anti-mouse/rat Foxp3 (eBioscience).

2.3 Reagents and Plasmids

1. Anti-sense (AS)-miR-34a and scramble control LNA oligonucleotides (Exiqon, Vedbaek, Denmark).
2. Pre-miR-34a mimic processor, scramble control oligonucleotides, and mirVana miRNA Isolation Kit (Ambion).
3. Poly(A) Polymerase Tailing Kit (EpiCentre).
4. Isolation Buffer for in vitro migration assay: Ca^{2+} and Mg^{2+} free phosphate buffered saline (PBS) supplemented with 0.1 % BSA and 2 mM EDTA.
5. MiR-34a and miR-34b/c expression vectors (constructed using the pTRIPZ lentiviral system from Thermo Scientific Open Biosystems).
6. Dynabeads Regulatory $CD4^+ CD25^+$ T cell kit (Invitrogen).
7. Reverse transcriptase Superscript III (Invitrogen).
8. Bioluminescent substrate (PerkinElmer).
9. 2.5× qPCR Mix (Eppendorf/Qiagen).

3 Methods

3.1 Quantification of MiRNA Expression Level in Response to TGF-β

We used the SYBR-based real-time quantitative PCR (qPCR) method to quantify mature miRNA expression.

Experimental Procedures

1. Culture hepatocellular carcinoma cells, including HepG2, HepG2.2.15, and PVTT-1 cells, at 75–85 % confluency in 10 % FBS DMEM with 5 % CO_2.
2. Treat cells with TGF-β (400 pM) for different durations (0, 3, 6, and 12 h).
3. Extract total RNA, including the small RNA fraction, from cultured cells using a mirVana miRNA Isolation Kit following the below procedures:
   (a) Remove the PBS wash, and add 600 μL Lysis/Binding Solution for 10^7 cells. Cells will lyse immediately upon exposure to the Lysis/Binding Solution. Collect the lysate with a rubber spatula, and pipet it into a tube (*see* **Note 1**).

(b) Vortex or pipet vigorously to completely lyse the cells and to obtain a homogenous lysate.

(c) Add 1/10 volume of miRNA Homogenate Additive to the cell lysate (or homogenate), and mix well by vortexing or inverting the tube several times.

(d) Add a volume of Acid-Phenol:Chloroform that is equal to the lysate volume before addition of the miRNA Homogenate Additive. Vortex for 30–60 s to mix (*see* **Note 2**).

(e) Centrifuge for 5 min at maximum speed (10,000 × *g*) at room temperature to separate the aqueous and organic phases (*see* **Note 3**).

(f) Carefully remove the aqueous (upper) phase without disturbing the lower phase, and transfer it to a fresh tube. At the end of this procedure, RNA can be eluted in nuclease-free water.

4. Add adenines at the 3′ end of RNA molecules lacking a polyA tail by using E. coli polyA polymerase. For 10 μL reaction, add:

   1.0 μL 10× buffer (500 mM Tris–HCl pH 8.0, 2.5 M NaCl, 100 mM $MgCl_2$).

   1.0 μL 10 mM ATP.

   0.5 μL SUPERase-In RNase Inhibitor (Invitrogen, AM2694).

   0.25 μL A-Plus Polymerase.

   1 μg Total RNA (Range at 0.1–5 μg).

   Add DEPC-treated ddH_2O to total volume into 10 μL.

   Incubate on heated top PCR machine at 37 °C for 30 min, then, 4 °C for over 2 min.

5. After oligo-dT annealing, use the reverse transcriptase Superscript III to generate cDNA. This process attaches an universal tag to the 3′ end of cDNAs (*see* **Note 4**). With this universal tag, qPCR can be performed with miRNA-specific forward primers and a reverse universal primer mix.

6. Detect mature miRNAs and U6 utilizing an RT-PCR-based array with gene-specific primers. For 10 μL reaction, add:

   1.0 μL cDNA.

   1.0 μL 10× Universal Primer Mix.

   1.0 μL 10× Gene-Specific Primer (10 μM).

   4.0 μL 2.5× qPCR Mix.

   Add ddH_2O to total volume into 10 μL.

Program the real-time cycler according to this cycling condition:

95 °C 5 min.

40 Cycles of the below three steps:

94 °C 15 s.

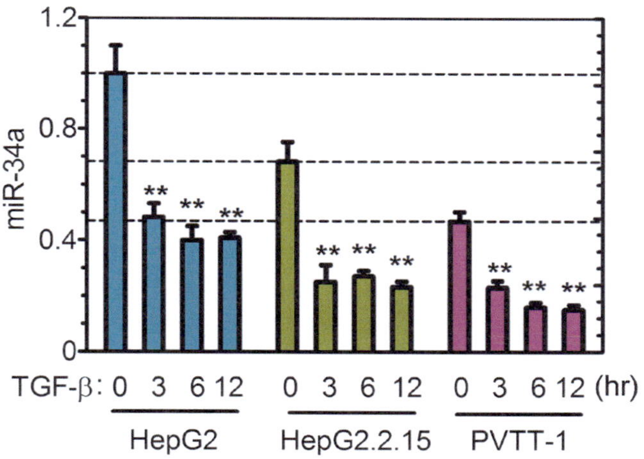

Fig. 1 The relative level of miR-34a in different cell lines mock-treated or treated with TGF-β1 for indicated time points was measured by qRT-PCR and normalized to U6. $N = 3$. ** $p < 0.01$ calculated by student's *t*-test

55 °C 30 s.

70 °C 15 s → Perform fluorescence data collection.

An example of the results produced is shown in Fig. 1, in which we found that TGF-β could downregulate miR-34a.

3.2 In Vitro Migration Assay

In this study, we found that TGF-β-regulated miR-34a can target the chemokine CCL22, which has been implicated in the tumorigenic process by binding to its receptor CCR4 on the surface of Treg cells, consequently recruiting those immunosuppressive cells to the tumor microenvironment. To test the hypothesis that the miR-34a-CCL22 axis can regulate Treg cell recruitment, we used an in vitro migration assay to examine the ability of conditioned culture media from PVTT-1 cells to mobilize $CD4^+CD25^+$Treg cells.

Experimental Procedures

1. Purify and separate $CD4^+CD25^+$ T cells and $CD4^+CD25^-$ T cells by using Dynabeads Regulatory $CD4^+CD25^+$ T cell kits (*see* **Note 5**) from human blood monocyte cells (PBMC).
   (a) Isolate $CD4^+$ T cells: Transfer 1×10^8 PBMC in 500 μL of Isolation Buffer to a tube. Add 200 μL of heat inactivated FBS/FCS and 200 μL of Antibody Mix for human CD4. Mix well and incubate for 20 min at 2–8 °C. Wash the cells by adding 4 mL of Isolation Buffer. Mix well by tilting the tube several times and centrifuge at $350 \times g$ for 8 min. Discard the supernatant. Resuspend the cells in 2 mL of Isolation Buffer. Add 1 mL of pre-washed and resuspended Depletion MyOne™ Dynabeads®. Incubate for 15 min

at 18–25 °C with gentle tilting and rotation. Add 3 mL of Isolation Buffer. Resuspend the bead-bound cells thoroughly by pipetting >10 times using a pipette with a narrow tip opening. Place the tube in the magnet for 2 min. Transfer the supernatant containing the untouched human $CD4^+$ T cells, to a new tube. Spin down the cells at $350 \times g$ for 8 min and resuspend the cells in Isolation Buffer to 1.5×10^7 $CD4^+$ T cells/mL.

(b) Isolate $CD4^+CD25^+$ Cells: Add 200 μL of pre-washed and resuspended Dynabeads® CD25 per 1.5×10^7 $CD4^+$ cells. Mix well and incubate for 25 min at 2–8 °C with rolling and tilting. Place the tube in the magnet for a minimum of 1 min. Carefully remove the supernatant containing the $CD4^+CD25^-$ (effector) cells. Remove the tube from the magnet and carefully resuspend the bead-bound cells in 5 mL of Isolation Buffer by gently shaking the tube instead of pipetting the cells. Place the tube in the magnet for a minimum of 1 min. Remove and discard the supernatant. Wash one more time by carefully resuspending the cells in 5 mL of Isolation Buffer and gently shaking the tube. Place the tube in the magnet for a minimum of 1 min. Remove and discard the supernatant. Resuspend the bead-bound cells in 500 μL RPMI with 1 % FBS.

(c) Release of $CD4^+CD25^+$ Regulatory T cells: Add 80 μL of DETACHaBEAD reagent and incubate for 45 min at room temperature with tilting and rotation. Place the tube in the magnet for a minimum of 1 min. Carefully remove the supernatant containing the $CD4^+CD25^+$ cells to a new tube.

(d) Wash the Dynabeads CD25 twice in 1 mL of RPMI with 1 % FBS to obtain the residual cells and collect the supernatant after separation on a magnet. Add 5 mL of RPMI with 1 % FBS to wash the cells, followed by centrifugation at $350 \times g$ for 8 min. Remove all visible liquid without disturbing the pellet.

(e) Repeat **step d**. Discard the supernatant and resuspend the cells in a preferred cell culture medium.

2. Collect the culture media from PVTT-1 cells, which are treated with different concentration of Anti-sense (AS)-miR-34a oligos for 24 h. Incubate 600 μL per well of threefold diluted culture media in the bottom chamber of a 24-well plate.
3. Resuspend $CD4^+CD25^+$ T cells and $CD4^+CD25^-$ T cells at 2×10^6 cells/mL. Add 200 μL of T cells to the upper chamber of the 3.0 μm trans wells.

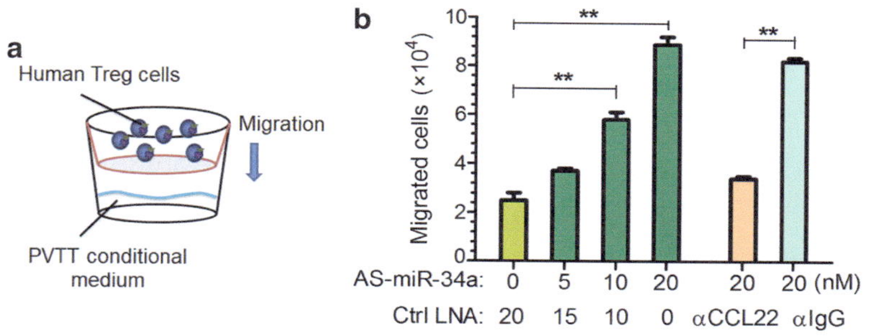

Fig. 2 The miR-34a-CCL22 pathway regulates $CD4^+CD25^+$ Treg cell migration. 1×10^5 human $CD4^+CD25^+$ Treg cells were tested by transwell assay (**a**) for their migration towards culture media of PVTT-1 cells harvested 24 h after the cells were transfected with different amounts of AS-miR-34a or control oligos as indicated. CCL22 neutralizing or control antibody was added to the culture media of AS-miR-34a oligos transfected PVTT-1 cells to determine the effect on Treg cell migration. Four hours after plating of the T cells in the upper chamber, migrated Treg cells were quantified by Cell Counter analysis (**b**). $N=3$, mean ± s.d. $^{**}p<0.01$ (Student *t*-test)

4. Incubate at 37 °C for 4 h. Remove the upper chambers and collect the cells from the bottom chamber. Add Cell staining buffer to ~2 mL and centrifuge at 350 × *g* for 5 min, discard supernatant.
5. Add reagents that block Fc receptors for reducing nonspecific immunofluorescent staining.
6. Add anti-human CD4-APC antibody in the T cells for 15–20 min on ice. Wash 2× with at least 2 mL of Cell Staining Buffer by centrifugation at 350 × *g* for 5 min, discard supernatant. Resuspend cell pellet in 0.5 mL of stain buffer and analyze cells in a FACSCanto flow cytometer. An example of the results is shown in Fig. 2.

3.3 An Animal Model of Hepatic Metastasis

To investigate the functional role of the miR-34a-CCL22-Treg link in liver cancer metastasis, we employed a model system utilizing a murine liver tumor cell line, Hepa1-6, originally derived from the C57BL/6J mouse strain. C57BL/6J mice are immune-competent with fully functional T cell lineages, which when combined with their syngeneic liver tumor cell line permit us to study the functional role of Treg cells in tumor metastasis.

Experimental Procedures

1. Infect Hepa1-6 cells that stably express a luciferase reporter with recombinant Lentivirus to overexpress pri-miR-34a.
2. Anesthetize male C57/BL6 mice at the age of 4 to 8 weeks with Ketamine (100 mg/kg) plus Xylazine (5 mg/kg)

administered by intraperitoneal injection. Administer sterile ophthalmic lubrication (without antibiotics) during anesthesia. Keep the mice on a heating pad throughout the procedure (before, during, and after the surgery), and monitor the mice until they recover from anesthesia (usually in ~30 min).

3. Shave the skin around the left upper abdomen of the mice, and clean with an alcohol derivative three times. Make an incision in the left side of the midsection about 1–1½ centimeters below the bottom of the ribcage of the mice to isolate the spleen.
4. Make a loop of 4-0 surgical suture and wrap the suture around the bottom of spleen. Gently close the loop around the spleen until it is loosely tied. Insert needle and inject cells into spleen; when finished, tighten and knot suture (*see* **Note 6**).
5. Close the skin with either wound clips or monofilament non-absorbable suture. Remove sutures and clips in 7–10 days. Following recovery from anesthesia and surgery, keep the animals in warm chambers with a 12-h light/12-h dark cycle within the mouse room, with access to rodent chow and water ad libitum.
6. Monitor the formation of abdominal metastases by the appearance of luciferase activity from the tumor cells stably expressing the luciferase reporter (Luc2 vector). Use a Xenogen Imaging System to monitor the bioluminescence every 3–4 days. After intraperitoneal injection of bioluminescent substrate at the dosage of 30 μg/mouse, anesthetize the mice and measure the luciferase signal in Xenogen Imaging monitor (*see* **Note 7**). An example is shown in Fig. 3.

4 Notes

1. Use the low end of the range (~300 μL) for small numbers of cells (hundreds), and use closer to 600 μL when isolating RNA from larger numbers of cells (thousands-millions).
2. For example: if the lysate volume is 300 μL, add 30 μL of miRNA Homogenate Additive, add 300 μL of Acid-Phenol: Chloroform.
3. After centrifugation, the interphase should be compact; if it is not, repeat the centrifugation.
4. The cDNA product should be tenfold diluted in ddH_2O. Alternatively, cDNA can be diluted in Tricine-EDTA buffer (10 mM Tricine, 1.0 mM EDTA, adjust pH to 8.5 with NaOH).
5. Using this kit, the purified $CD4^+CD25^+$ T cells actually should be $CD4^+CD25^{high}$ T cells, which were confirmed to be almost all $FoxP3^+$Treg cells.

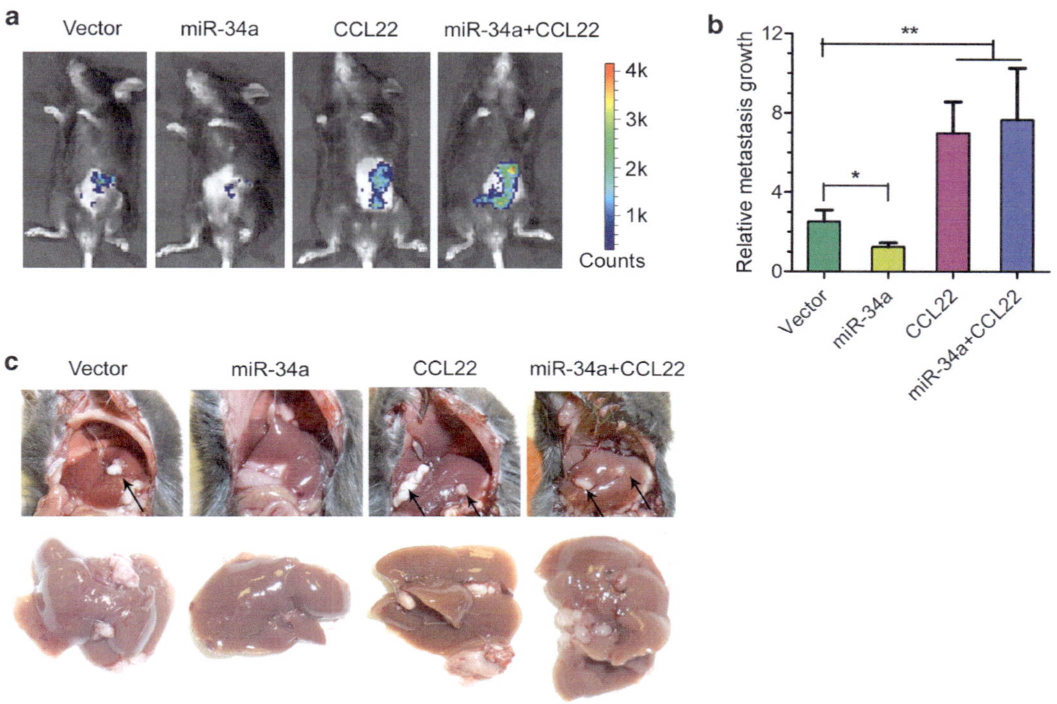

Fig. 3 Impact of miR-34a and CCL22 on HCC intrahepatic metastasis. (**a**) Four Hepa1-6 cell populations, pTRIPZ-miR-34a or pTRIPZ-mock together with pcDNA3-CCL22-ORF or pcDNA3 vector as indicated were stably transfected into Hepa1-6 cells, co-expressing a luciferase reporter were introduced into the C57BL/6J mice via intrasplenic injection. After 21 days, luciferase signals derived from abdominal metastatic tumor growth as shown in the representative images were determined by the Xenogen IVIS Lumina system. (**b**) Luciferase signal levels from the four groups of mice shown in panel E were normalized to that of control mice without tumor growth, and computed as the values for relative metastatic growth by each of the indicated Hepa1-6 cell populations. *$p<0.05$, **$p<0.01$ (Student *t*-test). $N=4$ per experimental group; experiment repeated three times. (**c**) Representative pictures showing the formation of tumor metastases in the mouse liver by each of the indicated Hepa1-6 populations. *Arrows* indicate metastatic tumors in the mouse liver

6. In this step, the tumor cells are very easy to leak out from the spleen. Make sure to tighten and knot suture to avoid cells exiting the spleen into the peritoneal cavity.
7. To get a better view of imaging, the skin can be shaved again before this monitoring procedure.

References

1. Ikushima H, Miyazono K (2010) TGFbeta signalling: a complex web in cancer progression. Nat Rev Cancer 10(6):415–424
2. Massague J (2012) TGFbeta signalling in context. Nat Rev Mol Cell Biol 13(10):616–630
3. Massague J (2008) TGFbeta in cancer. Cell 134(2):215–230, PMCID: 3512574
4. Bartel DP (2004) MicroRNAs: genomics, biogenesis, mechanism, and function. Cell 116(2):281–297

5. Siomi H, Siomi MC (2010) Posttranscriptional regulation of microRNA biogenesis in animals. Mol Cell 38(3):323–332
6. Farazi TA, Hoell JI, Morozov P, Tuschl T (2013) MicroRNAs in human cancer. Adv Exp Med Biol 774:1–20
7. Rogler CE, Levoci L, Ader T, Massimi A, Tchaikovskaya T, Norel R et al (2009) MicroRNA-23b cluster microRNAs regulate transforming growth factor-beta/bone morphogenetic protein signaling and liver stem cell differentiation by targeting Smads. Hepatology 50(2):575–584
8. Shan H, Zhang Y, Lu Y, Zhang Y, Pan Z, Cai B et al (2009) Down regulation of miR-133 and miR-590 contributes to nicotine-induced atrial remodelling in canines. Cardiovasc Res 83(3):465–472
9. Petrocca F, Vecchione A, Croce CM (2008) Emerging role of miR-106b-25/miR-17-92 clusters in the control of transforming growth factor beta signaling. Cancer Res 68(20): 8191–8194
10. Davis BN, Hilyard AC, Lagna G, Hata A (2008) SMAD proteins control DROSHA-mediated microRNA maturation. Nature 454(7200):56–61, PMCID: 2653422
11. Davis-Dusenbery BN, Hata A (2011) Smad-mediated miRNA processing: a critical role for a conserved RNA sequence. RNA Biol 8(1):71–76, PMCID: 3230544
12. Zhong X, Chung AC, Chen HY, Meng XM, Lan HY (2011) Smad3-mediated upregulation of miR-21 promotes renal fibrosis. J Am Soc Nephrol 22(9):1668–1681, PMCID: 3171938
13. Kim VN, Han J, Siomi MC (2009) Biogenesis of small RNAs in animals. Nat Rev Mol Cell Biol 10(2):126–139
14. Lee Y, Ahn C, Han J, Choi H, Kim J, Yim J et al (2003) The nuclear RNase III Drosha initiates microRNA processing. Nature 425(6956):415–419
15. Lund E, Guttinger S, Calado A, Dahlberg JE, Kutay U (2004) Nuclear export of microRNA precursors. Science 303(5654):95–98
16. Gregory RI, Yan KP, Amuthan G, Chendrimada T, Doratotaj B, Cooch N et al (2004) The microprocessor complex mediates the genesis of microRNAs. Nature 432(7014):235–240
17. Kato M, Putta S, Wang M, Yuan H, Lanting L, Nair I et al (2009) TGF-beta activates Akt kinase through a microRNA-dependent amplifying circuit targeting PTEN. Nat Cell Biol 11(7):881–889, PMCID: 2744130
18. Kato M, Zhang J, Wang M, Lanting L, Yuan H, Rossi JJ et al (2007) MicroRNA-192 in diabetic kidney glomeruli and its function in TGF-beta-induced collagen expression via inhibition of E-box repressors. Proc Natl Acad Sci U S A 104(9):3432–3437, PMCID: 1805579
19. Wang Y, Yu Y, Tsuyada A, Ren X, Wu X, Stubblefield K et al (2011) Transforming growth factor-beta regulates the sphere-initiating stem cell-like feature in breast cancer through miRNA-181 and ATM. Oncogene 30(12):1470–1480, PMCID: 3063856
20. Mani SA, Guo W, Liao MJ, Eaton EN, Ayyanan A, Zhou AY et al (2008) The epithelial-mesenchymal transition generates cells with properties of stem cells. Cell 133(4):704–715, PMCID: 2728032
21. Gregory PA, Bert AG, Paterson EL, Barry SC, Tsykin A, Farshid G et al (2008) The miR-200 family and miR-205 regulate epithelial to mesenchymal transition by targeting ZEB1 and SIP1. Nat Cell Biol 10(5):593–601
22. Wang B, Komers R, Carew R, Winbanks CE, Xu B, Herman-Edelstein M et al (2012) Suppression of microRNA-29 expression by TGF-beta1 promotes collagen expression and renal fibrosis. J Am Soc Nephrol 23(2):252–265, PMCID: 3269175
23. Wang J, Wang Y, Wang Y, Ma Y, Lan Y, Yang X (2013) TGF-beta regulated miR-29a promotes angiogenesis through targeting PTEN in endothelium. J Biol Chem 288(15): 10418–10426
24. Gebeshuber CA, Zatloukal K, Martinez J (2009) miR-29a suppresses tristetraprolin, which is a regulator of epithelial polarity and metastasis. EMBO Rep 10(4):400–405, PMCID: 2672883
25. Fabbri M, Garzon R, Cimmino A, Liu Z, Zanesi N, Callegari E et al (2007) MicroRNA-29 family reverts aberrant methylation in lung cancer by targeting DNA methyltransferases 3A and 3B. Proc Natl Acad Sci U S A 104(40):15805–15810, PMCID: 2000384
26. Pekarsky Y, Santanam U, Cimmino A, Palamarchuk A, Efanov A, Maximov V et al (2006) Tcl1 expression in chronic lymphocytic leukemia is regulated by miR-29 and miR-181. Cancer Res 66(24):11590–11593
27. Lopez-Novoa JM, Nieto MA (2009) Inflammation and EMT: an alliance towards organ fibrosis and cancer progression. EMBO Mol Med 1(6-7):303–314, PMCID: 3378143
28. Zhou L, Wang L, Lu L, Jiang P, Sun H, Wang H (2012) Inhibition of miR-29 by TGF-beta-Smad3 signaling through dual mechanisms promotes transdifferentiation of mouse myoblasts into myofibroblasts. PLoS One 7(3): e33766, PMCID: 3306299

29. Qin W, Chung AC, Huang XR, Meng XM, Hui DS, Yu CM et al (2011) TGF-beta/Smad3 signaling promotes renal fibrosis by inhibiting miR-29. J Am Soc Nephrol 22(8):1462–1474, PMCID: 3148701
30. Winbanks CE, Wang B, Beyer C, Koh P, White L, Kantharidis P et al (2011) TGF-beta regulates miR-206 and miR-29 to control myogenic differentiation through regulation of HDAC4. J Biol Chem 286(16):13805–13814, PMCID: 3077581
31. Noetel A, Kwiecinski M, Elfimova N, Huang J, Odenthal M (2012) microRNA are central players in anti- and profibrotic gene regulation during liver fibrosis. Front Physiol 3:49, PMCID: 3307137
32. Chung AC, Huang XR, Meng X, Lan HY (2010) miR-192 mediates TGF-beta/Smad3-driven renal fibrosis. J Am Soc Nephrol 21(8):1317–1325, PMCID: 2938591
33. Yao Q, Cao S, Li C, Mengesha A, Kong B, Wei M (2011) Micro-RNA-21 regulates TGF-beta-induced myofibroblast differentiation by targeting PDCD4 in tumor-stroma interaction. Int J Cancer 128(8): 1783–1792
34. Yang P, Li QJ, Feng Y, Zhang Y, Markowitz GJ, Ning S et al (2012) TGF-beta-miR-34a-CCL22 signaling-induced Treg cell recruitment promotes venous metastases of HBV-positive hepatocellular carcinoma. Cancer Cell 22(3): 291–303, PMCID: 3443566

Chapter 22

Epigenomic Regulation of Smad1 Signaling During Cellular Senescence Induced by Ras Activation

Atsushi Kaneda, Aya Nonaka, Takanori Fujita, Ryota Yamanaka, Mai Fujimoto, Kohei Miyazono, and Hiroyuki Aburatani

Abstract

Epigenomic modification plays important roles in regulating gene expression during development, differentiation, and cellular senescence. When oncogenes are activated, cells fall into stable growth arrest to block cellular proliferation, which is called oncogene-induced senescence. We recently identified through genome-wide analyses that Bmp2-Smad1 signal and its regulation by harmonized epigenomic alteration play an important role in *Ras*-induced senescence of mouse embryonic fibroblasts. We describe in this chapter the methods for analyses of epigenomic alteration and Smad1 targets on genome-wide scale.

Key words Epigenetics, Smad1, Bmp2, Histone, Chromatin immunoprecipitation (ChIP)

1 Introduction

Epigenome is modification of the genome, which does not involve alterations in the DNA sequence itself, but is heritable during cell division. The modifications include DNA methylation and histone modification, and play important roles in controlling gene expression during development, differentiation, and cellular senescence [1–3].

Replicative senescence is a state of stable growth arrest of cells, which is due to the limited capacity of proliferation in cultured primary cells [4, 5]. Stresses such as enforced expression of cancer-promoting genes are known to provoke premature form of cellular senescence, where cellular proliferation is irreversibly blocked [6, 7]. These stable states of proliferation arrest, the so-called oncogene-induced senescence, can act as a natural barrier to cancer progression, like cell death programs such as apoptosis and autophagy [8–10].

We recently performed genome-wide analyses of epigenomic and gene expression changes in *Ras*-induced senescence using mouse embryonic fibroblasts. The study revealed that Bmp2-Smad1

Xin-Hua Feng et al. (eds.), *TGF-β Signaling: Methods and Protocols*, Methods in Molecular Biology, vol. 1344, DOI 10.1007/978-1-4939-2966-5_22,

signal is critical in *Ras*-induced senescence, and the signal is regulated by coordinated epigenomic alteration [11]. The methods for genome-wide analyses of Smad1 targets and epigenomic alteration during Ras-induced senescence are presented here to illustrate alteration of epigenomic regulation of Bmp2-Smad1 signal. We hope these methods would provide researchers with necessary tools for epigenomic analysis of Smad1 signal.

2 Materials

2.1 Cells

Mouse embryonic fibroblast (MEF) is established from 13.5 embryonic day embryos [12], following the Subheading 3.1 described below. After cells are passed twice (MEFp2), cells are infected with retroviruses for 48 h. Then cells are exposed to 4 μg/mL puromycin during Day 0–Day 3 for selection. Materials to be prepared are as follows.

1. PBS/antibiotics (PBS containing 100 units/mL penicillin and 100 g/mL streptomycin sulfate):

   PBS with 1/100 volume of penicillin/streptomycin (Invitrogen 15140-122).

2. PBS/0.02%EDTA:

   PBS with 1/930 volume of 0.5 M EDTA.

3. Culture medium:

   Dulbecco's modified Eagle's medium (DMEM) containing 100 units/mL penicillin and 100 g/mL streptomycin sulfate, supplemented with 10 % (vol/vol) fetal bovine serum.

4. rBMP2 protein (R&D systems #355-BM).

2.2 Retroviruses

To induce senescence in MEF, we constructed retroviral vector by cloning mutated HRAS (RasV12) cDNA into a retroviral vector, e.g., pMX-puro (a kind gift of T. Kitamura, The University of Tokyo, Japan) [13], by using reverse-transcription PCR product from SK-BR3 cell (American Tissue Culture Collection) RNA. To prepare retroviruses, we transfected the RasV12 vector into plat-E packaging cells (a kind gift of T. Kitamura) using FuGENE 6 Transfection Reagent (Roche, Mannheim, Germany), but other packaging kits are also commercially available.

2.3 Buffers for Senescence-Associated β-Galactosidase (SA-βgal) Analysis

1. Fixation Buffer:

   2 % (w/v) formaldehyde, 0.2 % (w/v) glutaraldehyde in PBS.

2. Staining Solution:

   40 mM citric acid/sodium phosphate, pH 6.0, 150 mM NaCl, 2.0 mM $MgCl_2$, 5 mM $K_4[Fe(CN)_6]$, 5 mM $K_3[Fe(CN)_6]$, 1 mg/mL X-gal.

2.4 Buffers for Chromatin Immunoprecipitation (ChIP)

1. 100× Complete proteinase inhibitor.

   Dissolve one tablet of Complete proteinase inhibitor (Roche) in 10 mM Tris–HCl, pH 8.0, and keep at −20 °C.
2. 100× PMSF.

   Dissolve 34 mg PMSF in 1 mL DMSO and store aliquots at −20 °C.
3. SDS Lysis Buffer:

   10 mM Tris–HCl, pH 8.0, 150 mM NaCl, 1 % SDS, 1 mM EDTA, 1× Complete proteinase inhibitor (*see* **Note 1**).
4. ChIP Dilution Buffer:

   20 mM Tris–HCl, pH 8.0, 150 mM NaCl, 1 mM EDTA, 1 % Triton X-100, 1× Complete proteinase inhibitor (*see* **Note 1**).
5. Wash Buffer 1:

   20 mM Tris–HCl, pH 8.0, 500 mM NaCl, 2 mM EDTA, 1 % Triton X-100, 0.1 % SDS, 1 mM PMSF (*see* **Note 2**).
6. Wash Buffer 2:

   10 mM Tris–HCl, pH 8.0, 1 mM EDTA, 0.25 M LiCl, 0.5 % Na-Deoxycholate, 0.5 % NP-40, 1 mM PMSF (*see* **Note 2**).
7. Elution Buffer:

   50 mM Tris–HCl, pH 8.0, 10 mM EDTA, 1 % SDS.
8. 24:1 Chloroform/isoamyl alcohol:

   Mix chloroform and isoamyl alcohol at 24:1.
9. 25:24:1 Phenol/chloroform/isoamyl alcohol:

   Mix phenol, chloroform, and isoamyl alcohol at 25:24:1.
10. 20 mM DSG (Disuccinimidyl Glutarate, PIERCE, MW 326.26):

    Dissolve 6.5252 mg in 1 mL of DMSO.

2.5 Antibodies

1. Anti-H3K4me3.

   Abcam ab8580 1 μg/reaction, or ActivMotif #39159 2 μL/reaction.
2. Anti-H3K27me3.

   Millipore 07-449, 2 μg/reaction.
3. Anti-Smad1.

   BioMatrix, mouse monoclonal, 2.5 μg/reaction.

3 Methods

3.1 Establishment of MEF Cells

1. Cross C57/B6 mice or other strains by timed-mating, and sacrifice the pregnant mouse on embryonic day 13.5.
2. Wash uterus in PBS/antibiotics, in 10-cm dish, and wash once more in a new 10-cm dish.

3. Wash embryos in PBS/antibiotics in a new dish with uterus and placenta taken away.
4. Wash embryos in PBS/antibiotics in a new dish with amnion taken away.
5. Wash embryos in PBS/antibiotics in a new dish with head and red organs taken away, and wash once more in a new dish.
6. Approximately four embryos were minced together using scalpels, and collect them in a 50-mL tube with 16 mL of PBS/0.02 % EDTA.
7. Add 4 mL of 0.25 % trypsin/EDTA, and incubate for 15 min at 37 °C to digest (*see* **Note 3**).
8. Add 16 mL of PBS/0.02 % EDTA and 4 mL of 0.25 % trypsin/EDTA, and incubate for another 15 min at 37 °C (*see* **Note 4**).
9. Transfer to a cell culture flask containing 40 mL of culture medium. Pipette gently, and stand the flask for 5 min.
10. Collect 50–60 mL of supernatant into two 50-mL tubes, and centrifuge at $350 \times g$ for 5 min at RT.
11. Aspirate the supernatant, and suspend the cells with culture medium. Count the cells, seed them at 1.5×10^7 cells/10-cm dish, and incubate at 37 °C in 5 % CO_2. Change the medium after 4–6 h.
12. Collect the cells on the next morning and stock them as MEFp0. After passing twice (MEFp2), infect cells with retroviruses.

3.2 Senescence-Associated β-Galactosidase (SA-βgal) Analysis

MEFp2 and infected cells on days 7–10 underwent SA-βgal staining [14].

1. Cells were washed twice with PBS.
2. Cells were immersed in the Fixation Buffer for 5 min.
3. After two additional PBS washes, the cells were allowed to stain overnight in the Staining Solution.
4. After 12–16 h, check the blue color in senescent cells using a light microscopy (Fig. 1). If the blue color is detected, cells are washed with distilled water three times, and mounted.

3.3 ChIP for Histone Modification

MEF cells without viral infection (MEFp2 cells) and those with RasV12 infection at days 7–10 are cross-linked with 1 % formaldehyde and undergo ChIP analysis for histone modification. For ChIP analysis of Smad1 binding, stronger fixation protocol is necessary (*see* Subheading 3.4). ChIP'ed DNA samples were obtained as below, using anti-H3K4me3 and H3K27me3 antibodies.

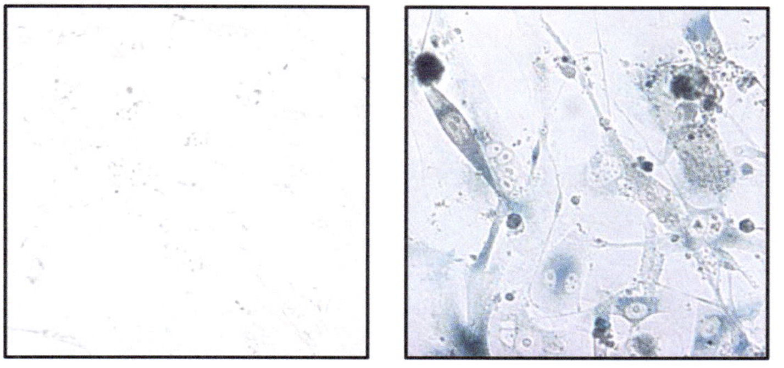

Fig. 1 SA-βgal staining. While growing MEFp2 cells are negative for SA-βgal staining (*MEFp2*), blue staining is observed in cells senesced by mutant Ras (*RasV12*)

1. *Cell Preparation:* cells are cultured at 70–80 % confluency on 15-cm dish.
2. *Day 1, cross-linking:* add formaldehyde to culture dishes at final concentration of 1 % (540 μL of 37 % HCHO for 20 mL of culture medium, 270 μL for 10 mL).
3. Incubate at room temperature for 10 min, gently swirling culture dishes at approx. 40 rpm on shaker.
4. Add 2.5 M Glycine to a final concentration of 0.2 M (1.8 mL for 20 mL, 0.9 mL for 10 mL) (*see* **Note 5**).
5. Incubate at room temperature for 5 min, gently swirling culture dishes at approx. 40 rpm on shaker.
6. Rinse cells three times with 5 mL of ice-cold PBS (*see* **Note 6**).
7. Harvest cells using a cell scraper.
8. Centrifuge cells at 700 × *g* for 5 min at 4 °C, and aspirate the supernatant.
9. Flash-freeze cells in liquid nitrogen. The frozen pellet can be stored at −80 °C.
10. *Day 1, cell sonication:* resuspend the cell pellet (from one 15-cm dish) in 300 μL of SDS Lysis Buffer with 3 μL of 100× Complete proteinase inhibitor (*see* **Note 1**).
11. Sonicate with Tomy UD-201 for 4 cycles (3–5 cycles) of 15 s at 50 % duty, output level 2 to obtain 300–700 bp DNA, or more cycles to obtain shorter fraction (*see* **Note 7**).
12. Centrifuge at 13,000 × *g* at 4 °C for 10 min.
13. Collect supernatant in 15-mL tube, and add 3 mL of ChIP Dilution Buffer with 30 μL of 100× Complete proteinase

inhibitor. Supernatant or diluted supernatant can be kept at −80 °C.

14. *Day 1, exclusion of nonspecific binding:* add 45 μL of protein A-Sepharose beads to 3 mL of diluted sonication sample. Use cell saver pipette tips to handle beads.
15. Rotate for 30 min at 4 °C.
16. Centrifuge at 700 × *g* for 2 min at 4 °C. Keep on ice for 5 min.
17. Collect 1 mL of the supernatant for each ChIP. (Three tubes could be prepared at the most.) Keep the remaining 100–300 μL of the supernatant at 4 °C for Input preparation. Add 1 mL of the supernatant to antibody-bound beads in a low-retention tube (**step 22**) and rotate at 4 °C for 4 h–O/N.
18. *Day 1, binding antibody to beads:* this step can be started before cell sonication. Equilibrate 20 μL protein A-Sepharose with 680 μL of ChIP Dilution Buffer with 6.8 μL of 100× Complete proteinase inhibitor (*see* **Note 1**). Use cell saver pipette tips, and 1.5-mL low-retention tubes.
19. Add antibody (e.g., anti-H3K4me3 and anti-H3K27me3), and rotate tubes at 4 °C for at least 1.5 h–O/N. Cell sonication (**steps 10–13**) can be performed during this step.
20. Centrifuge at 700 × *g* at 4 °C for 2 min.
21. Wash the Antibody-prebound beads three times with 500 μL of ChIP dilution buffer with 1× Complete proteinase inhibitor. Mix by inverting the tube several times, and centrifuge at 700 × *g* for 2 min at 4 °C. Discard the supernatant or eluate.
22. To the beads, add 1 mL of the sonicated cell lysate (**step 17**).
23. Rotate at 4 °C for 4 h–O/N.
24. *Day 2, washing, eluting, and reverse cross-linking:* spin the beads at 400 × *g* for 2 min at 4 °C, and waste the supernatant.
25. Add 550–600 μL of ChIP Dilution Buffer with 6 μL of 100× Complete proteinase inhibitor and transfer to Spin-X centrifuge tube (Costar #8162) (*see* **Note 8**).
26. Rotate for 5 min at RT.
27. Centrifuge at 350 × *g* at 4 °C for 2 min. Waste the eluate.
28. Wash with 600 μL of ChIP Dilution Buffer with 100× Complete proteinase inhibitor three times more (*see* **Note 9**). Rotate for 5 min at RT. Centrifuge at 350 × *g* at 4 °C for 2 min. Waste the eluate.
29. Add 600 μL of Wash Buffer 1 with 6 μL of 100× PMSF (*see* **Note 2**).
30. Rotate for 5 min at RT (*see* **Note 10**).
31. Centrifuge at 350 × *g* at 4 °C for 2 min. Waste the eluate.

32. Wash with 600 μL of Wash Buffer 1 with 6 μL of 100× PMSF once more (*see* **Note 11**).
33. Add 600 μL of Wash Buffer 2 with 6 μL of 100× PMSF (*see* **Note 2**).
34. Rotate for 5 min at RT.
35. Centrifuge at 350 ×*g* at 4 °C for 2 min. Waste the eluate.
36. Wash with Wash Buffer 2 with 6 μL of 100× PMSF once more (*see* **Note 12**).
37. Add 600 μL of TE.
38. Rotate for 1 min at RT.
39. Centrifuge at 350 ×*g* at 4 °C for 1 min.
40. Wash twice more with 600 μL of TE (*see* **Note 13**).
41. Add 300 μL of Elution Buffer to the column and transfer to new 1.5-mL tube.
42. Add Pronase to a final concentration of 1.5 mg/mL (*see* **Note 14**). Input samples should be reacted with RNaseA in prior (*see* **Note 10**), and Pronase should be added to the Input sample as well as ChIP'ed sample at this step.
43. Rotate at 42 °C for 2 h.
44. Incubate at 65 °C for 6 h–O/N (*see* **Note 15**).
45. *Day 3, purification:* spin at 700 ×*g*, 4 °C for 2 min and collect the supernatant. Add LiCl to a final concentration of 0.8 M (*see* **Note 16**).
46. Add 1× volume of 25:24:1 Phenol/chloroform/isoamyl alcohol, and vortex.
47. Centrifuge at 20,000 ×*g* for 5 min at 4 °C. Collect the water layer.
48. Add 1× volume of 24:1 Chloroform/isoamyl alcohol, and vortex.
49. Centrifuge at 20,000 ×*g* for 5 min at 4 °C. Collect the water layer.
50. Add 1 μL of 5 mg/mL glycogen. Add 2.5× volume of ethanol, vortex, and incubate for 20 min at −80 °C.
51. Centrifuge at 20,000 ×*g* at 4 °C for 20 min. Wash the pellet with 170 μL of 70 % EtOH.
52. Centrifuge at 20,000 ×*g* at 4 °C for 7 min. Dry the pellet and resuspend with 20–30 μL of 10 mM Tris–HCl, pH 8.0.
53. Measure DNA concentration.
54. Check the enrichment of control regions by ChIP-PCR (Fig. 2). For ChIP-PCR, dilute 4 μL of sample (*see* **Note 17**). ChIP-PCR protocol is described in Subheading 3.5.

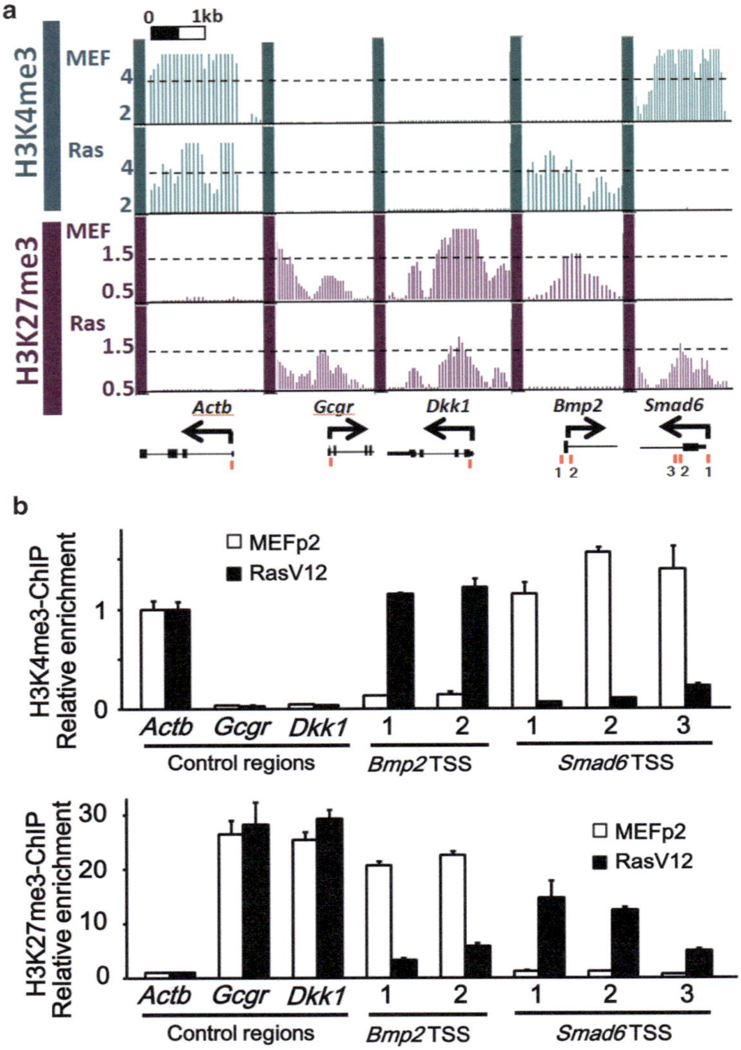

Fig. 2 ChIP for histone modification. (**a**) Epigenomic maps around TSS of *Actb*, *Gcgr*, *Dkk1*, *Bmp2*, and *Smad6*. Alteration of histone modification status in *Bmp2* and *Smad6* was shown. *Red bars*: regions for ChIP-PCR. *Y-axis*: the number of mapped reads for the region per total million reads. (**b**) Real-time ChIP-PCR for H3K4me3 and H3K27me3. In each sample of MEFp2 and RasV12 cells, relative enrichment compared to *Actb* was shown. These figures were modified from Figs. 3 and 4 of our previous report [11]

3.4 ChIP Using Anti-Smad1 Antibody

MEFp2 cells are starved for 16 h and exposed to rBMP2 protein at 25 ng/mL in serum-free medium for 1.5 h to activate Bmp2-Smad1 signal. Cells are fixed with 1 mM DSG and 1 % formalin as below.

1. *Day 1, cross-linking:* wash cells on 15-cm dish with PBS three times.

2. Expose cells to 1 mM DSG (by adding 1 mL of 20 mM DSG to 19 mL PBS). Incubate at room temperature for 20–30 min, gently swirling culture dishes at approx. 40 rpm on shaker.
3. Wash with PBS twice.
4. Add formaldehyde to culture dishes at final concentration of 1 % (by adding 540 μL of 37 % HCHO in 20 mL PBS, or 270 μL in 10 mL).
5. Incubate at room temperature for 10 min, gently swirling culture dishes at approx. 40 rpm on shaker.
6. Follow **step 4** of Subheading 3.3. For antibody, use anti-Smad1 antibody. As for beads, use protein G-Sepharose beads instead of protein A-Sepharose.
7. Check the enrichment of control regions by ChIP-PCR (Fig. 3). ChIP-PCR protocol is described in Subheading 3.5.

3.5 Quantitative Real-Time ChIP-PCR

Real-time PCR is performed to amplify ChIP samples by using SYBR Green and iCycler Thermal Cycler (Bio-Rad Laboratories) or other real-time PCR methods, and quantify them by drawing the standard curve using control samples such as 20, 2, 0.2, and 0.02 ng of sonicated genomic DNA of MEF. The quantity of ChIP'ed DNA of an analyzed region is compared to that of Input sample to calculate ratio against Input (Input%). The PCR primers and conditions are shown in Table 1.

3.6 ChIP-Sequencing

Sample preparation for ChIP-sequencing is performed according to the manufacturer's instructions (Illumina), and sequencing is performed using Solexa Genome Analyzer II. Briefly:

1. Extract the size-fractionated DNA, and add a single adenosine using Klenow exo- (3′ and 5′ exo minus; Illumina).
2. Add Illumina adaptors and amplify the DNA by 20 cycles of PCR according to the manufacturer's instructions.
3. Purify DNA and perform cluster generation and 36 cycles of sequencing on the Illumina cluster station following the manufacturer's instructions.
4. Map the 36-bp single end reads to the reference mouse genome such as NCBI Build #36 (UCSC mm8), using the Illumina pipeline software. Then, distribution of mapped reads in a certain region can be analyzed as below.
5. Obtain eland sorted file from the output of Illumina pipeline software. We count uniquely mapped reads only.
6. To count the mapped reads using "samtools" [15], convert the eland sorted file into BAM file.

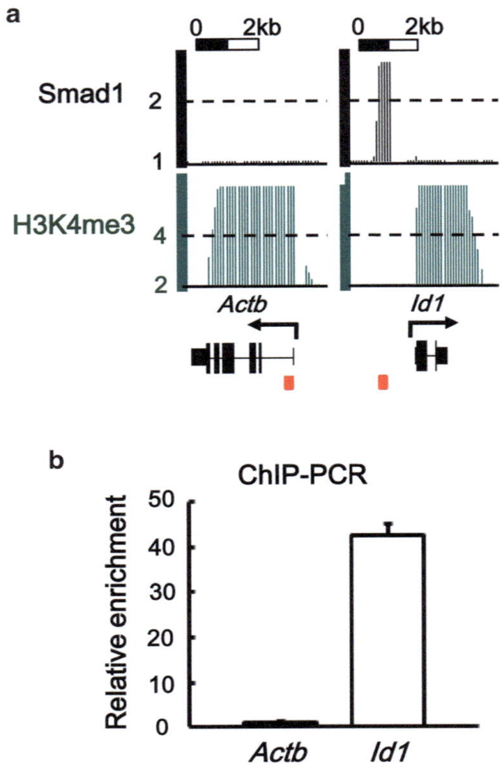

Fig. 3 ChIP using anti-Smad1 antibody. (**a**) Whereas no Smad1 binding was detected around *Actb*, Smad1 binding was detected at 1 kb upstream of TSS of *Id1*. *Red bars*: regions for ChIP-PCR. (**b**) Real-time ChIP-PCR for Smad1 binding. Relative enrichment compared to *Actb* was shown. Smad1 binding at 1 kb upstream of TSS of *Id1* was validated. These figures were modified from Fig. S11 of our previous report [11]

Table 1
ChIP-PCR primers

Genes	Primer sequences
Actb	TGAGGTACTAGCCACGAGAGAG and ACACCCGCCACCAGGTAAGCA
Dkk1	AGAGCCATCATTGTAAACACGG and ACCTTTGCCTGTTTGCGTCCT
Gcgr	TGCTGTCATGTCTGGTGAGTG and GGAGCTGTCAGCACTTGTGTA
Bmp2_1	CTTGGCTGGAGACTTCTTGAACT and TGGAGGCGGCAAGACTGGAT
Bmp2_2	ACTGGTGGAGTGGAGTGGAC and CTGGGGTTTGGAATGCCTAA
Smad6_1	CGCTTTGTGCTCGTGTACCA and CGATGCTAGAGACACCCTGC
Smad6_2	GTGAAACGGGATAGTAAGCCAT and CTAAAAGCTATGTACCGACTGAGG
Smad6_3	GCTGTCAGTAGGGAAATCACGC and GGCTAAAACTACAGAAAGGGACAA
Id1	CGGGTTTTATGAATGGGTGAC and GCGTCTGAACAAGCGGCTC

These primer sequences were also available in Table S8 of our previous report [11]

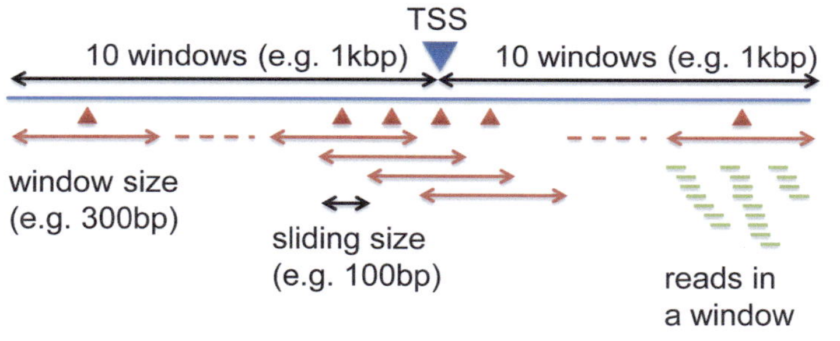

Fig. 4 Windows to count the number of mapped reads. We set the window size to 300 bp for ChIP-seq of H3K4me3 and Smad1 binding sites, and 500 bp for ChIP-seq of H3K27me3. Sequenced reads mapped within the window are simply counted and divided by total mapped reads. The calculated number represents the epigenomic status of the center position of the window. In this example, sliding size is set as 100 bp, and 10 windows are set in both the upstream and downstream of TSS

7. Make the list of genomic positions which we focus. We listed the positions of the transcription start sites (TSSs) of RefSeq genes to calculate the epigenetic status of the genes.
8. Set the size of the windows, where we count the number of mapped reads (Fig. 4). We set the window size to 300 bp for ChIP-seq of H3K4me3 and Smad1-binding sites, because they are distributed in narrow DNA regions and 300 bp is wide enough to cover each of those regions. Since H3K27me3 is rather broad modification than H3K4me3, wider window, 500 bp, is necessary to detect accurate read counts.
9. Set the number of windows and the sliding size. When we set the number of windows to 10 and the sliding size to 100 bp, 10 windows are put with equal sliding size of 100 bp both in front and back of the TSS.
10. For each TSS position in the list, obtain the list of reads which are mapped around the TSS with "samtools view" command.
11. For each TSS position in the list, calculate the start and end positions of all windows.
12. For each window, count the reads whose start positions are within the window.
13. Calculate the normalized score (reads per total million reads) dividing each read counts by total read counts and multiplying by a million. We assume that this normalized score shows the modification status for the center position of each window. We also assume that the maximum score within ±1 kb from TSS shows the modification status of the gene.

4 Notes

1. Buffer should be prepared without Complete proteinase inhibitor, and 1/100 volume of 100× Complete proteinase inhibitor should be added just before use.
2. Buffer should be prepared without PMSF, and 1/100 volume of 100× PMSF should be added just before use. 100× PMSF should be aliquoted rapidly in preparation, and use the aliquot immediately when thawed.
3. Invert the tube two to three times to mix well every 5 min.
4. Pipette two to three times when adding PBS/0.02 % EDTA and 0.25 % trypsin/EDTA.
5. The medium color changes to yellow.
6. Rinse quickly, and drain sufficiently.
7. Keep the samples at 4 °C during sonication. After each sonication, mix by inverting the tube several times and spin down.
8. Because the lids of Spin-X centrifuge tubes can be loosened, it is recommended that the columns of Spin-X centrifuge tubes be set in other types of 1.5-mL tubes. If 600 μL of buffer is too much depending on the types of tubes, the buffer volume could be reduced to 550 μL.
9. Repeating these steps, wash with ChIP Dilution Buffer four times in total.
10. In parallel, add 1 μL RNase A to Input sample around this step, and incubate for 30 min at 37 °C. Go to **step 42**.
11. Repeating these steps, wash with Wash Buffer 1 twice in total.
12. Repeating these steps, wash with Wash Buffer 2 twice in total.
13. Repeating these steps, wash with TE three times in total. At the last wash step, the column is recommended to be put in a new 1.5-mL tube.
14. Add 3 μL of 150 mg/μL Pronase or 11.25 μL of 40 mg/μL.
15. Though it is possible to rotate O/N, the lids of tubes could be loose by heating. To avoid this, rotate at 65 °C for 30 min, and incubate at 65 °C in a tube rack.
16. Add 57 μL of 5 M LiCl to 300 μL sample, for example.
17. As for ChIP'ed DNA derived from 1 mL, dilute to three to four times with 10 mM Tris–HCl, pH 8.0. Dilute Input samples to approximate 2 ng/μL with 10 mM Tris–HCl, pH 8.0.

Acknowledgements

We thank Prof. Toshio Kitamura, The University of Tokyo, for supply of retrovirus vector and packaging cells, and Prof. Mitsuo Oshimura, Tottori University, for technical advice on MEF establishment. This work was supported by a grant of Core Research for Evolutional Science and Technology (CREST), Japan Science and Technology Agency and Grants-in-Aids for Scientific Research from Ministry of Education, Culture, Sports, Science and Technology of Japan.

References

1. Feinberg AP (2007) Phenotypic plasticity and the epigenetics of human disease. Nature 447:433–440
2. Bracken AP, Kleine-Kohlbrecher D, Dietrich N, Pasini D, Gargiulo G et al (2007) The Polycomb group proteins bind throughout the INK4A-ARF locus and are disassociated in senescent cells. Genes Dev 21:525–530
3. Di Micco R, Sulli G, Dobreva M, Liontos M, Botrugno OA et al (2011) Interplay between oncogene-induced DNA damage response and heterochromatin in senescence and cancer. Nat Cell Biol 13:292–302
4. Hayflick L (1965) The limited in vitro lifetime of human diploid cell strains. Exp Cell Res 37:614–636
5. Kuilman T, Michaloglou C, Mooi WJ, Peeper DS (2010) The essence of senescence. Genes Dev 24:2463–2479
6. Serrano M, Lin AW, McCurrach ME, Beach D, Lowe SW (1997) Oncogenic ras provokes premature cell senescence associated with accumulation of p53 and p16INK4a. Cell 88: 593–602
7. Sharpless NE, DePinho RA (2005) Cancer: crime and punishment. Nature 436:636–637
8. Campisi J (2005) Senescent cells, tumor suppression, and organismal aging: good citizens, bad neighbors. Cell 120:513–522
9. Prieur A, Peeper DS (2008) Cellular senescence in vivo: a barrier to tumorigenesis. Curr Opin Cell Biol 20:150–155
10. Narita M, Lowe SW (2005) Senescence comes of age. Nat Med 11:920–922
11. Kaneda A, Fujita T, Anai M, Yamamoto S, Nagae G et al (2011) Activation of Bmp2-Smad1 signal and its regulation by coordinated alteration of H3K27 trimethylation in Ras-induced senescence. PLoS Genet 7:e1002359
12. Kaneda A, Wang CJ, Cheong R, Timp W, Onyango P et al (2007) Enhanced sensitivity to IGF-II signaling links loss of imprinting of IGF2 to increased cell proliferation and tumor risk. Proc Natl Acad Sci U S A 104: 20926–20931
13. Kitamura T, Onishi M, Kinoshita S, Shibuya A, Miyajima A et al (1995) Efficient screening of retroviral cDNA expression libraries. Proc Natl Acad Sci U S A 92:9146–9150
14. Dimri GP, Lee X, Basile G, Acosta M, Scott G et al (1995) A biomarker that identifies senescent human cells in culture and in aging skin in vivo. Proc Natl Acad Sci U S A 92: 9363–9367
15. Li H, Handsaker B, Wysoker A, Fennell T, Ruan J et al (2009) The sequence alignment/map format and SAMtools. Bioinformatics 25:2078–2079

Chapter 23

The Role of Ubiquitination to Determine Non-Smad Signaling Responses

Shyam Kumar Gudey and Marene Landström

Abstract

Ubiquitination is a posttranslational modification of proteins which acts as a key regulator of their function as well as fate. We have recently reported transforming growth factor β (TGFβ)-induced activation of non-Smad signaling responses through a specific Lys63-linked polyubiquitination of TGFβ type I receptor and TGFβ-associated kinase 1 (TAK1) that are utilized to specify cellular responses in cancer cells. This chapter gives a brief introduction of the biological importance of ubiquitination of proteins, the methods we have used for detecting new partners in the TGFβ signaling pathway and for performing ubiquitination assays.

Key words Non-Smad signaling, TAK1, TGFβ receptor I, TRAF6, Ubiquitination, Ubiquitination assays

1 Introduction

Ubiquitination is a dynamic, reversible, ATP-dependent posttranslational modification of proteins to determine their stability, subcellular localization, and function. In this process, an amide bond is formed between the terminal glycine of the activated ubiquitin protein and the ε-amine of the lysine in the acceptor protein (termed as acceptor lysine) or in the N-terminal amino group. Moreover, recent studies indicate that the ubiquitin moieties can also bind to cysteines, methionines, serines, and threonines of the acceptor protein [1]. The process of ubiquitination involves triple different steps. The first step in this process is initiated by a thioester bond formation between an E1-activating enzyme and ubiquitin in an ATP-dependent manner. The second step takes place by the transfer of ubiquitin to the E2-conjugating enzyme. The third and final step process is regulated by the family of E3 ubiquitin ligases, which transfer the ubiquitin from the E2-conjugating enzyme to the ε-amine of the lysine residue of the target protein [2]. Typically, the E3 ubiquitin ligases possess either Really Interesting New Gene (RING) or Homologous to the E6-AP Carboxyl

Xin-Hua Feng et al. (eds.), *TGF-β Signaling: Methods and Protocols*, Methods in Molecular Biology, vol. 1344,
DOI 10.1007/978-1-4939-2966-5_23, © Springer Science+Business Media New York 2016

Terminus (HECT) or U-box domains. The E3 ubiquitin ligases play an important role in the identification of substrates. TNF receptor-associated factor 6 (TRAF6) is one among such E3 ubiquitin ligases that catalyzes Lys63-linked polyubiquitin chains. TRAF6 was initially recognized for its crucial role in inflammatory processes, resulting in activation of the pro-inflammatory transcription factor nuclear factor-κB (NF-κB) [3]. Interestingly, TRAF6 was recently discovered to play a significant role downstream of the activated TGFβ receptor complex resulting in activation of the TGFβ-associated kinase 1 (TAK1), which is a member of the mitogen-activated protein kinase kinase kinase (MAPKKK) family [4] (Fig. 1).

The process of adding ubiquitin moieties to a target protein is termed ubiquitination or ubiquitylation. Proteins can undergo various posttranslational modifications (PTM) such as phosphorylation,

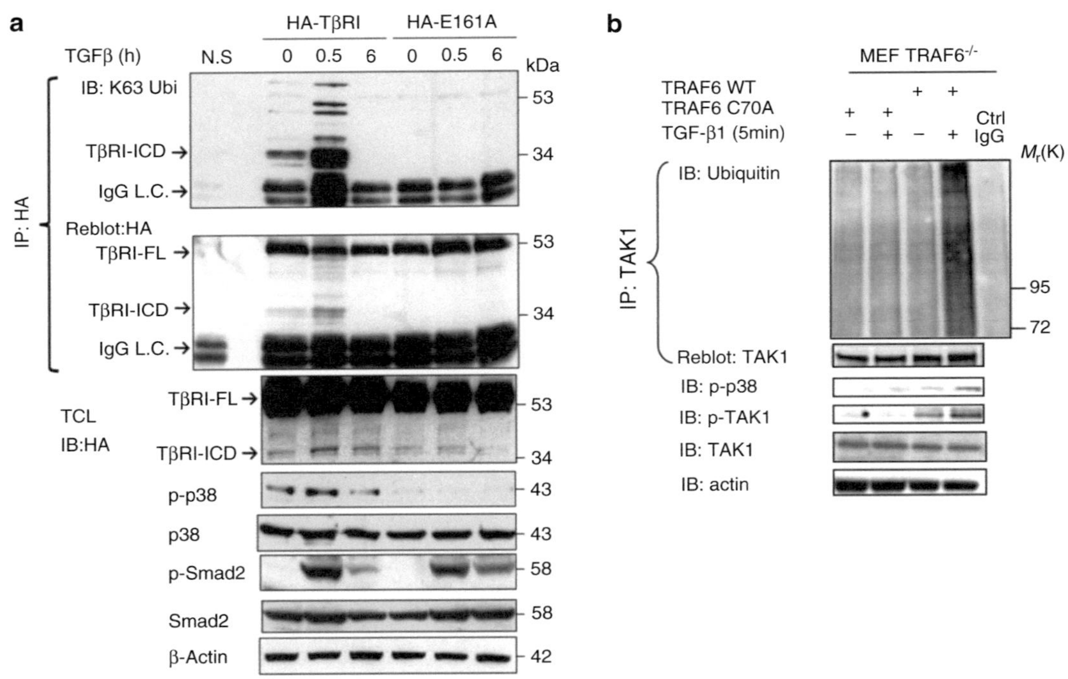

Fig. 1 (**a**) PC3U cells transiently transfected with C-terminally HA-tagged wild-type (wt) TβRI or E181A-mutant TβRI (that is unable to associate with TRAF6) were treated as indicated. Thereafter extracted cell lysates were conducted to an in vivo ubiquitination assay. Protein complexes were immunoprecipitated (IP) with anti-HA antibodies and then subjected to immunoblotting (IB) with the indicated antibodies. A light chain specific antiserum was used to avoid cross-reaction with IgG heavy chain. Molecular weight markers are indicated. A part of the corresponding total cell lysate (TCL) was also subjected to IB to visualize the TβRI-ICD and activation of p38 MAPK and Smad2, respectively. Actin served as internal control for equal loading of proteins. (**b**) TRAF6$^{-/-}$ mouse embryonic fibroblasts were transfected with WT- or C70A-mutant TRAF6 (lacking its E3-ligase activity) and treated as indicated. Extracted cell lysates were subjected to an in vivo ubiquitination assay and thereafter to IP and IB as indicated in the figure. Data presented in Fig. 1a is from Mu et al., Nature Comms 2011 [27] and in Fig. 1b from Sorrentino et al., Nature Cell Biology 2008 [20]

glycosylation, sumoylation, and ubiquitination, which determine the fate and function of the protein. The discovery of process of ubiquitination of proteins, by Ciechanover et al., in 1980, is one example of a PTM of key regulatory proteins that continues to attract great interest. The elucidation of the fundamental importance of the ubiquitination pathway by Aaron Ciechanover, Avram Hershko, and Irwin Rose was rewarded with the Nobel Prize in Chemistry in 2004 [5–7]. Ubiquitin is a small molecule of 76 amino acids with a molecular weight of 8.5 kDa. The name ubiquitin is commonly used rather than the original term "ubiquitous immunopoietic polypeptide." Ubiquitination is implicated in numerous and diverse array of cellular processes such as signal transduction, cell cycle regulation, membrane protein endocytosis, intracellular trafficking, DNA repair, assembly of signaling complexes, etc., depending on the type of ubiquitination (there are distinct ubiquitinations known today including mono-, multi-, Lys63-, Lys48-, and linear ubiquitination) [8–11]. Interestingly, if Lys63-linked chains are added to a kinase substrate, it often results in its activation. If Lys48-linked chains are added to the target protein, it is regarded as "kiss of the death" as it results in protein degradation through the proteasomal complex [12].

1.1 TGFβ Signaling

Transforming growth factor β (TGFβ) signaling plays a crucial role in cell survival, proliferation, differentiation, growth arrest, and apoptosis [13, 14]. TGFβ signaling is initiated upon ligand binding to the TGFβ receptor II (TβRII), which then heterodimerizes with TGFβ receptor I (TβRI). Upon heterodimerization, the TβRII phosphorylates multiple serine and threonine residues of the so-called GS domain located at the juxtamembrane region of the TβRI. Upon phosphorylation, the activated TβRI thereby transduces signals to the downstream partners, namely the canonical Smad or the non-Smad signaling cascades [15–18]. However, the molecular mechanisms underlying activation of the non-Smad signaling pathway have not been well understood until recently.

1.2 Ubiquitination to Determine Smad and Non-Smad Signaling Responses

It has been known that TGFβ elicits signaling responses also via non-Smad signaling pathways, through the activation of TGFβ-associated kinase 1 (TAK1), extracellular signal-regulated kinase (Erk), p38, mitogen-activated protein kinase (MAPK), and Akt, as described by Wakefield and Roberts [19]. Recently, our group identified that a specific non-Smad signaling cascade is operated by the E3 ubiquitin ligase, TRAF6. We reported that TRAF6 interacts with TβRI at a conserved consensus motif in TβRI. The TβRI-TRAF6 interaction leads to TRAF6 autoubiquitination in a TGFβ-dependent manner. The autoubiquitinated TRAF6 in turn activates TAK1 through Lys63-linked polyubiquitination and thereby activates protein kinase kinase 3/6 (MKK3/6), leading to p38 activation [20]. The importance of TRAF6 for TGFβ-induced activation of TAK1 was also shown to be required for transdifferentiation of epithelial cells in a process named epithelial to mesenchymal transition (EMT) [21]. EMT is today recognized as a phenomena occurring in normal and pathological conditions, such as embryogenesis and cancer, respectively.

Components of the TGFβ signaling pathway have been subjected to various types of ubiquitination, to promote either positive or negative regulation of TGFβ signaling. The E3 ligases Smurf1 and 2 are a classical example for negative regulation of TGFβ signaling. Smad7, an inhibitory Smad, associates with Smurf1 (a HECT type E3 ligase) in the nucleus; this complex of Smad7 and Smurf1 is later exported into the cytoplasm where it binds to the TβRI. Smurf1 then binds to the TβRI, causing degradation of TβRI [22]. On the other hand, the E3-ligase Arkadia positively regulates TGFβ signaling by targeting negative regulators of TGFβ signaling such as Smad7, c-Ski, and SnoN, for proteasomal degradation [22, 23].

Research on regulatory components of the non-Smad signaling pathway, and how they are induced or modified by PTMs induced by TGFβ, has been explored quite extensively in recent years. Some of these factors are described below. For instance, Kruppel-like factor 4 (KLF4) is a transcription factor, which is ubiquitinated and degraded through the proteasomal pathway upon TGFβ stimulation of cells [24, 25]. In the presence of active TGFβ ligand, TβRII phosphorylates Par6, a key regulator of cell polarity. The phosphorylated Par6 then binds to Smurf1, whereby Smurf1 leads to degradation of the small GTPase, RhoA, resulting in the loss of cellular tight junctions. This event is a hallmark of EMT, utilized in transformation of normal epithelial cells to cancer cells [26]. Our group has recently identified the importance for TRAF6 to cause polyubiquitination of TβRI in a Lys63-dependent manner. TβRI ubiquitination leads to its cleavage by TNF-α-converting enzyme (TACE) and thereby releases the TβRI intracellular domain (ICD), from the cell membrane, whereby the ICD translocates to the nucleus. Once the TβRI-ICD has entered into the cell nucleus, it binds to the promoter of the pro-invasive gene Snail1 to transcribe *Snail1* and *MMP2*, crucial regulators of invasion in cancerous cells [27].

2 Materials

RPMI-1640, penicillin, streptomycin and L-Glutamine were purchased from Sigma. Hereafter the various vendors given for each material are given in parenthesis.

Fugene6 transfection reagent (Promega).

Aprotinin and Pefabloc (Roche).

2.1 Western Blotting Ingredients

10 % Bis–tris gels, 20× MOPS Buffer, 10× Transfer Buffer, sample buffer, reducing agent from Life Technologies (Invitrogen).

Prestained protein ladder (Fermentas).

TBS and PBS (1 tablet respectively, in 500 mL of MQ water) (Medicago).

Nitrocellulose membrane, Protein G beads, ECL solution, ECL films (GE Healthcare).

2.2 Buffers

RIPA lysis buffer: 50 mM Tris pH 8.0, 150 mM NaCl, 10 % glycerol, 1 % NP40.

0.5 % Sodium deoxycholate.

1× MOPS Buffer: 50 mL of 20× MOPS Buffer + 950 mL of MQ water.

PBS-0.5 % NP40 buffer: 995 mL of 1× PBS + 5 mL of NP 40.

2.3 Antibodies

Mouse P4D1 and mouse IgG control were from Santa Cruz Biotechnology; rabbit K48 Ub, rabbit HA antibodies were from Cell Signaling; mouse K63 Ub was from Enzo Life Sciences, and rabbit TRAF6 was from Invitrogen.

Anti-Mouse HRP conjugated antibody, anti-Rabbit HRP conjugated antibody (DAKO).

Anti-Mouse Light chain specific HRP conjugated antibody, anti-Rabbit Light chain specific HRP conjugated antibody (Jackson Laboratories).

2.4 Plasmids

HA-ALK5ca was a gift from Peter ten Dijke.

Mutant HA-K63-Ub and HA-K48-Ub were kind gifts from Vishwa Dixit. Wild type HA-Ub was a kind gift from Ivan Dikic, respectively.

3 Methods

3.1 Cell Culturing

Day 1

1. The human prostate cancer (PC)-3U cell line, originated from the PC3 cell line derived from ATCC, was primarily used in the study.
2. PC3U cells were cultured for 24 h in RPMI-1640 growth medium constituting 10 % FBS, 1 % L-Glutamine, 1 % penicillin and streptomycin to reach a confluence of 60 %.
3. It is recommended to add two controls (1 10 cm^2 plate of cells as a "bead control" to rule out non-specific binding of proteins to beads and 1 10 cm^2 plate of cells as "IgG control" to rule out non-specific binding to IgG).

Day 2

4. PC3U cells were transiently transfected with HA-ALK5ca using Fugene 6 transfection reagent.

Day 3

5. Twenty-four hours post-transfection, cells were starved in RPMI-1640 growth medium constituting 1 % FBS, 1 % L-Glutamine, 1 % penicillin and streptomycin for no longer then 12–16 h.

3.2 Ubiquitination Assay (See Notes 1 and 2)

Day 4

6. PC3U cells were treated with TGFβ1 (10 ng/mL) for the indicated time periods.
7. After treatment, harvest the cells by removing the media and wash the cells in ice-cold 1× PBS (3 mL/plate).
8. Place the plate with cells on ice.
9. Add 1 mL of 1× PBS and scrape the cells with a new scrapper by keeping the plate on ice.
10. Transfer the scraped cells along with PBS into a 1.5 mL Eppendorf tube.
11. Transfer a small portion of the sample for total cell lysate and the remaining for Ub assay.
12. Centrifuge the sample at 0.4×*g* for 5 min at 4 °C with a microfuge.
13. Aspirate the supernatant with a suction pipet.
14. Place the samples with pellet on ice immediately.
15. Freshly prepare PBS/1 % SDS solution at room temperature (RT) (*see* **Note 3**).
16. Add 200 μL of PBS/1 % SDS solution to each sample. It is recommended to prepare a master mix for all the samples (*see* **Notes 4** and **5**).
17. Start the vortex and keep it in continuous on option at 400 RPM (*see* **Note 6**).
18. Boil the samples at 95 °C for 10 min.
19. Add 1.2 mL of PBS/0.5 % NP 40 solution containing Aprotinin (1:100) and Pefabloc (1:200).
20. Centrifuge at 16.2×*g* for 10 min at 4 °C with a microfuge.
21. After centrifugation, place the samples on ice.
22. Place 1 mL pipette tip to the bottom of the tube and pipette a little amount of the solution, you can observe a slime attached to the tip. Carefully pull out the whole slime, and the slime can then be discarded (*see* **Note 7**).
23. Now transfer the solution without the slime into a new Eppendorf tube. Label the tube and incubate on ice until all the samples are finished.

24. Add 2 μg of the antibody.
25. Incubate the sample in an end to end rotator for overnight at 4 °C by rotating.

Day 5

26. Take the samples from the end to end rotator and place the samples on ice
27. Add 50 μL of protein G beads to the sample. Follow the instructions of the manufacturer for washing prior to use (*see* **Notes 8** and **9**).
28. Rotate to mix the solution.
29. Incubate the sample in an end to end rotator for 1 h at 4 °C.
30. Prepare washing solution. Each sample is washed four times. Prepare master mix according to the number of samples.
31. Collect the samples from the end to end rotator.
32. Centrifuge the samples at 13,000 RPM for 1 min at 4 °C with a microfuge.
33. Aspirate the supernatant from the samples.
34. Continue washing four times with the washing solution.
35. Centrifuge the samples at 13,000 RPM for 1 min at 4 °C with a microfuge and aspirate the supernatant from the samples.
36. After washing samples four times with the washing solution, wash the samples finally with 1 mL of 1× PBS.
37. Centrifuge the samples at 13,000 RPM for 1 min with a microfuge and aspirate the supernatant after wash.
38. Add 30 μL of sample buffer and reducing agent mixture to each sample and keep the samples at 95 °C for 5 min.
39. Collect the samples and spin down at 13,000 RPM for 30 s with a microfuge.
40. The samples are then ready for running on an SDS-PAGE electrophoresis.
41. After completing electrophoresis, transfer the proteins to a nitrocellulose membrane at 4 °C for 1 h.
42. Take the membrane and keep it in a small box, block the membrane in 5 % BSA for 1 h at RT or as recommended by the antibody manufacturer.
43. Dilute primary antibody and add onto the membrane and incubate it overnight on a rocker.
44. Collect the membrane, remove the primary antibody, and wash it in 1× TBS-T for 10 min. Repeat three times.
45. Prepare secondary antibody solution. Dilute secondary antibody in TBS-T and mix it by vortexing.

46. Discard TBS-T and add secondary antibody solution, incubate for 1 h at RT.
47. Remove the secondary antibody solution and wash it in 1× TBS-T for 10 min. Repeat three times.
48. The membrane can be developed with ECL solution (follow manufacturer instructions) using photographic films or CCD camera system.

4 Notes

1. Follow the instructions of waste disposal according to the guidelines of respective laboratories.
2. We recommend placing all samples on ice unless mentioned, and operate quickly if possible.
3. Do not place the PBS/1 % SDS solution on ice. Placing on ice leads to formation of a white precipitate.
4. Pipette 200 μL of PBS/1 % SDS solution with a 200 μL pipette.
5. Vortex the sample while adding the PBS/1 % SDS solution dropwise to the pellet while vortexing the sample.
6. SDS gives a uniform charge to the proteins. Vortexing the cells will loosen the pelleted cells and dropwise addition of the solution will give efficient penetration of SDS into the cells.
7. Presence of slime may interfere with the immune-precipitation step when antibody is added.
8. Protein G beads are washed three times in 1× PBS to remove the ethanol from the beads, and leave little amount of PBS with the beads, which helps to pipette the beads easily.
9. Cut the pipette tip, so that it is easy to pipette out protein G beads.

Acknowledgements

The authors thank Susanne Grimsby, Alessandro Sorrentino, Reshma Sundar, and Noopur Thakur for their contributions to the original research. We thank Carl-Henrik Heldin and our colleges at the Ludwig Institute for Cancer Research, Uppsala Branch, for valuable discussions. This work was supported by the Swedish Medical Research Council, the Swedish Cancer Society, the Torsten and Ragnar Söderbergs Foundation, Umeå University, Cancer Research Foundation Norrland and a regional agreement between Umeå University and Västerbotten County Council on cooperation in the field of Medicine, Odontology and Health (ALF).

References

1. Hochstrasser M (2009) Origin and function of ubiquitin-like proteins. Nature 458(7237): 422–429, PMCID: 2819001
2. Hershko A, Ciechanover A, Rose IA (1981) Identification of the active amino acid residue of the polypeptide of ATP-dependent protein breakdown. J Biol Chem 256(4):1525–1528
3. Wertz IE, Dixit VM (2010) Signaling to NF-kappaB: regulation by ubiquitination. Cold Spring Harb Perspect Biol 2(3):a003350, PMCID: 2829959
4. Landstrom M (2010) The TAK1-TRAF6 signalling pathway. Int J Biochem Cell Biol 42(5):585–589
5. Ciechanover A, Heller H, Elias S, Haas AL, Hershko A (1980) ATP-dependent conjugation of reticulocyte proteins with the polypeptide required for protein degradation. Proc Natl Acad Sci U S A 77(3):1365–1368, PMCID: 348495
6. Hershko A, Ciechanover A, Heller H, Haas AL, Rose IA (1980) Proposed role of ATP in protein breakdown: conjugation of protein with multiple chains of the polypeptide of ATP-dependent proteolysis. Proc Natl Acad Sci U S A 77(4):1783–1786, PMCID: 348591
7. Goldstein G, Scheid M, Hammerling U, Schlesinger DH, Niall HD, Boyse EA (1975) Isolation of a polypeptide that has lymphocyte-differentiating properties and is probably represented universally in living cells. Proc Natl Acad Sci U S A 72(1):11–15, PMCID: 432229
8. Hershko A, Ciechanover A (1998) The ubiquitin system. Annu Rev Biochem 67:425–479
9. Haglund K, Dikic I (2005) Ubiquitylation and cell signaling. EMBO J 24(19):3353–3359, PMCID: 1276169
10. Haglund K, Di Fiore PP, Dikic I (2003) Distinct monoubiquitin signals in receptor endocytosis. Trends Biochem Sci 28(11): 598–603
11. Emmerich CH, Schmukle AC, Walczak H (2011) The emerging role of linear ubiquitination in cell signaling. Sci Signal 4(204):re5
12. Pickart CM, Fushman D (2004) Polyubiquitin chains: polymeric protein signals. Curr Opin Chem Biol 8(6):610–616
13. Heldin CH, Landstrom M, Moustakas A (2009) Mechanism of TGF-beta signaling to growth arrest, apoptosis, and epithelial-mesenchymal transition. Curr Opin Cell Biol 21(2):166–176
14. Massague J (2008) TGFbeta in cancer. Cell 134(2):215–230, PMCID: 3512574
15. Moustakas A, Heldin CH (2005) Non-Smad TGF-beta signals. J Cell Sci 118(Pt 16): 3573–3584
16. Mu Y, Gudey SK, Landstrom M (2012) Non-Smad signaling pathways. Cell Tissue Res 347(1):11–20
17. Roberts AB (1999) TGF-beta signaling from receptors to the nucleus. Microbes Infect 1(15):1265–1273
18. Derynck R, Zhang YE (2003) Smad-dependent and Smad-independent pathways in TGF-beta family signalling. Nature 425(6958):577–584
19. Wakefield LM, Roberts AB (2002) TGF-beta signaling: positive and negative effects on tumorigenesis. Curr Opin Genet Dev 12(1):22–29
20. Sorrentino A, Thakur N, Grimsby S, Marcusson A, von Bulow V, Schuster N et al (2008) The type I TGF-beta receptor engages TRAF6 to activate TAK1 in a receptor kinase-independent manner. Nat Cell Biol 10(10):1199–1207
21. Yamashita M, Fatyol K, Jin C, Wang X, Liu Z, Zhang YE (2008) TRAF6 mediates Smad-independent activation of JNK and p38 by TGF-beta. Mol Cell 31(6):918–924, PMCID: 2621323
22. Ebisawa T, Fukuchi M, Murakami G, Chiba T, Tanaka K, Imamura T et al (2001) Smurfl interacts with transforming growth factor-beta type I receptor through Smad7 and induces receptor degradation. J Biol Chem 276(16): 12477–12480
23. Koinuma D, Shinozaki M, Komuro A, Goto K, Saitoh M, Hanyu A et al (2003) Arkadia amplifies TGF-beta superfamily signalling through degradation of Smad7. EMBO J 22(24):6458–6470, PMCID: 291827
24. Hu D, Wan Y (2011) Regulation of Kruppel-like factor 4 by the anaphase promoting complex pathway is involved in TGF-beta signaling. J Biol Chem 286(9):6890–6901, PMCID: 3044944
25. De Boeck M, ten Dijke P (2012) Key role for ubiquitin protein modification in TGFbeta signal transduction. Ups J Med Sci 117(2):153–165, PMCID: 3339547
26. Ozdamar B, Bose R, Barrios-Rodiles M, Wang HR, Zhang Y, Wrana JL (2005) Regulation of the polarity protein Par6 by TGFbeta receptors controls epithelial cell plasticity. Science 307(5715):1603–1609
27. Mu Y, Sundar R, Thakur N, Ekman M, Gudey SK, Yakymovych M et al (2011) TRAF6 ubiquitinates TGFbeta type I receptor to promote its cleavage and nuclear translocation in cancer. Nat Commun 2:330, PMCID: 3113296

Chapter 24

Genome-Wide RNAi Screening to Dissect the TGF-β Signal Transduction Pathway

Xiaochu Chen and Lan Xu

Abstract

The *t*ransforming *g*rowth *f*actor-β (TGF-β) family of cytokines figures prominently in regulation of embryonic development and adult tissue homeostasis from *Drosophila* to mammals. Genetic defects affecting TGF-β signaling underlie developmental disorders and diseases such as cancer in human. Therefore, delineating the molecular mechanism by which TGF-β regulates cell biology is critical for understanding normal biology and disease mechanisms. Forward genetic screens in model organisms and biochemical approaches in mammalian tissue culture were instrumental in initial characterization of the TGF-β signal transduction pathway. With complete sequence information of the genomes and the advent of RNA interference (RNAi) technology, genome-wide RNAi screening emerged as a powerful functional genomics approach to systematically delineate molecular components of signal transduction pathways. Here, we describe a protocol for image-based whole-genome RNAi screening aimed at identifying molecules required for TGF-β signaling into the nucleus. Using this protocol we examined >90 % of annotated *Drosophila* open reading frames (ORF) individually and successfully uncovered several novel factors serving critical roles in the TGF-β pathway. Thus cell-based high-throughput functional genomics can uncover new mechanistic insights on signaling pathways beyond what the classical genetics had revealed.

Key words Growth factor-β (TGF-β), Embryonic development, Adult tissue homeostasis, *Drosophila*, Mammals

1 Introduction

Like in many other signaling pathways, the knowledge on TGF-β signaling mechanism was first acquired through genetic analyses in *Drosophila melanogaster*. Forward genetic screens uncovered several components in the TGF-β pathway, including Decapentaplegic (DPP, the *Drosophila* counterpart of TGF-β), the transmembrane Ser/Thr kinases Punt and Thickvein which are the TGF-β receptors, Mothers against Decapentaplegic (MAD), and others [1, 2]. Biochemical studies of these factors indicated that MAD (Smad in mammalian cells) is a transcription factor that is phosphorylated by the receptor kinases Punt/Thickvein (TβRI/II in mammals) upon binding of the ligand Dpp (TGF-βs in mammals) [3–5].

Xin-Hua Feng et al. (eds.), *TGF-β Signaling: Methods and Protocols*, Methods in Molecular Biology, vol. 1344,
DOI 10.1007/978-1-4939-2966-5_24,

Once phosphorylated/activated, MAD migrates into the nucleus and acts as transcription factors to elicit cellular responses to TGF-β [6, 7]. These key components of the TGF-β pathway, as well as the activation mechanism, are highly conserved in *Drosophila*, *c. elegans*, and vertebrates.

In spite of the success of classical genetics in delineating the TGF-β pathway, significant gaps remained. In particular, nuclear translocation of Smad is an essential step in transmitting the TGF-β signal into the nucleus, yet the molecules involved in this signal-regulated process were largely unknown. Some of this could be attributed to intrinsic limitations in whole organism-based genetics. For example, if a TGF-β pathway component is required for early embryonic development, classical genetic method may not uncover it because a homozygous null mutation may cause lethality and a heterozygous allele may not have any phenotype.

RNA interference (RNAi) is triggered by short 21–23 nucleotide short double-stranded RNAs (short interfering RNA, i.e., siRNA) that specifically recognize complementary sequences in the targeted mRNA, which consequently results in mRNA degradation [8]. The phenomenon of RNAi was first discovered in vivo in plants and *c. elegans*, and it was soon discovered that exogenous siRNA also triggers RNAi in tissue culture cells, thus providing a method for somatic loss-of-function genetic analysis [9]. Applied at the whole-genome scale, RNAi greatly complemented forward genetics and has been widely used in molecular dissection of many biological pathways in the past decade.

We have successfully carried out a whole-genome RNAi screening in *Drosophila* S2 cells that allowed us to genetically dissect the TGF-β pathway at a cellular level not achieved in whole animal-based traditional genetic screens [10, 11]. There are a number of advantages in using the *Drosophila* S2 cells in an RNAi screening. First, *Drosophila* S2 or S2R+ cells readily uptake double-stranded RNA (dsRNA) in the tissue culture media without any transfection reagents, making it technically very convenient [12]. Second, there is much less redundancy in the fly genome, avoiding the need to simultaneously knockdown several homologues in order to elicit a phenotype. Thirdly, there is a large collection of *Drosophila* mutant alleles or in vivo RNAi strains in academic centers (i.e., the Bloomington stock center, the Drosophila RNAi Screening Center at Harvard Medical School, and the Vienna Drosophila RNAi center), which makes it very convenient to test candidate hits in vivo. The initial proof of concept test, overall methodology, and data analyses of genome-wide RNAi screens in *Drosophila* cells can be found in these early publications [13–15]. The Drosophila RNAi Screening Center web page (www.flyrnai.org) provides an excellent up-to-date source for web-based tools, reagents, results of prior screens, and other information useful for designing RNAi screens.

The most important element in a whole-genome RNAi screening is the readout assay which needs to be robust and high throughput. In the TGF-β signaling pathway, nuclear translocation of Smad is an immediate early step. Image-based high-throughput screens demand automated, fast, high resolution and quantitative imaging instrumentations. For this, we used the Opera High Content Screening System from Evotech/Perkin Elmer and the nuclear/cytoplasmic signal quantification algorithm that accompanies the instrument. Our pilot experiments showed that there was sufficient dynamic range in the change of nuclear versus cytoplasmic signal of Smad in response to TGF-β. We imaged pilot screens in 384-well plates which contain positive (i.e., RNAi against the TGF-β receptor kinase) and negative (i.e., nontargeting) controls, and computed *Z*-scores based on the quantification of nuclear/cytoplasmic ratio of Smad (*see* Quantification of nuclear and cytoplasmic GFP signal in Methods). All the positive controls lied more than 2 standard deviations away from the mean, further demonstrating the robustness of the readout in the 384-well format (Fig. 1). Using this Opera-based imaging platform, we screened a whole-genome RNAi library covering >90 % of the *Drosophila* open reading frames, and successfully identified Msk (moleskin) as the nuclear transport factor of Smad [10].

2 Materials

2.1 Drosophila S2R+ Cell Culture

1. *Drosophila* S2R+ cells: the *Drosophila* RNAi Screening Center at Harvard Medical School. This cell line is more adherent than the more common S2 cell line. This property is important for screens that involve multiple washing and treatment of cells. Seeding density should be at least 5×10^6/10 cm plate. Passage cells when necessary and not allowing the culture density to exceed 5×10^7 cells/10 cm plate.
2. 24 °C Humidified incubator, no need for CO_2 control.
3. Culture media: Schneider's *Drosophila* medium (Invitrogen) supplemented with 10 % fetal bovine serum, penicillin (100 U/mL), and streptomycin (100 U/mL) (all from Invitrogen).
4. Transfection reagent: Effectene (Qiagen).
5. Hygromycin B (Invitrogen).
6. $CuSO_4$ (Sigma-Aldrich).

2.2 MAD Nuclear Translocation Reporter Cell Line

1. pSH-GFP-MAD: GFP-tagged MAD cloned into the metallothionein promoter driven pSH vector (inducible by $CuSO_4$) [16].
2. pMK-Punt and pMK-Tkv: Punt and Tkv (TGF-β receptor kinases) cloned into the pMK33 vector which is also driven by a metallothionein promoter.

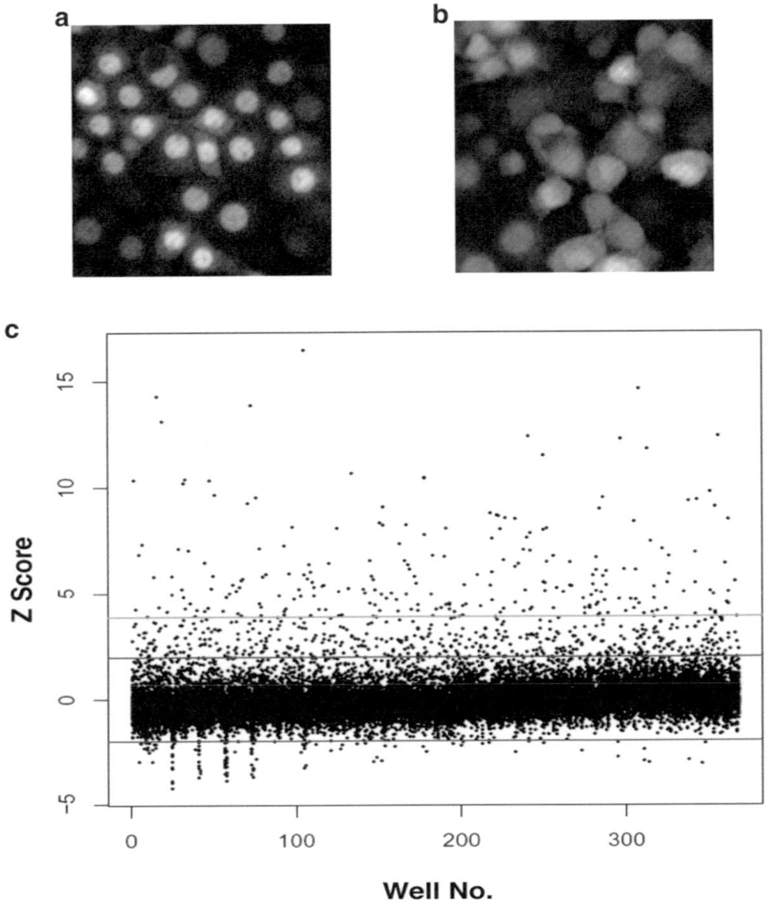

Fig. 1 Whole-genome RNAi screening to identify nuclear transport factors of Smad/MAD. (**a**) GFP-MAD distribution in a typical "non-hit" well. The image was acquired by the Opera High Content Screening System. (**b**) GFP-MAD distribution in a positive hit well (i.e., *msk* RNAi). (**c**) *Z*-score (*see* **step 3** of Quantification of nuclear and cytoplasmic GFP signal in Methods) distribution of the entire genome-wide screen (62 × 384-well plates). The *red lines* mark the 95 percentile. All points below the −95 % mark were considered preliminary hits (those required for Smad nuclear transport). All positive controls indeed fell into the 95 percentile

3. GFP-MAD + R cell line: S2R+ cells stably co-transfected with pSH-GFP-MAD, pMK-Punt, and pMK-Tkv. The resulted cell line—GFP-MAD + R—exhibited GFP signal exclusively in the nucleus upon $CuSO_4$ induction (0.5 mM, 3 h).

2.3 Fixation Solutions and Nuclei Staining

1. Fixing: paraformaldehyde (Electron Microscopy Sciences), 4 % in PBS.
2. Permeabilization: Triton X-100, 0.1 % in PBS (PBST).
3. Nuclei staining: 4′-6-Diamidino-2-phenylindole (DAPI, Sigma-Aldrich), 1 μg/mL in PBS.

2.4 Reagents and Kits to Generate Amplicons and Double-Stranded RNA (dsRNA)

1. Total RNA isolation: RNAeasy kit (Qiagen).
2. Reverse transcription: Superscript II Reverse Transcriptase kit (Invitrogen).
3. Genomic DNA isolation: Blood & Cell Culture DNA kit (Qiagen).
4. PCR primers: gene-specific primers with an additional T7 promoter sequence (5′-TAATACGACTCACTATAGGG) at the 5′ end. Primer design was based on recommendations by the *Drosophila* RNAi Screening Center web page (DRSC, www.flyrnai.org).
5. In vitro transcription: T7 MEGA script kit (Applied Biosystems).
6. Double-stranded RNA (dsRNA) purification: NucAway Spin Columns (Ambion/Life Technology).

2.5 High-Throughput Image-Based Screening

1. Cell culture: Black-walled 384-well plates with clear flat bottom (Perkin Elmer #6007460). The thin plastic bottom of the plate is optically clear and ensures best fluorescence reading and reduces focusing time.
2. Cell seeding: Matrix WellMate (Thermo Scientific).
3. Dispenser for nonsterile reagents: Multidrop 384 (Thermo Scientific).
4. High-throughput automated epifluorescent microscope: Opera High Content Screening System (Evotec/Perkin Elmer).
5. Plate-loading robotic arm: BioRobot Twister II (Qiagen).
6. The nuclear/cytoplasmic signal quantification software: Acapella (part of the Opera High Content Screening System).

2.6 Drosophila Whole-Genome RNAi Library 2.0

1. The DRSC 2.0 is a collection of 22,490 dsRNAs targeting about 13,900 genes which are 94.3 % of all annotated *Drosophila* genes. On average, there are one or two dsRNAs per gene (distributed by the *Drosophila* RNAi Screening Center, Harvard Medical School).

3 Methods

3.1 Generation of the GFP-MAD + R Reporter Cell Line

1. The day before transfection, seed 2×10^6 cells per 10 cm dish in 10 mL Schneider Medium containing serum and antibiotics.
2. The day of transfection, dilute 2 μg of pSH-GFP-MAD with the DNA-condensation buffer—Buffer EC (Qiagen)—to a final volume of 300 μL. Add 16 μL of Enhancer and mix well. Incubate at room temperature for 5 min and then add 20 μL of Effectene to the DNA-Enhancer mixture. Mix well and incubate at room temperature for 10 min.

3. During the 10 min incubation, change the cells into 8 mL of fresh culture media.
4. Add 2 mL of complete media to the transfection complexes. Mix well and immediately add the mixture dropwise onto the cells. Gently swirl the dish to ensure uniform distribution of the transfection complexes.
5. Incubate the cells with the transfection complexes at 24 °C without changing media.
6. Forty-eight hours after transfection, passage cells 1:5 into selective medium containing 200 μg/mL of Hygromycin B (Invitrogen). Maintain cells in selective medium until colonies appear, usually within 2 weeks. Change media and passage cells if necessary.

3.2 Generation of Amplicons

1. Isolate genomic DNA from cultured S2R+ cells using the Qiagen Blood & Cell Culture DNA kit.
2. Set up 50 μL PCR reactions as below, in a 96-well plate containing 96 different specific primer pairs.

Component	Amount per reaction (μL)	Final
10× PCR buffer	10	1×
10 mM dNTP mixture	2	0.2 mM each
50 mM $MgCl_2$	3	1.5 mM
Genomic DNA (0.35 mg/mL)	1	350 ng
Forward primer (10 μM)	2	20 μM
Reverse primer (10 μM)	2	20 μM
Taq (Invitrogen) DNA pol.	0.5	2.5 U
Dnase/Rnase-free water	79.5	100 μL

3. Run the following touchdown PCR program.

94 °C, 3 min	
94 °C, 30 s	
60 °C (−1/cyc), 45 s	Repeat 10 cycles
72 °C, 1 min	
94 °C, 30 s	
50 °C, 45 s	Repeat 23 cycles
72 °C, 1 min	
72 °C, 10 min	
4 °C, ∞	

4. Run 5 μL of PCR product on a 0.8 % agarose TAE (Tris-Acetate-EDTA) gel to check for size and quantity. PCR products can be stored at −20 °C for up to 1 year.

3.3 Generation of dsRNA

1. In vitro Transcription. Use the T7 MEGAscript kit (Applied Biosystems) to set up in vitro transcription (IVT) reactions as detailed in the table below.

Component	Amount per reaction (μL)
ATP (75 mM)	2
UTP (75 mM)	2
CTP (75 mM)	2
GTP (75 mM)	2
T7 enzyme	2
T7 reaction buffer	2
Amplicon from 3.2	8
Total volume	20

Incubate at 37 °C for 16 h or overnight.

2. Dilute the IVT reaction to 100 μL with nuclease-free TE buffer (pH 8.0). Sometimes the IVT reaction is highly viscous, pipette up and down a few times to dissolve.
3. Annealing. Heat the samples at 95 °C for 3 min and let cool gradually to room temperature. Place another heat block on top of the tubes to minimize condensation in the inside of the cap.
4. Purification. Hydrate the NucAway Spin Column (Applied Biosystems, AM10070) with 650 μL of nuclease-free TE buffer. Tap the column gently to let out air bubbles from the gel bed. Let sit at room temperature for 5 min.
5. Place the NucAway Spin Column in a 2 mL collection tube and spin the column at 750 × *g* for 2 min to remove excess interstitial fluid. Keep track of the orientation of the column in the rotor.
6. Discard the Collection Tube and immediately apply the sample (100 μL) to the center of the gel bed at the top of the column.
7. Place the NucAway Spin Column in a new 1.5 mL elution tube and place in the rotor, keeping the same orientation as in the last centrifugation.
8. Spin the NucAway Spin Column at 750 × *g* for 2 min and collect the sample in the elution tube. Discard the column.

9. Take an 1 μL aliquot, dilute in 100 μL of TE buffer and measure RNA concentration: Concentration (μg/μL) = OD 260 × 45 × 100 (dilution factor)/1000.
10. Run 10 μL of the above diluted sample on a 1 % TAE agarose gel with 0.5 μg/mL ethidium bromide (Sigma-Aldrich). Use DNA markers with known concentration. Check for proper size and verify concentration of the dsRNA by comparing ethidium bromide staining intensity. dsRNA should migrate only slightly differently from double-stranded DNA of the same size. dsRNA bands on 1 % TAE should appear somewhat fussy but not overly diffused.
11. Once passed the quality control, dilute the dsRNA to the working concentration (0.05 μg/μL) with nuclease-free water. For 384-well plates, add 250 ng dsRNA per well. In pre-designated wells, add control dsRNAs, i.e., GFP, *Rho1*, and *Thread*. Also leave some designated wells empty for investigators to put in their preferred controls. Seal the plate and store immediately in −80 °C.
12. The plates are provided to the investigators without disclosing the contents in each well, so the screens will be done in a double-blinded fashion.

3.4 RNAi Screening

1. Collect S2R+ cells by trypsinization and count live cells using a hemocytometer.
2. Resuspend cells in serum-free and antibiotics-free media at 1.0×10^6/mL.
3. The dsRNA library DRSC 2.0 is arrayed in a total of 66 384-well plates in a randomized manner (250 ng of dsRNA per well). Each plate is bar coded and contains both positive and negative control wells. For positive controls, dsRNAs targeting Punt and Tkv, which are known to be required for MAD nuclear translocation, were plated into well 13A, 13H and 14I, 14P respectively. Other controls include GFP, Rho1, and Thread dsRNAs. The layout for the genome screening plates can be found at: www.flyrnai.org/DRSC-LAY.html.
4. Plate 10 μL (1×10^4) cells per well in the 384-well genome library plates as above using WellMate (Thermo Scientific). Centrifuge the plates at 290 × *g* for 2 min.
5. Incubate the cells at room temperature for 1 h. Add 30 μL of complete serum-containing media to each well. Centrifuge plates at 290 × *g* for 2 min.
6. Leave the plates in a 24 °C humidified incubator for 3 days.

3.5 Smad (MAD) Nuclear Translocation Assay

1. Dispense 5 μL of 5 mM $CuSO_4$/PBS (final concentration 0.5 mM) per well using WellMate. Centrifuge the plates at 290 × *g* for 2 min. Continue to incubate for 3 h.

2. Remove media by aspiration. Add 100 μL 4 % paraformaldehyde/PBS per well using Multidrop and fix the cells for 8 min at room temperature.
3. Aspirate and add 100 μL of 0.2 % Triton X-100/PBS, permeabilize the cells for 8 min at room temperature.
4. Remove the permeabilization solution and add 100 μL of 0.1 % Triton X-100/PBS containing 1 μg/mL DAPI.
5. Seal and store the plates at 4 °C before imaging. GFP signal is usually stable for weeks. However it is highly recommended to image the plates as soon as possible.

3.6 High-Throughput Imaging

1. Load the plates onto the staging tray of the Opera High Content Screening System.
2. Calibrate the chosen wavelengths and focus plane in a test run. Image at 20× magnification, four fields per well at two wavelengths: a confocal laser at 488 nm (for GFP) and a non-confocal UV beam (for DAPI). The scanning speed at this particular setting is about 3 s per well. Allow enough time for the microscope to image without interruption.

3.7 Quantification of Nuclear and Cytoplasmic GFP Signal

1. Run the Acapella program bundled with the Opera High Content Screening System to analyze the acquired confocal images. The software recognizes the boundary of the whole cell or nucleus (DAPI positive region) and quantifies the GFP signal from the nucleus and cytoplasm. The program also counts the number of GFP positive cells from the four fields. Averaging the quantification from all cells in the four fields each well, the program calculates an average nuclear/cytoplasmic signal ratio (N/C ratio) for each dsRNA/well. Submit the results, including the plate bar code, to DRSC for data analysis.
2. Log onto the DRSC website and use the Heatmap tool to visualize numeric N/C ratio data generated for each well in each plate. Check the degree of statistical normality of each plate (non-control wells) on the Heatmap page. Select the "Q-Q Plot" button for a Quantile-Quantile plot of the data against a normal curve. Visually inspect the Q-Q Plot to evaluate the distribution of the data. Normally distributed data should form a straight diagonal line across the graph. A probability of normality score (*P* value) is also provided. Any plate with *P* value below 1.0×10^{-5} should be re-evaluated.
3. As experimental conditions vary from plate to plate, it is impossible to directly compare raw data across plates. The Heatmap page provides normalization solutions (*Z*-scores) for comparison between plates. *Z*-score is a measure of distance, in standard deviation, from the mean. Select the *Z*-score method to

re-display the Heatmap data. Each well now has a *Z*-score value which can be compared meaningfully to the *Z*-score values on other plates.

4. Choose a cut-off *Z*-score that would include the positive controls while minimizing the noises. In our case, we chose the *Z*-score cut-off as <−1.4. This corresponds to about 95 percentile (*Z*-score 1.6), and ~100 potential hits.
5. Identify the wells with *Z*-scores below the cut-off, which would be considered candidate hits. To minimize false negative rate, we visually inspect wells with the 20 lowest *Z*-scores in each plate.

3.8 Secondary Screening to Validate Hits

1. To avoid off-target artifacts. Validate the hits by testing additional dsRNAs that do not overlap with the ones in the whole-genome library. DRSC maintains a validation set of amplicons that do not overlap with dsRNAs in the genomic dsRNA library. One can order these amplicons from DRSC for in vitro transcription to generate dsRNAs.
2. Repeat the **steps 3** to 7 in Generation of dsRNA (Methods) but on a much smaller scale. At least two independent dsRNAs must show the same phenotype in order for that gene to be considered a validated hit.
3. Submit to DRSC the list of hits that passed the secondary screen. DRSC will then release the decoding keys to unveil the dsRNA identity.

4 Critical Issues

1. *Hit selection.*
   One of the biggest challenges in whole-genome RNAi screen is to prioritize the hits for further in-depth analysis. Once the identities of the hits are known, the decision will rest not only on the *Z*-score. For example, many of the genes are covered by more than one dsRNA/well in the library, so if there are multiple hits corresponding to the same gene, this will be a high confidence hit. Similarly, if several genes in the same pathway or protein complexes score as hits, it is likely they are real. Another sign that the screen and the scoring algorithm have been robust is that factors known to participate in the biological process are identified.
2. *The off-target issue.*
   False positive hits due to off-target RNAi are frequently encountered in whole-genome screens [13, 17, 18]. The first step to address off-target effect is to repeat the experiment with a second or more dsRNA against the same gene but targeting

an entirely non-overlapping region. If the same phenotype is observed, then it is likely not due to an off-target effect. The most convincing validation, however, is to use dsRNAs targeting the 5′ or 3′ untranslated region and test whether the cDNA could rescue the phenotype. In designing dsRNAs, certain principles could be applied to minimize the chance of an off-target effect, such as avoiding repeat sequences and short stretches (i.e., 19 bp), that matches perfectly to other genes. SnapDragon in the DRSC web page provides a convenient tool to evaluate dsRNA constructs and identify genes that have significant uninterrupted sequence similarities (i.e., >16 bp) which may be knocked down due to off-target effects.

3. *RNAi efficiency.*
   RNAi is usually robust in S2R+ cells using the direct feeding method. However the RNAi efficiency sometimes could be inconsistent, especially when a different batch of media or serum is used. Carefully test media and serum lots before launching a large-scale screening. Keep S2R+ cells healthy by seeding at least $5 \times 10^6/10$ cm plate but never reaching more than $5 \times 10^7/10$ cm plate. Feeding is substantially less effective in S2 and some other *Drosophila* cell lines. However, in most cases RNAi is effective when the dsRNA is transfected into cells with reagents such as DharmaFect 4 (Thermo Scientific).

4. *Expected results in control wells.*
   The various positive and negative controls serve different purposes and are expected to exhibit different phenotypes. RNAi against Punt or Tkv substantially reduced nuclear accumulation of MAD (*Z*-score ranged from −1.4 to −2.0 in our screen), because they blocked MAD activation/phosphorylation which is prerequisite for nuclear translocation of MAD. Even though Punt and Tkv are being induced in the GFP-MAD + R cells upon $CuSO_4$ treatment, the dsRNA appeared to be still highly effective. GFP RNAi resulted in much reduced GFP signal but not DAPI. Cells treated with dsRNA against Rho1 failed to complete cytokinesis and displayed much enlarged cell body. The *thread* gene is essential for S2R+ cell viability, so the RNAi would greatly reduce the number of GFP and DAPI positive cells.

5. *Potential issues in quantifying nucleus versus cytoplasmic GFP signal.*
   A high percentage of dead cells in the field may sometimes eschew the results, especially if the nucleus is highly condensed and could be mistaken by the algorithm as very small nucleus. The cell shape also affects the quantification. Therefore it is best to test run the several different scripts of the algorithms which are tailored to different types of cell shape and determine which one fits the cell line under study the best.

In addition, it is always necessary to confirm the candidates by visual inspection of the images.

6. *Z-score distribution.*
When the *Z*-score for each dsRNA in the entire screen is plotted, the data is slightly skewed to the right, indicating the tendency to have more dsRNAs that caused an increase of the nucleus/cytoplasmic ratio of GFP-MAD. The reason behind this phenomenon is not immediately obvious and could be due to the nature of the readout or the quantification algorithm. Nevertheless, the *Z*-scores passed the statistical normality test (the Q-Q plot, *P* value $>1.0\times10^{-5}$).

7. *Edge effect.*
Using the HeatMap tool to visualize *Z*-scores across the plates, we sometimes see clear signs of edge effects, i.e., the wells in the outermost rows and columns have much higher frequencies of outliers. Like all high-throughput screens, this is often caused by excessive evaporation of media in the edge wells. One could put 384-well plates in Tupperware with wetted paper towel laid at the bottom, and make sure the incubator is properly humidified. Check for signs of edge effects using HeatMap and exclude these wells to avoid distortion of the statistical analyses.

Acknowledgements

We would like to thank the DRSC at the Harvard Medical School for providing critical reagents, instrumentations, technical advice, and other resources that greatly facilitated our screening. The authors work was funded by the NIH (RO1 CA108509).

References

1. Raftery LA, Twombly V, Wharton K, Gelbart WM (1995) Genetic screens to identify elements of the decapentaplegic signaling pathway in Drosophila. Genetics 139:241–254
2. Sekelsky JJ, Newfeld SJ, Raftery LA, Chartoff EH, Gelbart WM (1995) Genetic characterization and cloning of *Mothers against dpp*, a gene required for *decapentaplegic* function in Drosophila melanogaster. Genetics 139: 1347–1358
3. Macias-Silva M, Abdollah S, Hoodless PA, Pirone R, Attisano L, Wrana JL (1996) MADR2 is a substrate of the TGFß receptor and phosphorylation is required for nuclear accumulation and signaling. Cell 87: 1215–1224
4. Kretzschmar M, Liu F, Hata A, Doody J, Massagué J (1997) The TGF-ß mediator Smad1 is directly phosphorylated and functionally activated by the BMP receptor kinase. Genes Dev 11:984–995
5. Souchelnytskyi S, Tamaki K, Engstrom U, Wernstedt C, ten Dijke P, Heldin CH (1997) Phosphorylation of Ser465 and Ser467 in the C terminus of Smad2 mediates interaction with Smad4 and is required for transforming growth factor-beta signaling. J Biol Chem 272(44): 28107–28115
6. Raftery LA, Sutherland DJ (1999) TGF-beta family signal transduction in Drosophila development: from Mad to Smads. Dev Biol 210(2):251–268

7. Shi Y, Massagué J (2003) Mechanisms of TGF-beta signaling from cell membrane to the nucleus. Cell 113(6):685–700
8. Sharp PA, Zamore PD (2000) Molecular biology. RNA interference. Science 287(5462): 2431–2433
9. Hannon GJ (2002) RNA interference. Nature 418(6894):244–251
10. Xu L, Yao X, Chen X, Lu P, Zhang B, Ip YT (2007) Msk is required for nuclear import of TGF-{beta}/BMP-activated Smads. J Cell Biol 178:981–994
11. Chen X, Xu L (2010) Specific nucleoporin requirement for Smad nuclear translocation. Mol Cell Biol 30(16):4022–4034, PMCID: 2916443
12. Clemens JC, Worby CA, Simonson-Leff N, Muda M, Maehama T, Hemmings BA et al (2000) Use of double-stranded RNA interference in Drosophila cell lines to dissect signal transduction pathways. Proc Natl Acad Sci U S A 97(12):6499–6503
13. Armknecht S, Boutros M, Kiger A, Nybakken K, Mathey-Prevot B, Perrimon N (2005) High-throughput RNA interference screens in Drosophila tissue culture cells. Methods Enzymol 392:55–73
14. Ramadan N, Flockhart I, Booker M, Perrimon N, Mathey-Prevot B (2007) Design and implementation of high-throughput RNAi screens in cultured Drosophila cells. Nat Protoc 2(9): 2245–2264
15. Flockhart I, Booker M, Kiger A, Boutros M, Armknecht S, Ramadan N et al (2006) FlyRNAi: the Drosophila RNAi screening center database. Nucleic Acids Res 34(Database issue):D489–D494, PMCID: 1347476
16. Bunch TA, Grinblat Y, Goldstein LS (1988) Characterization and use of the Drosophila metallothionein promoter in cultured Drosophila melanogaster cells. Nucleic Acids Res 16(3):1043–1061
17. Echeverri CJ, Beachy PA, Baum B, Boutros M, Buchholz F, Chanda SK et al (2006) Minimizing the risk of reporting false positives in large-scale RNAi screens. Nat Methods 3(10):777–779
18. Perrimon N, Mathey-Prevot B (2007) Matter arising: off-targets and genome-scale RNAi screens in Drosophila. Fly (Austin) 1(1):1–5

Chapter 25

Measuring TGF-β Ligand Dynamics in Culture Medium

Zipei Feng, Zhike Zi, and Xuedong Liu

Abstract

TGF-β plays an important role in a myriad of cell activities including differentiation, proliferation, and growth arrest. These effects are influenced by the concentration of TGF-β in the surrounding milieu, which is interpreted by mammalian cells and subsequently translated into meaningful signals that guide their proliferation, survival, or death. To predict cellular responses to TGF-ß signaling based on molecular mechanisms, it is important to consider how cells respond to different ligand doses and how variations in ligand exposure impact Smad signaling dynamics and subsequent gene expression. Here we describe methods to measure TGF-β concentration in the environment and approaches to perturb cellular TGF-β exposure to gain a quantitative understanding of signaling dynamics of this pathway.

Key words TGF-β, Signaling, SMAD, Switch-like response

1 Introduction

Transforming growth factor β (TGF-β) is a cytokine that regulates a diverse group of cellular behaviors such as proliferation, differentiation, and growth arrest [1]. These influences in a broader perspective lead to the implication of TGF-β in cancer, cardiac and pulmonary diseases, congenital syndromes such as Marfan and Loeys-Dietz, and diabetes [2]. The classic TGF-β signaling cascade begins as TGF-β ligand binds to TGF-β receptor type 2 (TβRII), bringing it in close proximity to TGF-β receptor type 1 (TβRI) [1]. Next, the constitutively active kinase domain on TβRII transphosphorylates serine and threonine residues of TβRI, which then becomes active and passes the signal downstream to the SMAD proteins [1]. These SMAD proteins include receptor-regulated SMADs or R-SMADs (SMAD2 and SMAD3 in the TGF-β pathway), the common mediator SMAD (Co-SMAD, i.e., SMAD4), and the inhibitory SMAD (I-SMAD, i.e., SMAD6 and SMAD7) [3]. SMAD2 and SMAD3 directly interact with activated TβRI. They are phosphorylated by TβRI at the two distal serines of their C-terminal SSXS motif [4–6]. The phosphorylation event then

Xin-Hua Feng et al. (eds.), *TGF-β Signaling: Methods and Protocols*, Methods in Molecular Biology, vol. 1344,
DOI 10.1007/978-1-4939-2966-5_25, © Springer Science+Business Media New York 2016

triggers the homo-oligomerization of SMAD2 and SMAD3 and their hetero-oligomerization with SMAD4, resulting in the nuclear accumulation of these complexes followed by transcriptional regulation of various genes [7].

Similar to many other cell processes, TGF-β signaling is highly regulated. The cell is able to sense minor changes in the concentration of signal in the surrounding environment and translate them into largely different corresponding actions. In cell culture studies, exposing cells to different concentrations of TGF-β is able to elicit a switch-like response in transcription and translation of different genes such as PAI-1 [8]. Similar behavior is observed in human and animal models. For example, bone morphogenic protein (BMP), a member of TGF-β superfamily, is crucial in the development of dorsal part of spinal cord and brainstem [9]. The continual release of BMP by non-neural ectodermal cells during development leads to the formation of a concentration gradient in the adjacent dorsal neurotube [9]. The part of the neurotube that receives the highest amount of BMP is signaled to differentiate into somatic sensory neurons, while the cells farther away from the roof plate receiving less BMP are signaled to become visceral afferent neurons [9]. Another well-studied example is Dpp, also a member of the TGF-β superfamily, whose concentration gradient is again absolutely essential in the proper wing development of *Drosophila spp* [7].

The presence of TGF-β ligand is one way that cell response could be affected [8]. When the ligands are withdrawn from the environment, cells lose the stimulation and the phosphorylation of SMAD2 is shut down in a matter of hours [8]. Here we describe a method to efficiently remove most of the TGF-β molecules in the surrounding medium and induce a transient phospho-SMAD2 response, which is lower in amplitude and duration compared to the response elicited when there is sustained TGF-β in the extracellular milieu (Figs. 1 and 2).

Additionally, it seems that cells can also regulate the response to TGF-β signaling through ligand depletion [8, 10]. When the ligand depletion is blocked or slowed down, cells lose the switch-like behavior and respond in a Michaelis-Menten-like fashion to a TGF-β concentration gradient [8]. Ligand depletion can be accomplished experimentally by completely blocking endocytosis, which is often difficult due to off-target effects of certain drugs [8]. Alternatively, one can decrease the rate of ligand depletion simply by increasing the medium volume while keeping the absolute number of ligand molecules constant. This method is effective in slowing down the cellular uptake of TGF-β molecules (Fig. 3) and is able to reduce the switch-like behavior of the TGF-β signaling pathway (Fig. 4).

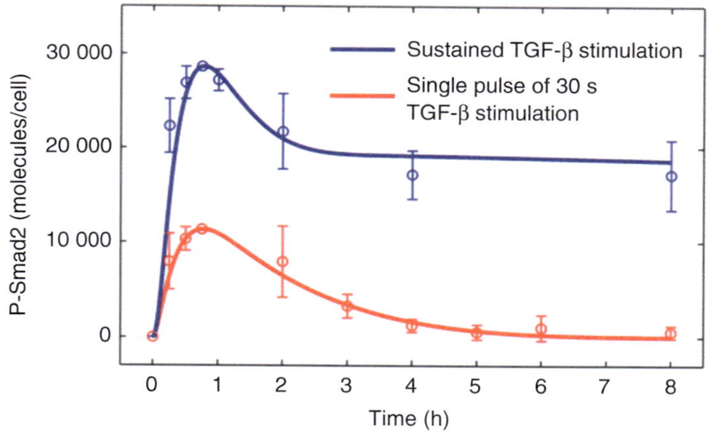

Fig. 1 Phospho-SMAD2 signal following TGF-quantitative analysis. Notice the decrease in amplitude and duration of signal with washout compared to continual stimulation. (Taken from [8])

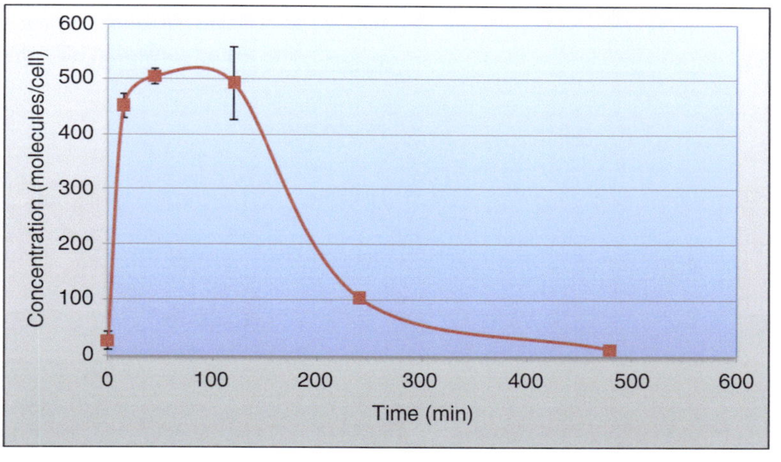

Fig. 2 Numbers of TGF-β molecule per cell after 30 s TGF-β stimulation followed by washout. Minimum amount of TGF-β is left following the washout, and 500 molecules of TGF-β per cell are insufficient to elicit a significant response when exposed to nonstimulated HaCaT cells. (Original data taken from [8])

2 Materials

2.1 Cell Culture

1. Dulbecco's Modified Eagle Medium (DMEM) (GIBCO 12800-082).
2. Fetal Bovine Serum (FBS) (SAFC Biosciences, 12303C).
3. GlutaMAX L-Glutamine Supplement (L-Glu) (GIBCO 35050).
4. Dulbecco's Phosphate Buffered Saline (D-PBS) (GIBCO, Invitrogen 14190-136).

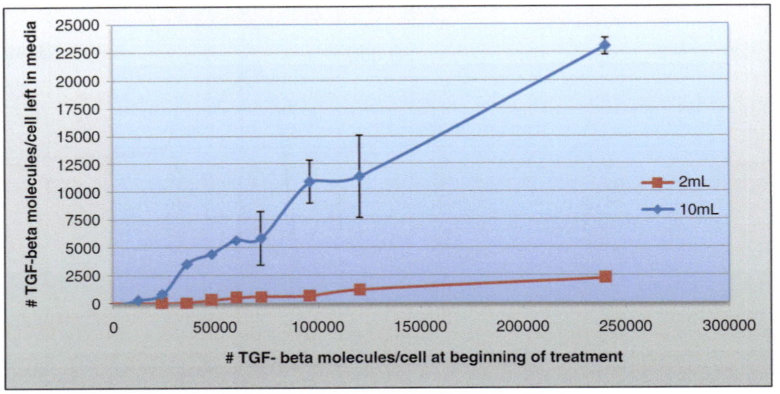

Fig. 3 Number of TGF-β molecules per cell left in media after 24 h treatment. Compared to Fig. 4 we can see that there is a direct relationship between number of TGF-β molecules left in medium and volume of medium. This can be interpreted as slowing down of endocytosis

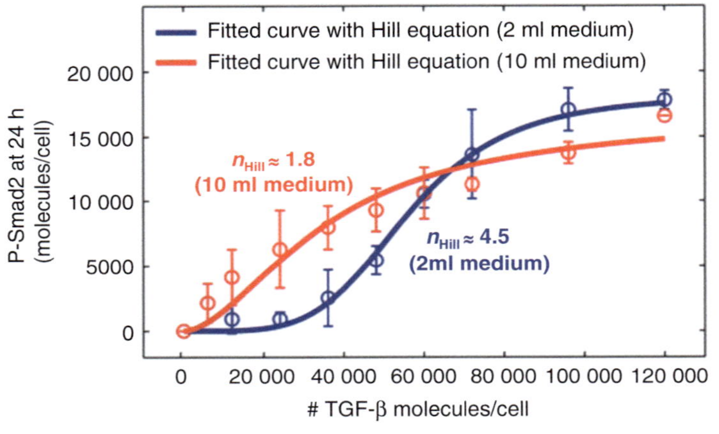

Fig. 4 The effect of increasing volume on phospho-SMAD2 levels. Notice how the switch-like response (described mathematically by Hill's coefficient) is greatly reduced when the volume increased. (Taken from [8])

5. 100× Penicillin G Potassium Solution (ICN Biomedicals, 100548, dissolved in ddH_2O to 10,000 U/mL).
6. 100× Streptomycin Sulfate solution (Sigma S-6501, dissolved in ddH_2O to 10,000 U/mL).
7. Trypsin (0.25 %) (GIBCO, Invitrogen 15050-057).

2.2 Live Cell Treatment

1. Transforming Growth Factor β1 (TGF-β1, stock solution 100 nM) (R&D Systems, 240-B-002).

2.3 Cellular Collection and Analysis

1. RIPA buffer (50 mM Tris–HCl, pH 7.4, 150 mM NaCl, 1 % v/v Triton X-100, 0.1 % w/v SDS, 1 % w/v sodium deoxycholate, 1 mM EDTA, 1 mM PMSF (added before use), 25 mM β-glycerophosphate (added before use), and 1 mM sodium orthovanadate (added before use)).
2. Cell Lifter (Costar, Corning 3008).
3. Ponceau S Staining Solution (0.5 g Ponceau S Dye, 5 % Glacial Acetic Acid to 100 mL with deionized water).
4. Protogel 30 % Acrylamide/Bisacrylamide Solution (National Diagnostics, EC-890).
5. Protogel 4× Resolving Gel Buffer (National Diagnostics, EC-892).
6. Protogel Stacking Gel Buffer (National Diagnostics, EC-893).
7. 10 % Sodium Dodecyl Sulfate Solution (10 % w/v SDS in distilled water).
8. Rabbit anti-pSMAD2 antibody (Cell Signaling #3108S).
9. Mouse anti-SMAD1/2/3 antibody (Santa Cruz Biotechnology SC-7960).
10. Mouse anti-β-Actin antibody (Abcam AB-8226100, Cell Signaling #4967).
11. Horse Radish Peroxidase conjugated Sheep anti-mouse antibody (GE Healthcare NA931-1 mL).
12. Horse Radish Peroxidase conjugated Goat anti-rabbit antibody (GE Healthcare RPN4301-1 mL).
13. SuperSignal West Dura Extended Duration Substrate chemiluminescent substrate (Pierce Biotechnology 34076).
14. Protran Nitrocellulose Membrane (Whatman 104024 BA83).
15. Whatman Chromatography Paper (Whatman 3030-917).
16. Semi-dry transfer apparatus (Hoefer TE70).
17. BCA protein assay kit (Thermo Scientific #23225).
18. Spectra broad range multicolor protein ladder (Fermentas SM184).
19. Powerwave X Scanning Spectrophotometer Plate Reader (Bio-Tek).
20. 4× SDS-Gel Loading Buffer (0.08 % w/v Bromophenol Blue, 50 mM EDTA, 40 mM DTT, 40 % v/v glycerol, 8 % w/v SDS, 200 mM Tris–HCl, pH 6.8 in deionized water).
21. Transfer Buffer (5.8 g Glycine, 11.6 g Tris Base, 0.72 g Sodium Dodecyl Sulfate, 400 mL Methanol, to 2 L with deionized water).
22. Tris Buffered Saline supplemented with 0.05 % Tween-20 detergent (TBS-t) (150 mM NaCl, 2 mM KCl, 25 mM Tri-HCl, pH 7.4, 0.05 % Tween 20, in deionized water).

23. 3 % BSA in TBST, filtered with 45 μm filter.
24. Western Blot Film (ISC BioExpress F-9023-8X10).

2.4 Luciferase Assay

1. Promega Dual Luciferase Kit (#E1910).
2. Beckman Coulter Multimode Detector DTX880 (Beckman).

3 Methods

3.1 Cell Culture and Reducing TGF-β Ligand Depletion

1. Working DMEM medium consisting of 10 % FBS, 2 mM L-Glu, 100 U/mL Penicillin G, and 100 U/mL Streptomycin is prepared and stored at 4 °C until use.
2. One 15 cm tissue culture plate is seeded with 8×10^6 adherent human keratinocytes (HaCaT) cells in 20 mL of the working DMEM medium. The cells are allowed to grow for 48 h at 37 °C in a 100 % humidified environment with 5 % CO_2. Medium is changed every 24 h (*see* **Note 1**).
3. Four 6-well plates are labeled as in Fig. 5. Cells are subsequently counted and reseeded into the 6-well plates at a density of 8.5×10^5 cells in 1.5 mL medium per well. Plates are then incubated overnight at 37 °C in a 100 % humidified environment with 5 % CO_2.
4. Next morning, twenty 15-mL conical tubes are prepared. 3 mL of pre-warmed medium is transferred into tubes 1–10 while 11 mL is transferred into tubes 11–20. Cells in the well

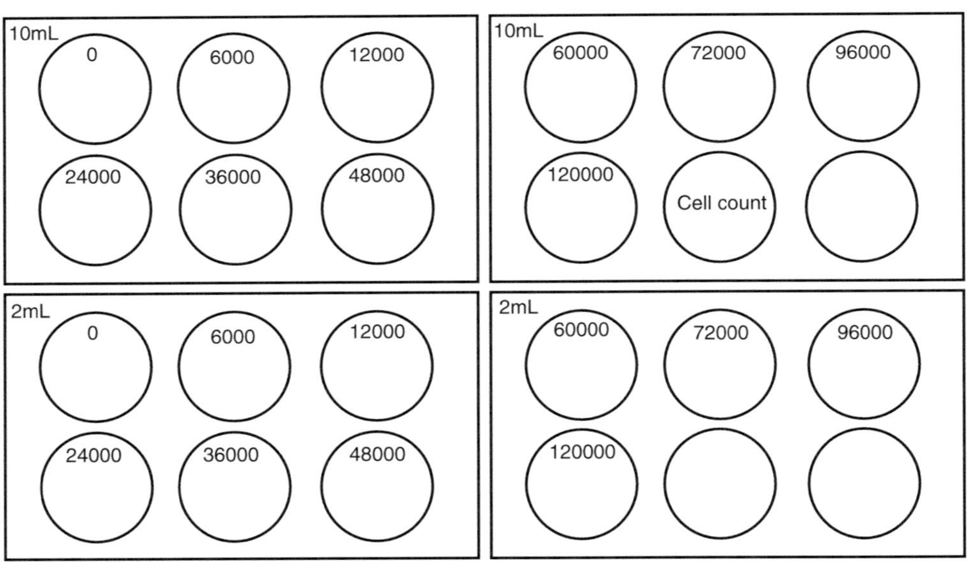

Fig. 5 Plate schematic. The labeled numbers correspond to different number of TGF-β molecules per cell

labeled as "cell count" in Fig. 5 are trypsinized and counted. The cell number is assumed to be the same in all the wells and is used to calculate the amount of TGF-β to be added into each conical tube. The final concentration of TGF-β in each tube is 0, 6000, 12,000, 24,000, 36,000, 48,000, 60,000, 72,000, 96,000, and 120,000 molecules of TGF-β per cell, respectively (Fig. 5). Medium in the remaining 20 wells is aspirated off and replaced with media from the conical tubes containing the corresponding volume and concentration of TGF-β. The amount of media will be 10 mL or 2 mL according to the labels in Fig. 5. Plates are then incubated for 24 h at 37 °C, 100 % humidity and 5 % CO_2 (*see* **Note 2**).

5. Plates are collected from incubator; each well containing roughly 8.5×10^5 adherent HaCaT cells are rinsed three times with 2 mL of pre-chilled D-PBS. D-PBS is subsequently removed by titling the plate vertically at an 80° angle in a bucket full with ice for 30 s and aspirating all liquid from the bottom corner of the plate using a P-200 pipette. 75 μL of RIPA buffer is added to each well, and the cells are scraped with a cell scraper and transferred into pre-chilled 1.7 mL Eppendorf microcentrifuge tubes. The tubes are rotated for 1 h at 4 °C (*see* **Note 3**).
6. Fractions are spun for 10 min at 13,200 × *g* and 4 °C. Supernatants are transferred to new labeled tubes and stored on ice.

3.2 Determining Protein Concentration and Performing Western Blot Analysis

1. Protein concentration is determined according to manufacturer's instructions using a BCA assay kit. In short, serial dilutions of bovine serum albumin (BSA) stock solution (2 mg/mL) are made in distilled water to produce final concentrations of 1.0, 0.5, and 0.25 mg/mL. A blank is also prepared from distilled water alone (*see* **Note 4**).
2. 1:10 Dilutions of cell lysates are made in separate tubes for a total of 20 μL.
3. In a 96-well plate, 5 μL of each unknown or standard is mixed with 100 μL of 1× BCA solution (50:1 mixture of solutions A and B, from the BCA kit) in triplicates. Plates are then completely sealed with Parafilm and incubated at 37 °C for 30 min.
4. The 96-well plate is read for absorbance at 562 nm using a 96-well plate scanning spectrophotometer. The raw data is plotted on Excel and the unknown is calculated using standard BSA concentrations. The yield is roughly 200 μg of protein per 75 μL of the lysate.
5. Cell lysate containing 60 μg of total protein is mixed with 4× SDS loading buffer and incubated at 95 °C for 5 min. Tubes are inverted and liquid contents are briefly spun at 13,200 × *g* for 10 s.

6. Two 1.5 mm, 12 % polyacrylamide gels (SDS-PAGE Protogel reagents from National Diagnostics) are polymerized with a 10-well comb according to manufacturer's instructions. All lanes are loaded with samples and 5 μL of Spectra protein ladder is added to the same lane as cell lysate without TGF-β treatment. Each gel is run at 30 mA until the bromophenol blue dye reaches the bottom of the gel (*see* **Note 5**).
7. Each gel is transferred in a semi-dry western blot horizontal transfer unit, using a sandwich from cathode (bottom piece) to anode (top piece) with the following scheme: three pieces of 0.5-cm (thick) chromatography paper, one piece of nitrocellulose membrane, SDS-PAGE gel containing samples, three pieces of thick chromatography paper. This sandwich is assembled in 100 mL of transfer buffer with gel above the nitrocellulose membrane, and transferred to the transfer apparatus in the same configuration. The sandwich is then lightly ironed with a test tube to ensure no air bubbles are trapped between the membrane and the gel. For each sandwich containing one gel, 45 mA (corresponding to 6 V) of current is applied to the apparatus for 2 h.
8. After the transfer is finished, the nitrocellulose membrane is removed from the apparatus and stained with 10 mL of Ponceau S staining solution at room temperature for 30 s. The staining solution is removed and the membrane is rinsed with distilled water until nonspecific background of Ponceau S stain is removed. Each lane should have even and bright red staining bands.
9. Each membrane is blocked with 5 mL of 3 % (w/v) BSA in TBS-t at room temperature for 30 min.
10. Blocking buffer is completely removed and both membranes are rinsed 1× with TBS-t prior to addition of 3.5 mL of 1:500 diluted rabbit anti-pSMAD2 in 3 % (w/v) BSA in TBS-t.
11. Both blots are incubated for 3 h at room temperature on a tabletop rocker.
12. Membranes are rinsed twice and washed 1× for 5 min with 10 mL of TBS-t.
13. Each membrane is then incubated with 3.5 mL of 1:2000 diluted anti-rabbit HRP conjugated secondary antibody solution in 3 % (w/v) non-fat dry milk in TBS-t for 50 min at room temperature on a tabletop rocker.
14. Each membrane is rinsed once, and washed 3× with 10 mL of TBS-t for 5, 10, and 10 min, respectively.
15. Membranes are removed from wash buffer and allowed to drip for 10 s before being placed protein side up on a piece of clear plastic(cut three-ringed binder sheet protector). 200 μL of

West Dura solution (a mixture of 100 μL solution A and 100 μL of solution B) is added to each membrane and the membranes are lightly tilted to ensure even spread. Another clear plastic sheet is then covered over the membranes and the sandwich is allowed to sit for 30 s prior to ironing out excess liquid from inside the sandwich with a paper towel.

16. The sandwich is then taped to the inside of an imaging cassette, and exposed with X-ray developing film in a dark room for 5 s, 15 s, 30 s and 1 min to produce various exposures. The film is then developed using an automatic film developer.
17. The membranes are taken out of the sandwich, rinsed with TBS-t, and placed in box with TBS-t.
18. **Steps 9–17** are repeated with mouse β-actin and mouse SMAD1/2/3 (primary antibody is incubated overnight instead of 3 h at room temperature).

3.3 Cell Culture and TGF-β Ligand Washout

1. Working DMEM media consists of 10 % FBS, 2 mM L-Glu, 100 U/mL Penicillin G, and 100 U/mL Streptomycin is prepared and stored at 4 °C until use.
2. Cells are seeded to 11 wells of 6-well plates at a density of 8.5×10^5 cells in 1.5 mL medium. Plates are then incubated overnight at 37 °C, 100 % humidity and 5 % CO_2.
3. Next morning, two 50 mL conical tubes each with 25 mL of pre-warmed DMEM are prepared. Cells in 1 well of the 11 wells are trypsinized and counted. The cell number is assumed to be the same in every well and is used to calculate the amount of TGF-β needed to make a final concentration of 60,000 molecules per cell. TGF-β is added to one of the 50 mL conical tubes. Medium in the remaining 10 wells is aspirated off and replaced with 2 mL DMEM containing 60,000 molecules per cell of TGF-β. Plates are then incubated for 30 s at room temperature and the medium is immediately aspirated off. The plates are rinsed 3× with D-PBS and fresh pre-warmed DMEM media is added to each well. Plates are incubated for 0, 15, 30, 45, 60, 120, 180, 240, 300, 360, 420, and 480 min (*see* **Note 6**).
4. Plates are individually collected from incubator, rinsed three times with 2 mL pre-chilled D-PBS. D-PBS is subsequently removed by titling the plate vertically at an 80° angle in a bucket full with ice for 30 s and aspirating all liquid from the bottom corner of the plate using a P-200 pipette. Plates are then snap-frozen with liquid nitrogen and placed in −80 °C incubator until all plates are collected.
5. Plates are allowed to thaw on ice. Seventy-five microliters of RIPA buffer is added to each well. Plates are scraped with a cell scraper and the mixture is transferred into corresponding

pre-chilled 1.7 mL Eppendorf microcentrifuge tubes. The tubes are rotated for 1 h at 4 °C.

6. Fractions are spun for 10 min at 13,200 × *g* and 4 °C. Supernatants are transferred to new labeled tubes and stored on ice.
7. Protein concentration, western blotting, is done using the same protocol described above.

3.4 Luciferase Assay on Washout Experiment

1. The assay is performed according to manufacturer's protocol (Promega E1910). Briefly, HaCaT cells with TGF-β responsive CAGAx12 luciferase reporter is seeded in 30 wells of 12-well plates at a density of 200,000 cell/well in 750 μL of medium. The cells are allowed to settle down for 12 h in 37 °C in a 100 % humidified environment with 5 % CO_2.
2. A TGF-β standard is prepared at concentrations of 0, 0.195, 0.39, 0.78, 1.56, 3.12, 6.25, 12.5, 25, and 50 pM (corresponding to 0, 282, 565, 1130, 2261, 4523, 9047, 18,094, 36,190, 72,380 molecules/cell, with cell number being estimated at 832,000).
3. The medium (750 μL) is aspirated off from each well and replaced with either 2 mL of standard, or 2.0 mL saved conditioned medium for each of the conditions in the 10 mL versus 2 mL TGF-β ligand depletion experiment (2.0 mL conditioned media from "10 mL" set and 2.0 mL conditioned media from the "2 mL" set). The plates are incubated for 10 h at 37 °C, 100 % humidity and 5 % CO_2.
4. The plates are taken out of incubator, rinsed 2× with pre-chilled D-PBS. D-PBS is subsequently removed by titling the plate vertically at an 80° angle in a bucket full with ice for 30 s and aspirating all liquid from the bottom corner of the plate using a P-200 pipetor.
5. 200 μL of 1× cell lysis buffer is added to each well, and the cells are scraped off the plate with a mini cell scraper and transferred to corresponding pre-chilled 1.5 mL Eppendorf microcentrifuge tubes. The tubes are then rotated for 45 min at 4 °C and spun down for 10 min at 13,200 × *g*.
6. 50 μL of the supernatant is added to each well of a white opaque 96-well plate in duplicates.
7. With a timer in hand, the luciferase reagent is added to each well in sequence at exactly 10 s from the time at which reagent is added to the previous well. When all the wells are finished, the plate is placed in the Beckman Coulter Multimode Detector DTX880 and each well is read at 10 s intervals.
8. The standard is fitted with a curve and is used to extrapolate the amount of TGF-β from the depletion experiment.

4 Notes

1. Cells reached roughly 80 % confluency by day 2.
2. Actual cell number is counted for more accurate calculation of number of TGF-β molecules per cell.
3. The term "rinse" is used here to refer to the brief adding, mixing, and removal of solutions.
4. As long as no more than 2 mM of diethylthreitol (DTT) or a similar reducing equivalent is added to RIPA buffer, there is minimal effect on the final absorbance read. Therefore diluting BSA standards in RIPA buffer is not necessary.
5. No effect on signal of p-SMAD2, SMAD1/2/3, and β-actin is observed with addition of Spectra protein ladder.
6. The media added are pre-warmed to 37 °C, based on the assumption that there is minimal change of temperature in 30 s even when plates are placed at room temperature. No significant differences in pSMAD2 signal are observed when the plate is incubated at 37 °C incubator with 100 % humidity and 5 % CO_2.

References

1. Massague J (1998) TGF-β signal transduction. Annu Rev Biochem 67:753–791
2. Akhurst RJ, Hata A (2012) Targeting the TGFβ signalling pathway in disease.Nat Rev Drug Discov. 11:790-811
3. Heldin CH, Miyazono K, ten Dijke P (1997) TGF-β signalling from cell membrane to nucleus through SMAD proteins. Nature 390(6659):465–471
4. Abdollah S, Macias-Silva M, Tsukazaki T, Hayashi H, Attisano L, Wrana JL (1997) TβRI phosphorylation of Smad2 on Ser465 and Ser467 is required for Smad2-Smad4 complex formation and signaling. J Biol Chem 272(44): 27678–27685
5. Liu X, Sun Y, Constantinescu SN, Karam E, Weinberg RA, Lodish HF (1997) Transforming growth factor β-induced phosphorylation of Smad3 is required for growth inhibition and transcriptional induction in epithelial cells. Proc Natl Acad Sci U S A 94(20): 10669–10674
6. Souchelnytskyi S, Tamaki K, Engstrom U, Wernstedt C, ten Dijke P, Heldin CH (1997) Phosphorylation of Ser465 and Ser467 in the C terminus of Smad2 mediates interaction with Smad4 and is required for transforming growth factor-β signaling. J Biol Chem 272(44): 28107–28115
7. Shi Y, Massague J (2003) Mechanisms of TGF-β signaling from cell membrane to the nucleus. Cell 113(6):685–700
8. Zi Z, Feng Z, Chapnick DA et al (2011) Quantitative analysis of transient and sustained transforming growth factor-β signaling dynamics. Mol Syst Biol 7:492
9. Hornbruch A, Ma G, Ballermann MA, Tumova K, Liu D, Cairine Logan C (2005) A BMP-mediated transcriptional cascade involving Cash1 and tlx-3 specifies first-order relay sensory neurons in the developing hindbrain. Mech Dev 122(7–8):900–913
10. Clarke DC, Brown ML, Erickson RA, Shi Y, Liu X (2009) Transforming growth factor β depletion is the primary determinant of smad signaling kinetics. Mol Cell Biol 29(9): 2443–2455

INDEX

Xin-Hua Feng et al. (eds.), *TGF-β Signaling: Methods and Protocols*, Methods in Molecular Biology, vol. 1344, DOI 10.1007/978-1-4939-2966-5_2, © Springer Science+Business Media New York 2016

V

W

X

Z

MIX
Papier aus verantwortungsvollen Quellen
Paper from responsible sources
FSC® C105338

If you have any concerns about our products,
you can contact us on
ProductSafety@springernature.com

In case Publisher is established outside the EU,
the EU authorized representative is:
Springer Nature Customer Service Center GmbH
Europaplatz 3, 69115 Heidelberg, Germany

Printed by Libri Plureos GmbH
in Hamburg, Germany